INTRODUCTION
MECHANICS

Mahendra K Verma

Indian Institute of Technology
Kanpur

CRC Press
Taylor & Francis Group
Boca Raton London New York

CRC Press is an imprint of the
Taylor & Francis Group, an **informa** business

First published 2008 by Universities Press (India) Private Limited

Published 2018 by CRC Press
Taylor & Francis Group
6000 Broken Sound Parkway NW, Suite 300
Boca Raton, FL 33487-2742

First issued in paperback 2018

© 2008 by Taylor & Francis Group, LLC
CRC Press is an imprint of Taylor & Francis Group, an Informa business

No claim to original U.S. Government works

ISBN 13: 978-1-138-11677-1 (pbk)
ISBN 13: 978-1-4398-0127-7 (hbk)

Visit the Taylor & Francis Web site at
http://www.taylorandfrancis.com

and the CRC Press Web site at
http://www.crcpress.com

Distributed in India, China, Pakistan, Bangladesh, Sri Lanka, Nepal, Bhutan, Indonesia, Malaysia, Singapore and Hong Kong by
Orient Blackswan Private Limited

Distributed in the rest of the world by
CRC Press LLC, Taylor and Francis Group

Set in Times Roman 10/12 by
OSDATA, Hyderabad, India

To the unnamed heroes and heroines

who believed in their convictions

and struggled because they were ahead of their times

Contents

Preface

Introduction to Mechanics grew out of a one-semester course we offer at the Indian Institute of Technology Kanpur to the first year Bachelor (B. Tech. and Integrated M. Sc.) students. The book covers Newtonian dynamics and the basics of special relativity. A natural question arises: why must we have another book on mechanics when there are so many excellent textbooks written by eminent physicists like Kleppner and Kolenkow, Knudsen and Hjorth, Kittel, Feynman etc.? We believe that this book has something new to offer. Along with a discussion on standard topics—Newton's laws of motion, energy, linear and angular momentum, rigid body motion, special relativity—we have introduced modern topics like symmetries, phase space, Newton's laws in the framework of differential equations, nonlinear dynamics and chaos.

We have attempted to keep the focus on the main ideas of Newtonian dynamics—the key issues (e.g., *absolute space and absolute time*) and its basic assumptions have been emphasised throughout the text. Special relativity is introduced as an alternative to the Newtonian framework. Newton's equation of motion is presented as a differential equation that helps us predict the future of a mechanical system. This kind of treatment brings out key issues like phase space and determinism in mechanical systems. This approach also helps us introduce modern research topics like chaos theory in a natural way. We hope that the students will appreciate these core issues rather than just getting entrenched in various technical aspects.

In addition to the standard topics of mechanics, we also investigate symmetries of Newton's laws (Chapter 7). Symmetries are very important in physics, and they are the basis of fundamental theories like quantum mechanics and quantum field theory. We have also introduced phase space in this book. Chaos theory, which is one of the frontier areas of physics now, is the study of interesting dynamics in phase space. One of the most important tools in modern science and engineering, dimensional analysis and estimation, has been introduced in Appendix B. We have included a brief history of mechanics (Chapter 1) and touched on what we call the 'paradigm of physics' (Appendix A) with the intention to highlight the key questions and issues of mechanics. Answer to these questions gave birth to modern science. We have attempted to put forth Galileo's and Newton's emphasis on experiments or observations to validate a theory.

Presently many intermediate (class XII) students in Indian schools learn physics at the level of Halliday and Resnick's book. This creates a necessity for a modern yet simple and standard textbook for the first year students of B. Sc. that could act as a stepping stone to more advanced level courses. We hope that this book would help students achieve this target.

We have observed that many students segregate learning into theory and numerical problems. Learning science is not memorising the laws, or solving numerical problems; it is a method of understanding nature by observations and logical arguments. The objective of a book or a teacher is to introduce the students to existing theories along with their assumptions and limitations. We should keep in mind that these theories are quite complex, and that the ramifications and assumptions of these theories must be studied from various angles. Good exercises assist students to achieve these objectives and they are an integral part of a course. We urge students to work out the exercises and projects given at the end of each chapter with the above motivation. The students and instructors can choose which problems to solve from the given set. We believe that doing too many problems mechanically is harmful; it does not achieve the objective of understanding a theory.

When I taught mechanics to the first year students of IIT Kanpur, I covered almost all the topics in the present book except Chapter 16 (waves) and Appendix E (tensors). I found that most of the students did not have the requisite background for the discussion on tensors, hence the discussion on tensors was skipped in the course. However interested students could refer to the discussions on tensors in Appendix E as a supplement if they have the requisite background and motivation. The history of mechanics (Chapter 1), the paradigm of physics and science (Appendix A), and dimensional analysis and estimation (Appendix B) were covered in the beginning of the course, after which we introduced Newtonian mechanics. Overall, I believe most of the material in the present book can be covered in a semester course with some additions and deletions depending on the interest of the instructor and students. I believe a book can only serve as a guide to a student who has to learn the material on her/his own. Hence I urge the students to explore the subtleties of mechanics from various sources. The references listed at the end of the book could be useful in this exploration.

The sequence of topics in this book is somewhat different from many classical texts on mechanics. We introduced Newton's laws in terms of differential equations, and wanted to expose the students to solve reasonably hard problems related to the differential equation formulation of Newton's laws. This was the reason why we introduced Kepler's problem rather early as an application of Newton's equation of motion in radial–polar coordinate. Also, we have tried to minimise the usual problems related to blocks, pulleys, and inclines since the students must have already solved them in schools. Noninertial reference frames contrast inertial reference frames, hence they have also been introduced rather early—before the discussion on rotation. These topics require a very basic definition of angular velocity and angular momentum that the students have already learnt in their school physics. If instructors choose to follow this book as a text, we strongly urge them to follow the topics in the sequence given in the book.

During our course we emphasised on plotting and finding the numerical solution of differential equations using the software Matlab or Octave. Most of the plots in the present book have been done using Matlab. Plotting programs and several other programs can be downloaded from my website *http://home.iitk.ac.in/~mkv*. Students are encouraged to learn Matlab; it is very simple, yet very useful for this course and other courses.

As mentioned earlier, this book is a polished form of various lecture notes of our mechanics course in IIT Kanpur. Hence, the ideas and problems in this book really belong to all the instructors and tutors, whom I thank for sharing their material. Particularly, I gratefully acknowledge the very fruitful interactions I had with A. K. Majumdar, P. Jain, H. Wanare, S. A. Ramakrishna, R. Prasad, A. K. Mallik, Y. N. Mohapatra, Z. Hussain, T. Sarkar,

G. Sengupta, A. Dutta, and A. Sahay. Among the textbooks on mechanics, I specially benefitted from the presentation style of the books by Knudsen and Hjorth, and Feynman, and I record my grateful acknowledgements to these wonderful teachers and authors. S. A. Ramakrishna did the necessary but painstaking task of editing and gave many useful suggestions to improve the presentation.

I thank the Quality Improvement Program (QIP) of IIT Kanpur for providing financial assistance for writing this book. I am grateful to Mayank Goel for producing most of the attractive figures. The initial typesetting was done using Lyx and Latex, and some of the figures were drawn using Octave and Matlab. I wish to acknowledge the developers of these software and Linux OS. I thank Wikipedia for the free use of the images of the scientists in the first chapter. I also thank Mr. Khan and Mr. Arvind Mishra for assistance in printing and photocopying.

I owe a great deal to my publishers, Universities Press, Hyderabad. I thank their director, Mr Madhu Reddy for his guidance and support. I am grateful to Ms Jebah S. David and the whole publication team for their help throughout the publication process from copyediting to designing the cover.

Finally I wish to acknowledge my appreciation and gratitude to the thinkers, named and unnamed, who challenged the paradigm of their days and thought of creative ways to understand the motion of objects around us. History shows that some of these scientists faced an uphill challenge and sacrificed a great deal (recall Bruno who was burnt at the stake more than four centuries ago for believing that the Sun and the not the Earth was the centre of the Solar System). The known heroes are at least vindicated. However the unknown ones have struggled and still struggle for the cause silently. This book is dedicated to these unknown heroes and heroines of science and other fields.

I have made all attempts to make the text free of errors and ambiguities. However I am sure that many errors are still present in the text. I would be grateful if you could point out these errors to me by email (mkv@iitk.ac.in) or by post.

Notation

Attempts were made to follow the following notation throughout the book.

Position vector	$\mathbf{r}$
Velocity vector	$\mathbf{v}$
acceleration vector	$\mathbf{a}$
Linear momentum	$\mathbf{p}$
Centre of mass	CM
Centre of mass coordinate	$\mathbf{R}_{\mathrm{CM}}$
Energy	E
Angular velocity	ω, Ω
Torque	$\mathbf{N}$
with respect to	wrt
Left-hand side	LHS
Right-hand side	RHS
Kinetic energy	KE
Differential equation	DE
Special relativity	SR

1

History of Mechanics

The study of motion of physical bodies is called *mechanics*. This subject dates back to ancient times. Many wise and curious people have wondered about the motion of heavenly bodies, often in preference to terrestrial bodies. The debate whether the Sun goes around the Earth, or the Earth goes around the Sun was one of the most heated issues in the history of science.

A study of history reveals that Aristotle formulated the first theory of mechanics. Aristotle's theory of motion was rather philosophical and relied heavily on logic. Many centuries later it was observed that Aristotle's predictions did not match with experimental (or empirical) observations. Around the sixteen century, Galileo, Newton, and other physicists formulated a completely new theory of mechanics which forms the basis for modern science. In this new theory, observations and experiments have a very important role as they are used to verify or reject a physical theory.

At the turn of the twentieth century it was found that the predictions of Newton's laws do not hold for particles moving with speeds close to the speed of light, and for microscopic particles, e.g. electrons. To understand these phenomena, two new theories were proposed: Einstein's theory of relativity and quantum mechanics. The present book will focus on Newton's laws, and it will only touch slightly on the above modern theories.

In the following discussion we will outline major discoveries in mechanics from the ancient to the present time. Due to lack of space we will be very brief. The reader can always refer to Ferris (1988), Spangenburg and Moser (1993) and Wikipedia for further discussions on the historical issues.

While studying the history of mechanics, it will become evident that mechanics was the front-runner in founding modern scientific ideas. Let us now start our discussion about the history of mechanics.

1.1 Ancient thoughts

Among ancient thinkers, Greeks philosophers contributed the most in the field of science.[1] They discovered that the Earth is round. They estimated the radius of the Earth fairly well.

---

[1] Due to lack of extensive research, the overall contributions of other ancient civilisations like the Chinese, Indian, Mayans, and Egyptians have not been recorded satisfactorily.

They also attempted to measure the Earth–Moon and Earth–Sun distances. They thought up many interesting ideas about the planets, the stars, and the universe as a whole. Due to lack of time and space, we cannot describe all these ideas here. In the following paragraphs we will only discuss the ideas of Eudoxus, Aristotle, and Ptolemy. Interested readers can refer to Ferris (1988), Spangenburg and Moser (1993), and Wikipedia for details.

Many Greek thinkers addressed the movement of stars and planets. The most prominent among them was Eudoxus (410 or 408 BC to 355 or 347 BC). He belonged to Plato's academy in Athens. Deeply influenced by Plato's appeal for perfect geometrical forms (sphere, regular solids etc.), Eudoxus proposed that the universe is composed of 27 concentric spheres surrounding the Earth as shown in Fig. 1.1. These spheres rotate with constant velocity, with the axes of some of the spheres inclined to each other. The stars were carried by the outermost sphere. Each planet moves in four spheres (20 for 5 planets), and the Sun and the Moon move in three spheres each. By adjusting the rates of rotation and the inclination of the axes of the spheres, Euxodus could explain the movements of heavenly bodies. Since the Earth is at the centre of the universe in this model, it is called *the geocentric model of the universe.*

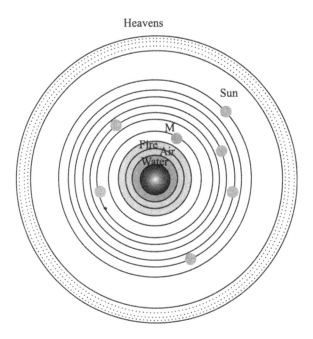

Figure 1.1   *The cosmos according to Eudoxus*

The most complex motion in the sky is that of the outer planets. When we observe these planets from the Earth at a fixed time in the night, they move westward, then eastward, and again westward as shown in Fig. 1.2. This is called retrograde motion of the planets. Eudoxus explained this phenomena by introducing epicycles as shown in Fig. 1.3(a). The planet moves on a smaller sphere that itself rotates around a larger sphere. The apparent motion of the planet from the Earth in the epicycle model is depicted in Fig. 1.3(b). Note that a single epicycle produced the trajectory reproduced in Fig. 1.3(b); more than a single epicycle is required to generate the retrograde motion seen in the sky.

Figure 1.2 *Observed motion of Mars and Saturn. These planets move westward, eastward, and again westward. This phenomenon is called retrograde motion.*

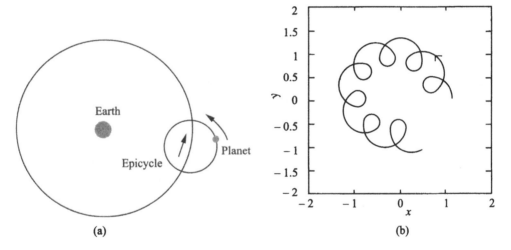

Figure 1.3 *(a) The retrograde motion of an outer planet explained using epicycles. The planet moves on a sphere that itself rotates around a sphere whose centre is the Earth. (b) The apparent motion of the planet in Earth's frame of reference with a single epicycle.*

Aristotle (384–322 BC) introduced more spheres to take the count up to 55. The results from later refined observations however did not match with the Aristotelian theory. To keep the geocentric picture intact, more complex set of spheres were introduced. Claudius Ptolemy (85–165 AD) constructed the most refined geocentric model that lasted for nearly 1400 years. Note that the heliocentric picture, much simpler than the geocentric one, can explain all the observations. Unfortunately, prevalent prejudices typically hindered thinkers from attempting simple alternate ideas.

Aristotle set out to explain why the heavenly spheres rotated. He postulated that the *prime mover* rotated the spheres containing the heavenly bodies. The planets, the Moon, and the Sun are regulated by several spheres. For Aristotle, the heavenly cosmos was imperishable and perfect, and the heavenly spheres moved in circular orbits which is the most perfect trajectory. On the contrary, everything on the Earth was changeable and corrupt. **The motion on the Earth is linear, not circular.**

To explain the motion in the terrestrial environment, Aristotle argued that each object is made up of four elements—earth, water, air, and fire in various proportions (see Fig. 1.1). Each object moves in such a way as to return to its natural state, that is, one of the spheres on the Earth as shown in Fig. 1.1 (sphere of earth, water, air, and fire). Solid objects like stones would tend to fall to the Earth, their natural place. Air and fire are above the ground; objects composed mainly of air or fire would go upward towards their natural place. A consequence of this theory is that heavier bodies containing more earth will fall to the Earth faster than lighter ones.

Aristotle's arguments for the vertical motion on the Earth are logically consistent, but they do not work for the projectile motion. He postulated a rather awkward argument to explain projectile motion, theorising that air rushes into the vacuum created by the projectile's trail, and the incoming air pushes the projectile. Compare the complexity of Aristotelian theory with the relatively simpler Newton's laws of motion.

According to Aristotle, an unforced object comes to rest when it reaches its natural place. Hence a block sliding on a horizontal plank will come to rest unless forced, since rest is the natural state of the block.[2] Aristotle also gave a theory for the origin of force that can be illustrated using an example. For a projectile motion, thrower's arm provides force to the projectile; the thrower to his/her arm; and so on, forming an infinite chain of causes. Aristotle argued that there must be an *unmoved mover*, something which can initiate motion without itself being set in motion. This view was preserved by the medieval Church during the Dark Ages, and it became a ruling paradigm.

In hindsight, Aristotle's theory of cosmos and causation were constructed by pure logic without resorting to proper observational and experimental tests. So they tended to be metaphysical and philosophical. Around two thousand years later, many inconsistencies were found in the Aristotelian theory, the resolution of which led to the birth of modern science.

1.2   Renaissance—Heliocentric theory

1.2.1   Nicolaus Copernicus

In medieval times, astronomers performed very accurate observations of planetary orbits. Ptolemy's system of planetary motion could not explain the refined observations on the retrograde motion of Mars and Saturn. *Copernicus* (1473–1543), a Polish astronomer, discovered that the retrograde motion of the planets could be explained quite easily if one assumes that the planets go around a stationary Sun. The time period of Earth's revolution around the Sun is smaller than that of Mars. In the heliocentric picture, for the configuration shown in Fig. 1.4(a), Mars exhibits retrograde motion when seen from the Earth. At instants 1,2,3 and 6,7 Mars is moving westward, and in between at instants 4 and 5, it moves eastward. These observations led Copernicus to propose the *heliocentric model*.

This was the beginning of heliocentric theory.

---

[2]Contrast this argument with Newton's first law of motion according to which no force is needed to sustain constant velocity of a moving body.

EXAMPLE 1.1   In the heliocentric picture, describe the motion of Mars in Earth's sky. Assume the orbits of Mars and Earth around the Sun to be circular.

SOLUTION   In the heliocentric model, the motion of Mars and Earth are as shown in Fig. 1.4(b). We assume circular orbits for both these planets. We also assume that the Earth, Mars, and the Sun are collinear at $t = 0$. In a coordinate system fixed to the Sun, the coordinates of Mars and Earth are

$$x_{Mars} = R_{Mars}\cos(\omega_{Mars}t); \quad y_{Mars} = R_{Mars}\sin(\omega_{Mars}t);$$

$$x_E = R_E\cos(\omega_E t); \quad y_E = R_E\sin(\omega_E t);$$

where R_{Mars} and R_E are the distances of Mars and the Earth from the Sun respectively, and ω_{Mars} and ω_E are the angular velocities of Mars and the Earth respectively around the Sun. The relative coordinates of Mars with respect to the Earth is

$$x_r = x_{Mars} - x_E = R_{Mars}\cos(\omega_{Mars}t) - R_E\cos(\omega_E t),$$

$$y_r = y_{Mars} - y_E = R_{Mars}\sin(\omega_{Mars}t) - R_E\sin(\omega_E t).$$

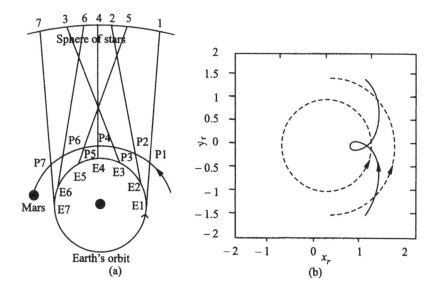

Figure 1.4   (a) The apparent retrograde motion of Mars as seen from the Earth in the heliocentric model. (b) In the heliocentric model, the apparent path of Mars as seen from the Earth is depicted as a solid line. The path of the Earth and Mars are also shown as inner and outer dashed lines respectively. The trajectories shown here cover $t' = t/T_E = -0.5 : 0 : 0.5$. The retrograde motion takes place when the Sun, Earth, and Mars are in an approximate straight line.

If we measure time in units of one terrestrial year, and the distance in the units of R_E, then

$$t = t'T_E = t'\frac{2\pi}{\omega_E}; \quad r = R_E r'$$

where t', r' are the non-dimensional time and distance respectively. In these units

$$x'_r = \left(\frac{R_{\text{Mars}}}{R_E}\right) \cos\left[\left(\frac{T_E}{T_{\text{Mars}}}\right) 2\pi t'\right] - \cos\left(2\pi t'\right)$$

$$y'_r = \left(\frac{R_{\text{Mars}}}{R_E}\right) \sin\left[\left(\frac{T_E}{T_{\text{Mars}}}\right) 2\pi t'\right] - \sin(2\pi t')$$

The ratios are $\dfrac{R_{\text{Mars}}}{R_E} = \dfrac{2.27 \times 10^{11}}{1.5 \times 10^{11}} = 1.51$ and $\dfrac{T_E}{T_{\text{Mars}}} = \dfrac{365.25}{686.93} = 0.53$. The angle made by Mars will be

$$\tan\theta = \frac{y'_r}{x'_r}.$$

The plot of y'_r vs. x'_r is shown in Fig. 1.4(b) for $t' = -0.5 : 0 : 0.5$. The retrograde motion takes place when the Sun, Moon, and Earth are approximately in a straight line.

1.2.2   Tycho Brahe

Tycho Brahe (Denmark, 1546–1601) can be considered to be the greatest astronomer before the advent of the telescope. In an island near Copenhagen, Brahe built the ultimate astronomical observatory of his day.[3] Brahe observed that some new bright objects suddenly appeared in the sky. This dispelled Aristotle's idea that the heavens are unchanging. These bright objects were comets and supernova.

Brahe did not believe in Copernicus' heliocentric picture. He thought of a composite model in which the planets moved around the Sun, but the Sun itself moved around the Earth. To get observational support for his model, Brahe made very precise astronomical observations of planets (up to 1 minute of an arc) from 1576 to 1597, which provided him a huge data bank. Brahe was a brilliant astronomer, but not a very good theorist. He could not analyse his data; this task was done by his student Kepler after his death.

1.2.3   Johannes Kepler

Kepler (Germany, 1571–1630) inherited Tycho Brahe's data and analysed them with great care. The analysis of the Mars data itself took 8 years. He summarised his findings in the following three laws of planetary motion:

---

[3]These instruments were very ingenious. The coordinates of stars and planets were measured by accurate sextants. Shadows were used to measured time. You can get a fairly good idea about the nature of these instruments by visiting Jantar Mantar in Delhi, Banaras, or Ujjain.

1. The orbit of a planet lies in a fixed plane containing the Sun. The planet moves in the plane in an elliptical orbit with the Sun as one focus.

2. The radius vector from the Sun to the planet sweeps out equal areas in equal amount of time.

3. $T^2 = kR^3$, where T is the period of revolution, R is the semi-axis of the elliptical path of the planet, and k is a constant.

Recall that all the earlier thinkers believed that the planetary orbits were circular. This prejudice came from Plato's fondness for symmetries in nature. Kepler claimed Mars' orbit to be elliptical. This claim was quite courageous because the eccentricity of Mars' orbit ($e = (r_{max} - r_{min})/(r_{max} + r_{min})$) is 0.093, that is very close to zero (eccentricity of a circle). Kepler strongly believed in Brahe's data and his own calculations.

Kepler's laws played a very important role in the development of mechanics. Newton used Kepler's laws to arrive at the law of universal gravitation.

1.3   Birth of modern science—Galileo and Newton

Around the fifteen century, the Renaissance brought in many new ideas to Europe. In this vibrant period, modern science was born due to the contributions of many thinkers like Descartes, Galileo, Newton, Laplace, etc. Here we will focus on Galileo and Newton who formulated the theory of mechanics.

1.3.1   Galileo Galilei

Galileo Galilei (Italy, 1564–1642) is considered to be the father of modern science. He placed a firm emphasis on observation. For example, he rolled balls down an inclined plane and measured the time elapsed using a water clock (similar to an hour glass). He recognised this motion to be equivalent to free fall. He observed that all the balls, independent of its constituents, took the same time. This observation clearly disproved Aristotle's idea that heavier bodies fall faster than lighter ones.[4]

Galileo made numerous important observations. For the first time he showed that the time period of a pendulum is independent of its amplitude, which led to the invention of clocks. He also got hold of a then-recently invented instrument called the telescope and started observing the sky. He discovered the moons of Jupiter. The moons revolving around Jupiter rather than the Earth provide further credence to Copernicus' heliocentric theory.

---

[4]Galileo also constructed a beautiful logical argument to prove that all masses fall at the same rate. Tie up a heavy body with a light body. By Aristotle's theory that heavier bodies fall faster than lighter ones, the composite body should fall faster than both its constituents, as it is heavier than both. But if we look at it another way, the lighter body should impede the motion of the heavier body, hence the speed of the composite body should be intermediate between the speeds of its constituents. This is a contradiction! Hence Aristotle must be wrong. Thus Galileo concluded that all the bodies fall under gravity at the same rate, a hypothesis that does not suffer from the above contradictions.

One of the most important contribution of Galileo is the *law of inertia* according to which a body retains its velocity unless a force acts on the body. This law contradicts Aristotle's theory that an object slows down and stops unless some force acts on it. Galileo's law of inertia is valid only for a free particle, which is an abstraction. Galileo also gave a *law of relativity* according to which the laws of nature are the same in all reference frames moving with constant velocity.

Galileo provided a new paradigm for science in which the laws are constructed and refuted on the basis of observations. Einstein said the following in his homage to Galileo, "Pure logical thinking cannot yield us any knowledge of the empirical world; all knowledge of reality starts from experience and ends in it.... Because Galileo saw this, and particularly because he drummed it into the scientific world, he is the father of modern physics—indeed, of modern science altogether."

For a brief discussion on the present paradigm of physics, refer to Appendix A.

1.3.2   Isaac Newton

Isaac Newton (England, 1642–1727) was born the year Galileo died. Newton synthesised the ideas of Copernicus, Kepler, and Galileo, and produced compact and abstract laws of motion and the law of universal gravitation. These theories had an enormous impact on science. Newton was very strong in mathematics; he is the first physicist who wrote the laws of physics in precise mathematical forms. Note that Newton is also the discoverer of calculus.

You would have been aware of Newton's laws of motion from your school days onwards. So we will not delve into it except to remind you that Newton's laws are universal, i.e., they are applicable to all systems except those that move with speeds comparable to the speed of light.

There is a story that Newton discovered the idea of gravitation when an apple fell on his head. This is a story for children and definitely untrue. Newton used Kepler's laws to discover the law of universal gravitation. Huygens (Netherlands, 1629–1695) first derived an expression for the acceleration of an object going around in a circular path. Using Huygens' formula, the acceleration of a planet going around the Sun is

$$a = \omega^2 R = \frac{4\pi^2}{T^2} R, \tag{1.1}$$

where ω is the frequency of the circular motion, and R is the distance of the planet from the Sun. According to Kepler's third law, $T^2 = kR^3$ (k is a constant), substitution of which in Eq. (1.1) yields

$$a = \frac{4\pi^2}{k} \frac{1}{R^2}. \tag{1.2}$$

From Newton's second law of motion, the force acting on the planet is

$$F = ma = m\frac{4\pi^2}{k} \frac{1}{R^2},$$

which is inversely proportional to R^2. So Newton concluded that the Sun attracts planets by a gravitational force that varies as $1/R^2$.

Newton felt that the same force must be making the Moon go around the Earth. He also noted that the moon falls towards the Earth all the time while still following a circular path. Newton cleverly used Kepler's third law for the Earth–Moon system. The acceleration of the Moon towards the Earth is

$$a_M = \frac{4\pi^2}{T_M^2 R_{EM}},$$

where R_{EM} is the distance between the Earth and the Moon, and T_M is the time period of the Moon's orbit around the Earth.

After this Newton used Eq. (1.2) to deduce the acceleration of an object on the surface of the Earth:

$$a_E = a_M \left(\frac{R_{EM}}{R_E}\right)^2 = \frac{4\pi^2}{T_M^2}\left(\frac{R_{EM}}{R_E}\right)^3 R_E$$

where R_E is the radius of the Earth. Substitution of $T_M = 27$ days 7 hours and 43 minutes, $R_{EM}/R_E = 60.1$, and $R_E = 6400$ km yields

$$a_E = 9.84 \text{m/s}^2.$$

Newton measured the acceleration due to gravity by rolling various masses on an inclined plane, and found the experimental value to be in general agreement with the above theoretical prediction. Using this result, Newton concluded that the forces acting between the Sun and planets, the Earth and Moon, and the Earth and objects on the Earth have the same origin, i.e. gravitational force. A remark is in order here. Newton proved the inverse square law of gravitational force for elliptical orbits. The proof is geometrical, and somewhat involved.

From the above calculations, it is also clear that the acceleration due to gravity on the surface of the Earth is independent of the mass of the object. Newton verified this theoretical observation by measuring the acceleration of objects made of various materials on an inclined plane. Thus Aristotle's theory of motion was proven to be incorrect.

This is the genius of Newton. He used Kepler's third law of planetary motion, and some mathematics to deduce the universal gravitational force between objects at disparate scales. The universality of gravitation indicates that the laws of physics on the Earth and in the heavens is the same, contrary to Aristotle's belief. Newton deduced the universal law of gravity according to which any two masses m_1 and m_2 separated by a distance r exert an attractive force on each other whose magnitude is

$$f = G\frac{m_1 m_2}{r^2},$$

where G is Newton's constant of gravitation, and whose direction is towards each other.

Although Newton constructed general abstract laws in the language of mathematics, these laws are based on observations. Newton made various predictions based on his laws, e.g., the orbits of comets, tidal effects, the oblate nature of the Earth, perturbations of planetary orbits from the ellipse due to other astronomical objects etc. Many of these predictions could be tested from observations. This method, started by Newton and Galileo, is the method of modern science. Newton's ideas had far-reaching consequences in the field of science. For his wide-ranging contributions, many people consider Newton to be the greatest scientist of all time.

Newton wondered very deeply about space and time; he conjectured them to be absolute, i.e. the same for all observers. Many of Newton's predictions based on absolute space and time proved to be correct. But problems surfaced when the speed of the particle is close to the speed of light. That's where Einstein comes to rescue.

1.4   Fundamental revision of spacetime—Einstein

At the turn of the twentieth century, Michelson and Morley observed a very puzzling experimental result. They measured the speed of light to be the same while an observer is moving towards the source or away from the source. In addition, the Maxwell's equations too indicated that electromagnetic waves move with a constant velocity independent of the motion of the source or the observer. These results contradict the addition of velocities given by Newton and Galileo. Many scientists tried to explain the constancy of the speed of light using various tricks within Newton's framework. However iconoclast Einstein (Germany and USA, 1879–1955) had another idea. He noticed that the constancy of the speed of light and the postulate of absolute space and absolute time are contradictory, and abandoned the concepts of absolute space and time. He proposed a new theory called the *special theory of relativity* in which the time difference between any two events could be different for different observers.

The spacetime structure of special relativity differs completely from that of Newton. The time difference between two events could be measured differently in two inertial frames; the length of a metre stick also is different in two inertial frames. The theory of relativity has many other startling predictions such as it is possible to convert mass to energy and vice versa. All the predictions of relativity have been found to be consistent with experiments, and hence it is considered to be a correct theory. For particles moving with small speeds, the kinematics and dynamics of the special theory are consistent with Newton's theory of motion.

Einstein went further and discovered an equivalence between accelerating frames and gravity. Using this idea, he proposed a new theory of gravity that is more general than Newton's theory of gravity. For example, Einstein predicted that light bends near a star, but no such conclusion can be drawn from Newton's theory of gravity. Einstein's theory of gravity is called the general theory of relativity. Einstein's theory of relativity has had a revolutionary impact in physics.

Figure 1.5 showcases the main thinkers and scientists through the ages who have made far-reaching contributions to the field of mechanics.

Aristotle (384–322 BC)   N. Copernicus (1473–1543)

T. Brahe (1546–1601)   J. Kepler (1571–1630)

Galileo (1564–1642)   I. Newton (1642–1727)   A. Einstein (1879–1956)

Figure 1.5   *Scientists who have made contributions to the field of mechanics. [Source: Wikipedia]*

1.5   Mechanics and modern physics

As discussed in the previous sections, modern science essentially starts with Galileo and Newton's theory of motion. New areas like fluid mechanics, elasticity and wave theory were directly inspired by Newton's theory. Electrodynamics and statistical mechanics took birth around mid-nineteenth century. Two revolutionary theories, relativity and quantum mechanics, were born in the twentieth century when Newton's theory was observed to fail at high speeds and at small length scales. These theories form modern physics. See Fig. 1.6 for an illustration of different branches in physics.

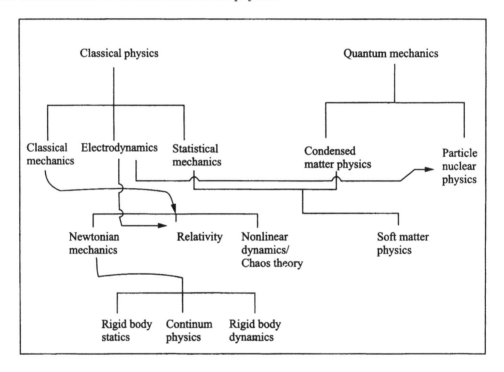

Figure 1.6   *Different areas of physics*

Quantum mechanics differs fundamentally from Newton's and Einstein's conception of dynamics. For example, in quantum mechanics we cannot simultaneously specify the position and the velocity of a particle precisely—unlike in Newton's formalism. The ideas of force and trajectory are abandoned in quantum mechanics.

In the major part of the twentieth century, the study of elementary particles was one of the main areas of research. A successful theory of elementary particles called *quantum field theory*, was discovered by marrying electrodynamics, relativity and quantum mechanics. The quantum field theory, which is still incomplete, is presently the best known physical theory of nature. One of the major unsolved problems in quantum field theory is the quantum theory of gravity. Quantum field theory is also used to study properties of matter like metals, semiconductors etc.

Another connected area of research which has come up in recent times is nonlinear dynamics and chaos. We will explain this field using an example. Newton solved the two-body problem exactly—bound particles under gravitational potential have elliptical orbits. Way back in 1888, Poincaré showed that no such solution exists for a three-body problem. In fact, if a test particle starts from two nearby initial conditions, the future for these two initial conditions could be very different. This kind of sensitivity to initial conditions is called *chaos*. A similar feature has been found for weather systems—this property of the weather makes long-term prediction of the weather impossible. Chaos theory and related fields are active areas of physics research.

Presently, it appears that we know all the basic equations needed to describe nature. For example, physicists are quite confident that quantum chromodynamics describes the physics of the nucleus; Navier–Stokes equation describes turbulent flows; certain quantum mechanical equations describe the physics of superconductors and other complex materials. The major difficulty in physics, however, is how to solve these equations. Physicists have been trying very hard to solve them using analytical methods and powerful computers, yet complete success has eluded them so far. Maybe, we need to change the whole paradigm! Only the future can tell!

Exercises

1. Galileo rolled various balls through a groove in an inclined plane. He found that the time taken by balls of different materials and different radii was approximately the same. Are these results consistent with Newton's laws of motion? Would the result have changed if some balls were hollow and some were solid?

2. Describe the motion of Saturn in the sky, assuming that Saturn and Earth orbit around the Sun in circular orbits. The time periods of Earth's and Saturn's orbit are 365.25 and 10756.2 terrestrial days respectively. The distances of the Earth and Saturn from the Sun are 1.5×10^{11} and 14.3×10^{11} metres respectively. Does retrograde motion take place? If yes, for how long?

3. Do the above problem for other solar planets. The data can be obtained from Appendix G.

4. When will the next retrograde motion take place for Mars and Saturn? Current coordinates of the planets can be obtained from the Internet.

5. Historically how were the following measurements made? Suggest improvements using modern gadgets.

    (a) Radius of the Earth

    (b) Distance of the Moon from the Earth

    (c) Distance of the Sun from the Earth

    (d) Distance of planets from the Sun

(e) Mass of the Earth

(f) Mass of the Sun

Projects

1. Repeat Galileo's experiment to show that all objects fall with the same acceleration on the surface of the Earth (Exercise 1).

2. Repeat Newton's experiment to compute acceleration due to gravity.

3. Construct a Sun dial.

4. Locate the positions of planets in the sky and observe their motion.

5. Use a sextant to record the position of Mars.

6. Observe the motion of stars, solar planets, and the Moon.

7. How did Kepler compute the distance between Mars and the Earth?

2

Newton's Laws of Motion

Around three hundred years ago, Newton formulated his laws of motion. His theory is truly remarkable and revolutionary. For the first time, many terrestrial (on the Earth) and heavenly (in the sky) motion could be explained by his theory. Newton's laws of motion are quantitative, hence they have great predictive power. Scientists (including Newton himself) could compute planetary and lunar orbits, tides, motion of projectiles and comets quite precisely. Newton's laws explain ordinary day-to-day things very well, and they are the foundations of engineering science and machine design. They form the basis for statics, dynamics, elasticity, fluid mechanics, strength of material etc.

Newton's laws are a combination of empirical observations, and certain major abstractions and assumptions. Newton associated certain properties to space and time. He introduced abstractions of *point mass, force* etc. Galileo had already introduced the abstraction of an *isolated body* before Newton. These ideas seem very common now, but they are very profound and path-breaking. Some of these conceptions were subsequently found to be incorrect at very small scales, and at very high speeds. Einstein modified Newton's ideas of space and time. At very small scales, Newton's laws are replaced by Schrdinger's equation of quantum theory. Newton's laws are approximate laws that work very well for particle motion with reasonably small speeds on macroscopic scales.

Now let us state Newton's laws of motion along with their assumptions.

2.1 Statements of Newton's laws of motion

2.1.1 First law or the law of inertia

In school textbooks, the first law of motion or the law of inertia is usually stated as *"A body remains in its state of rest or in uniform linear motion as long as no external force acts on it to change that state"*.

But the above statement is not quite correct. For example, in an accelerated frame of reference, an isolated object accelerates even if no force is acting on it. So, a more refined statement is that the first law holds in an *inertial frame*. Now what is an inertial frame? It is a reference frame in which the first law holds. This is a circular argument. To break the cycle, the first law is stated as:

It is always possible to find a coordinate system with respect to which isolated bodies move uniformly. This coordinate system is called an inertial reference frame.

Hence, the first law is a statement about the existence of inertial frames. In the above definition, we have a key word, *isolated body.* An isolated body is an abstraction for objects on which no force is acting. An example of an isolated body is an object in free space. Note that the law of inertia was first discovered by Galileo.

2.1.2   Second law

The velocity of a body changes on an application of force. According to the second law, the acceleration of the body is proportional to the force applied to the body, and is in the direction of the applied force. The proportionality constant is called *inertial mass.* It turns out that the law of motion is valid in more general situations (e.g., variable mass) if it is expressed in terms of the rate of change of linear momentum rather than acceleration. The linear momentum, $\mathbf{p}$, of a body is defined as the product of its inertial mass (m) and velocity ($\mathbf{v}$), i.e., $\mathbf{p} = m\mathbf{v}$. Newton stated his second law of motion as

The change in the linear momentum of a body is proportional to the external force acting on the body, and is in the direction of the external force.

If $\mathbf{p}$ is the linear momentum of the body, and $\mathbf{F}$ is the force applied on the body, then according to the second law

$$\frac{d\mathbf{p}}{dt} = \mathbf{F}.$$

The computation of the linear momentum of an extended object involves the knowledge of some more concepts. To simplify, Newton assumed the body to be a *point particle.* No real object is a point particle. However if the size of a body is much smaller than the length scale of the system under consideration, then we can treat the body to be a point particle. For example, the size of the Earth is much smaller than the Earth–Sun distance, hence the Earth can be treated as a point particle while studying its orbital motion around the Sun.

Further, the velocity of a body is the rate of change of its position, or,

$$\mathbf{p} = m\mathbf{v} = m\frac{d\mathbf{r}}{dt}$$

where $\mathbf{r}$ is the position vector of the particle. Hence, Newton's second law $d\mathbf{p}/dt = \mathbf{F}$ yields

$$m\frac{d^2\mathbf{r}}{dt^2} = \mathbf{F}. \tag{2.1}$$

The quantity $d^2\mathbf{r}/dt^2 = \ddot{\mathbf{r}}$ is the *acceleration* of the particle. Single dot and double dots are the short-hand notation for first and second derivatives respectively. The above equation, called the *equation of motion,* is the most important equation of classical physics. It is the basis for statics, dynamics, fluid mechanics, elasticity etc. Most of the discussion in this book revolves around this equation.

The inertial mass m of an object is a measure of inertia of that object, and it can be calculated by computing the ratio of the force applied and the acceleration of the particle ($m = \mathbf{F}/\ddot{\mathbf{r}}$). It can also be deduced that an object with a larger inertial mass will require a larger force to achieve the same amount of acceleration.

In the later part of the book, we will discuss the *gravitational mass* or the *gravitational charge* that couples with gravity. It has been found that gravitational mass and inertial mass are equal (Einstein's *principle of equivalence*). However we should keep in mind that the inertial mass and gravitational mass of a body are two different quantities that happen to be equal in nature. These ideas will be discussed in detail later in the book.

Equation (2.1) can be rewritten in terms of the Cartesian components as

$$m\frac{d^2x}{dt^2} = F_x, \tag{2.2}$$

$$m\frac{d^2y}{dt^2} = F_y, \tag{2.3}$$

$$m\frac{d^2z}{dt^2} = F_z. \tag{2.4}$$

These are three differential equations. Note that a differential equation is an equation involving an unknown function and its derivative. All the forces acting on a particle found till date are functions of the position and/or the velocity of the particle. Therefore, the equations of motion of a point particle have at the most second derivatives of the particle's position. Hence they are second-order differential equations. The coordinates of the particle can be found by solving these differential equations. The solution of the equations however requires knowledge of

1. The forces acting on the body at all times. The forces are determined by force laws that will be discussed in Chapter 3.

2. The initial conditions of the particle. Since Eqs. (2.2)–(2.4) are of the second order, we need to know the initial position and the initial velocity of the particle: $[x(0), v_x(0)]$ for Eq. (2.2), $[y(0), v_y(0)]$ for Eq. (2.3), and $[z(0), v_z(0)]$ for Eq. (2.4).

Given the above two ingredients, we can solve the equations of motion and in principle determine the future of the particle (its position and velocity). In the subsequent chapters we will solve the equations of motion for simple problems in one, two, and three dimensions.

Newton's laws of motion suggest that the future of a mechanical system is completely determined once its initial conditions are provided. Early success of Newton's laws in predicting planetary motion, tides, eclipses, etc. led scientists to believe in a completely deterministic universe. However, it was later found that the above properties do not hold for quantum systems and many nonlinear systems. See Appendix A for some discussions on *determinism*.

After discussing the second law of motion, we move on now to the third law.

2.1.3   Third law

Newton's third law connects action and reaction between two bodies. This law reads as

If a given body A exerts a force on another body B, then body B also exerts a force on body A with equal magnitude but opposite in direction.

Clearly the net force acting on the combined system of A and B is zero. The reader is familiar with various examples illustrating the third law. We will not discuss them here. Note however that the third law of motion in the form stated here does not always hold. For example, if the Sun is moved away abruptly from its present position, the gravitational forces on the Earth and the Sun will keep pointing to the original directions until the signal reaches the Earth that the Sun has been displaced. After the signal reaches the Earth, the forces will point towards the new position. Since the signal propagates with a speed equal to that of light, which is finite, the action and reaction are not opposite to each other in the intermediate time. Another example where Newton's third law is violated is when two current-carrying wires are placed at right angles to each other. As shown in Fig. 2.1, the action and reaction (magnetic forces) are perpendicular to each other.

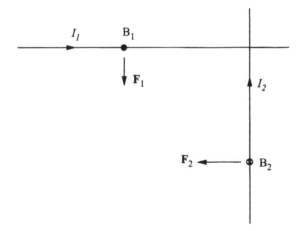

Figure 2.1   *The forces acting on two current carrying wires. The forces $\mathbf{F}_1$ and $\mathbf{F}_2$ are not opposite to each other.*

We will later show that Newton's third law implies conservation of linear momentum for the combined system of bodies A and B. In the above two examples, Newton's third law is violated. Note however that the conservation of linear momentum is valid for these systems if we include the linear momentum of gravitational and electromagnetic fields. The important point to note is that the Newton's third law may not hold for some systems, but the conservation of total linear momentum (including that of fields) is always valid. In quantum mechanics, one does not talk about forces or Newton's third law; yet the conservation of linear momentum is valid. Conservation laws are sacred in physics. The linear momentum of fields will not be discussed in this course; you will learn some of these ideas in your electrodynamics course.

2.1.4   Absolute time

In dynamics, the motion of a particle is quantified by its position as a function of time. Newton thought that *time is absolute*, i.e., it is the same for all observers. According to Newton[1],

Absolute, true, and mathematical time, of itself, and from its own nature flows equably without regard to anything external, and by another name is called duration: relative, apparent, and common time, is some sensible and external (whether accurate or unequable) measure of duration by the means of motion, which is commonly used instead of true time; such as an hour, a day, a month, a year.

The implication of the above postulate is that if the clocks of two observers (in the same or different inertial frames) are synchronised at some point, they will remain synchronised forever. Another important consequence of the above assumption is the time ordering of events is the same for all observers. That is, if event A occurs before event B for one observer, then event A occurs before event B for all observers. Also, if an observer reports that an event C causes another event D, then this cause–effect sequence is maintained for all the observers.

2.1.5   Absolute space

Newton pondered quite deeply about space in which motion takes place. He advanced the idea of absolute space. According to Newton,

Absolute space, in its own nature, without regard to anything external, remains always similar and immovable. Relative space is some movable dimension or measure of the absolute spaces; which our senses determine by its position to bodies; and which is vulgarly taken for immovable space; such is the dimension of a subterraneaneous, an aereal, or celestial space, determined by its position in respect of the earth. Absolute and relative space, are the same in figure and magnitude; but they do not remain always numerically the same. For if the Earth, for instance, moves, a space of our air, which relatively and in respect of the Earth remains always the same, will at one time be one part of the absolute space into which the air passes; at another time it will be another part of the same, and so, absolutely understood, it will be perpetually mutable.

The implication of the above postulate is that the physical distance between two events is the same for all observers. Let us imagine a fixed inertial frame I, and another inertial frame I' that is moving with a velocity V relative to I as shown in Fig. 2.2. The x-axis of the two inertial frames are coincident, while the y- and z-axes are parallel. Let us assume that the origin of both the inertial frames coincide at $t = 0$. If an event takes place at point

---

[1]http://members.tripod.com/~gravitee/

A whose coordinates in I and I' are (x, y, z) and (x', y', z') respectively, then according to Newton's postulate of absolute space

$$x = x' + Vt'$$

$$y = y'$$

$$z = z'$$

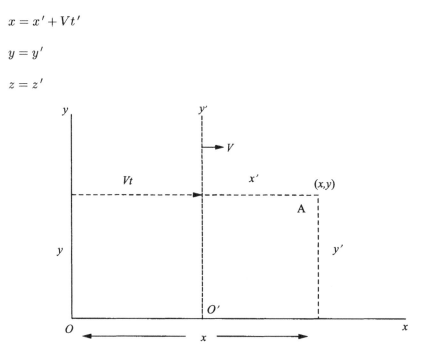

Figure 2.2   *Two inertial reference frames I and I' record an event at point A. The coordinates in the reference frames are (x, y, z) and (x', y', z') respectively.*

The postulate of absolute time yields

$$t = t'.$$

Using these four formulas, we can convert (x, y, z, t) to (x', y', z', t'), and vice versa. This transformation is called the *Galilean transformation*.

We can represent events occurring on the x-axis by a point in (x, t) or (x', t') coordinate systems as shown in Fig. 2.3. Note that x' and t' axes are obtained by setting $t' = 0$ and $x' = 0$ respectively. An event occurring at position x at time t is denoted by point P whose coordinates are (x, t). The same point is (x', t') in the other coordinate system with $t' = t$ and $x = x' + Vt'$.

Note however that the Galilean transformation is valid only when the velocities of particles and reference frames are much smaller than the speed of light. For speeds close to the speed of light, Einstein found that the idea of absolute space and absolute time are no more valid, and he provided a new transformation rule called the *Lorentz transformation*.

Typically, we find that several forces act on an object rather than a single force. Newton provided a procedure to apply the equations of motion under these situations.

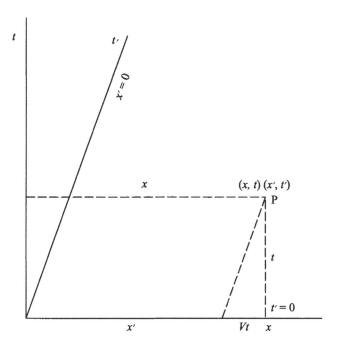

Figure 2.3   *Galilean transformation: the coordinate system (x', t') is moving with respect to the laboratory frame of reference with velocity V.*

2.2   Superposition principle

The superposition principle is one of the major assumptions made by Newton. Suppose the forces $\mathbf{F}_1$ and $\mathbf{F}_2$ induce accelerations $\mathbf{a}_1$ and $\mathbf{a}_2$ respectively on a body, that is,

$$m\mathbf{a}_1 = \mathbf{F}_1,$$

$$m\mathbf{a}_2 = \mathbf{F}_2.$$

We want to study the effect on the body when both the forces $\mathbf{F}_1$ and $\mathbf{F}_2$ act on the same body. By adding the above two equations we obtain

$$m(\mathbf{a}_1 + \mathbf{a}_2) = \mathbf{F}_1 + \mathbf{F}_2.$$

The resultant force $\mathbf{F}_1 + \mathbf{F}_2$ is the vector addition of the individual forces, and the resultant acceleration is $\mathbf{a}_1 + \mathbf{a}_2$. This principle is called the *Superposition principle*. It is valid because the relationship between the force and acceleration is linear. Note that the superposition principle will breakdown if $a \propto F^2$.

The superposition principle is very useful in physics. We often encounter situations when a body is under the influence of many forces. In these situations, we compute accelerations of the body due to individual forces, and then add all of them. Recall how you compute electric field due to many charges using the superposition principle of electrodynamics.

A word of caution: suppose in time t, forces $\mathbf{F}_1$ and $\mathbf{F}_2$ induce displacements $\mathbf{r}_1$ and $\mathbf{r}_2$ respectively, then the net force $\mathbf{F}_1 + \mathbf{F}_2$ does not induce a displacement $\mathbf{r}_1 + \mathbf{r}_2$. Why not?

Newton's laws were very successful in calculating the motion of planets, tides etc. Some ideas of Newton, however, were incorrect. Mach was one of the first critics of Newton's ideas. We will briefly discuss Mach's objections in the next section.

2.3   Newton vs. Mach

Let us explore some more consequences of the postulate of absolute space. If a particle is at rest in one inertial frame, then it will move with constant velocity in another reference frame that is moving with constant velocity relative to the former inertial frame. Both these reference frames are equivalent. Hence, if there is an *absolute space* we cannot determine whether a particle is at rest with respect to the absolute space or moving with respect to the absolute space. However one can determine whether a particle is accelerating with respect to absolute space by identifying the pseudo forces. So Newton set out to measure acceleration with respect to the *absolute space* using the pseudo forces. Newton devised the famous bucket experiment for this purpose.

Newton hung a bucket of water from a roof, and rotated it about the vertical axis. He found that the surface of water becomes a parabola. Newton argued that the parabolic shape of the water surface in the rotating bucket is due to the motion of the bucket relative to *absolute space* (derivation in Chapter 10). He then asked what will happen if the bucket did not rotate, but the whole universe (except the bucket) rotated in the reverse direction. The later configuration is kinematically equivalent to the former experiment. Newton conjectured that the water surface will remain flat because the bucket is not rotating with respect to the absolute space. This appears logical if the assumption of absolute space is correct.

In 1880, Mach countered Newton's idea of absolute space. He argued that in science we must avoid unverifiable assumptions. The idea of rotating the whole universe except the bucket is unverifiable. Mach argued that the centrifugal force on the water in Newton's bucket arises due to the acceleration-dependent interaction of water with the masses of the universe, not because of its motion with respect to absolute space. Hence inertia (mass) of a particle will change if the mass distribution of the universe changed, or a big mass comes near the bucket. These are difficult issues, and they have not been fully resolved yet.

The above discussions show that Newton's laws of motion involve several definitions (inertial frames, mass etc.), certain assumptions about space and time, and mathematical abstractions. We have been reading these laws since school days, and they appear to be simple. However they are far from obvious (recall that Aristotle's thinking was completely contrary to Newton's).

2.4   Newton's *Principia*

Newton published his discovery in the field of mechanics and other branches of physics in *Philosophiae Naturalis Principia Mathematica* (Latin) or *Mathematical Principles of Natural Philosophy*, often called the *Principia* or the *Principia Mathematica* in brief. These volumes contain the laws of motion, the discussion on space and time, the law of gravitation, and many other important aspects of dynamics (tides, motion in fluids, frictional forces, etc.). This book is considered to be one of the greatest works in science and philosophy. The reader can browse through some portions of the *Principia* in Chandrashekhar (2003), Wikipedia, and Stanford Encyclopedia of Philosophy, and http://members.tripod.com/gravitee/.

Newton's laws of motion provide a quantitative framework to predict the future of mechanical systems. Note however that to obtain the solution we require the initial conditions of the system and the forces acting on it. In this book we will study various aspects of Newton's formalism. Specifically, in the next chapter we will discuss various forces found in nature.

Exercises

1. If we set the force acting on a particle to zero in Newton's second law, we obtain the acceleration of the object to be zero. Does it imply that Newton's first law of motion is included in the second law?

2. To construct your own theory of motion, would you like to add any new law, or delete any of Newton's laws? Confine yourself to small velocities (non-relativistic regime).

3. Identify which among the following are inertial reference frames.

    (a) A spaceship moving with constant velocity in free space.

    (b) An accelerating spaceship

    (c) Earth's surface

    (d) Solar system

4. If a particle moves with velocity **v** in an inertial frame, what is its velocity in another reference frame that is moving with velocity **V** relative to the first reference frame. Derive your result using Galilean transformation.

5. In Newton's theory, with what speed does the signal of gravitational interactions propagate? Reason out your answer.

6. List all the assumptions of Newton's laws of motion. Do you know some situations where they fail?

3

Forces

According to Newton, the equation of motion of a point particle of mass m that is experiencing force $\mathbf{F}$ is

$$m\ddot{\mathbf{x}} = \mathbf{F}.$$

The determination of the solution $\mathbf{x}(t)$ requires knowledge of the forces acting on the particle. Hence, the study of forces become imperative for solving the equation of motion of a particle.

In physics we label some forces as fundamental, and others as derived. Gravity, electromagnetism, weak nuclear, and strong nuclear forces are called fundamental forces. All other forces in nature can be derived from these forces. For example, van der Waal's force can be derived from electrostatic forces. Other examples of derived forces are surface tension, friction, normal forces etc. In this chapter we will give a brief review of some of these forces.

3.1 Fundamental forces of nature

1. **Gravitational force**
   Newton discovered the *universal law of gravitation*, according to which the force on a test particle due to an object situated at a distance r from the test particle is

$$\mathbf{F} = -\frac{GM_Gm_G}{r^2}\hat{\mathbf{r}} = -m_G g\hat{\mathbf{r}}, \tag{3.1}$$

   where m_G and M_G are the *gravitational charges* of the test particle and the object respectively, G is Newton's constant of gravitation, and $g = GM_G/r^2$. If the inertial mass of the test particle is m, then the acceleration of the test particle is

$$a = \frac{GM_Gm_G}{r^2m} = \frac{m_G}{m}g. \tag{3.2}$$

   Experiments show that bodies with different inertial masses fall to the Earth with the same acceleration whose numerical value is 9.8 m/s². That is,

$$a = \frac{m_G}{m} g = 9.8 \ \text{m/s}^2.$$

The above calculation implies that $m_G/m = $ constant. We take $m_G = m$ or $(a = g)$ by choosing $G = 6.67 \times 10^{-11} \ \text{Nm}^2/\text{kg}^2$. However if we choose the ratio of the gravitational charge and the inertial mass (m_G/m) to be equal to k, then G in this system would be $G = 6.67 \times 10^{-11}/k^2 \ \text{Nm}^2/\text{kg}^2$. The equality of gravitational charge and inertial mass is the reason why all masses fall to the Earth with the same acceleration (g). Often gravitational charge is also referred to as *gravitational mass*. Galileo performed the first experiment to demonstrate the above equality by rolling balls of different masses and different material down an inclined plane. A more accurate experiment was performed by Eötvös in 1890 using a torsional pendulum. Modern experiments show that $m_G = m$ to within a precision of 1 in 10^{10}. Einstein used this idea to discover a new theory of gravity.

2. **Electromagnetic force**
Two charges either attract or repel each other depending on their relative signs. This force is called electrostatic force. The force law for electrostatic interactions was first given by Coulomb. Two current-carrying wires interact with each other through magnetic force. Later it was discovered that a changing magnetic field produces an electric field (Faraday), and a changing electric field produces a magnetic field (Maxwell). Maxwell discovered a comprehensive set of equations that include both electric and magnetic forces. Note that the magnetic forces between bar magnets can be understood in terms of microscopic currents inside the magnets. These topics will be covered in your electromagnetics course.

3. **Strong nuclear force**
A nucleus contains protons and neutrons. Naturally, the electric repulsion between protons can tear the nucleus apart due to the short distances between them. It was found that protons and protons, protons and neutrons, and neutrons and neutrons attract each other with a much stronger force called the *strong nuclear force*. The origin of this force is rather complex; here we will only provide a very sketchy picture.

According to the modern theory of nuclii, protons and neutrons are not fundamental particles. They are made of quarks. The quarks carry flavour (whose names are *up, down, charm, strange, up, bottom*) and colour charge (whose names are *red, green, blue*). The quarks are held together by colour forces between the colour charges. The colour force is transmitted or mediated by the exchange particle called *gluon*. The nuclear force between the protons and neutrons inside the nucleus is the resultant of the forces among the constituent quarks. Strangely, the force between two quarks increases with distance. There are many interesting issues in this field which are beyond the scope of this book. You could refer to Wikipedia for more details.

4. **Weak nuclear force**
The weak nuclear force is active inside the nucleus, but it is many orders of magnitude weaker than the strong nuclear force. This force is responsible for converting a neutron to a proton in a *beta decay*. An electron and a neutrino is emitted in the process. Even

though the weak nuclear force is also between quarks, their nature is very different. One major difference between the strong nuclear force and the weak nuclear force is the violation of mirror symmetry in the weak nuclear force that will be discussed in Chapter 7.

An important point to keep in mind is that in modern physics, interactions are typically expressed in terms of potentials rather than in terms of forces. For example, Schrödinger equation contains potential, not force. We will discuss potentials in Chapter 11.

In mechanics the force on a particle is induced either through a force field or through contact. Gravity and electromagnetic forces are transmitted through force fields, while frictional force, normal force, tension etc. are transmitted through contact.

3.2   Some examples of derived forces

In our daily lives we mostly come across gravity and derived forces of electromagnetic origin. The muscular forces in our body originate from chemical reactions that arise due to the electromagnetic forces between atoms and molecules. In fact all the forces inside an animal body are essentially electromagnetic. Surface tension and capillary forces are due to the electromagnetic forces between molecules. When two surfaces are pressed against each other, molecules at the surfaces try to repel each other by electromagnetic and quantum interactions; this is the *normal force*. The list is endless.

We encounter frictional force everywhere. Its origin is again electromagnetic. Frictional force plays a major role in mechanics. Therefore, we will discuss this force in some detail.

3.3   Frictional forces

Frictional force arises when a solid body moves on a solid surface or in a fluid. We call the former dry friction, and the latter wet friction. Some of the properties of these forces are discussed below.

3.3.1   Dry friction

When a solid body A slides on another solid body B, both the bodies experience a force called *frictional forces* that opposes the relative motion of the bodies. This force is not well understood. The irregularities on the surfaces get locked with each other; the lockings break when the bodies move relative to each other. The frictional force between the sliding blocks supposedly originate due to these interactions.

Many experiments have been done to understand frictional force. Empirical laws describing the dry frictional force are

1. The frictional force on a surface is proportional to the normal force acting on that surface. The proportionality constant is called the coefficient of friction. Curiously, for a given normal force, the frictional force is independent of the surface area of contact.

2. When we apply a force on a block resting on a horizontal surface, the frictional force opposes the movement of the block. The maximum value the frictional force can take is $\mu_s N$, where μ_s is a constant called the *static frictional coefficient*, and N is the normal force on the block. If the applied force is less than this value, the block does not move. When the applied force is more than $\mu_s N$, the block starts moving and the frictional force is $\mu_k N$, where μ_k is another constant called the *kinetic frictional coefficient*. Experiments show that typically $\mu_k < \mu_s$.

3. A rougher surface experiences larger frictional force than a smoother one.

4. Lubricants such as oil and grease decrease the frictional force between two surfaces.

5. When we slide a block of pure substance on a surface of the same substance, such as copper on copper, the frictional force is typically due to various oxides and impurities sticking on the surface. But if we clean and polish the surfaces very carefully, contrary to expectation, the frictional force increases. This is because copper atoms try to fuse together and form a crystal. This phenomena is called *cold weld*.

We will now study typical examples of frictional force. Observe that frictional force opposes the relative motion between contact surfaces. However the frictional force is not necessarily in the direction opposite to the direction of net motion of the body; the frictional force only opposes the relative motion of the body on the surface (See Fig. 3.1(b), (g)). Some illustrative examples are

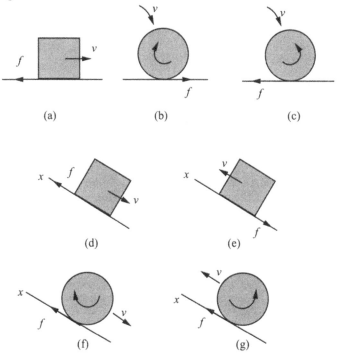

Figure 3.1   *Examples of frictional forces*

1. A block is moving on a horizontal surface as shown in Fig. 3.1(a). The direction of frictional force is to the left, opposing the motion of the block.

2. A ball, which is spinning clockwise, is thrown to the ground as shown in Fig 3.1(b). The frictional force will oppose the relative slip, and it will be in the forward direction. Hence the ball will move faster in the forward direction after the impact. This is the principle behind the top spin used by cricketers and tennis players. If the ball was spinning in the counter-clockwise direction, then the frictional force will be in the backward direction (Fig. 3.1(c)), and the ball will slow down after impact. This is the underspin used in tennis. Note that in case (b) the frictional force is in the direction of motion of the ball.

3. A block is sliding down an incline as shown in Fig. 3.1(d). To oppose the relative motion, the frictional force is along $\hat{x}$. If the block was moving upward, the frictional force will point downward $(-\hat{x})$ (Fig. 3.1(e)).

4. A ball is rolling down an incline. The ball has a tendency to slip downward at the contact, hence the frictional force points upward along $\hat{x}$ (Fig. 3.1(f)). If the ball rolls upward, then too the ball has a tendency to slip downward at the contact making the frictional force again point upward (Fig. 3.1(g)). Hence, the frictional force points upward for both cases. Contrast this with the sliding block on an incline (item 3) where the frictional force is in the direction opposite to the direction of motion.

5. When we walk, we push the ground backward. Consequently the frictional force is in the forward direction, and it helps us walk.

We will now work out a problem involving dry friction.

EXAMPLE 3.1   A block of mass M_1 rests on a block of mass M_2 that lies on a frictionless table. The coefficient of friction between the blocks is μ. What is the maximum horizontal force that can be applied to the blocks for them to accelerate without slipping on one another if the force is applied to (a) the block of mass M_1 and (b) the block of mass M_2? Describe the motion of the blocks as a function of magnitude of horizontal force.

SOLUTION   **Case (a): The force is applied to the upper block**
The free-body diagrams of the blocks for this configuration are shown in Fig. 3.2(a). There is no motion in the vertical direction ($a_y = 0$). Therefore, Newton's laws yield

$$N_1 = M_1 g$$

$$N_2 - N_1 = M_2 g.$$

For frictional force f below $f_{\text{max}} = \mu N_1$, both the blocks move together with acceleration a:

$$M_2 a = f,$$

$$M_1 a = F - f.$$

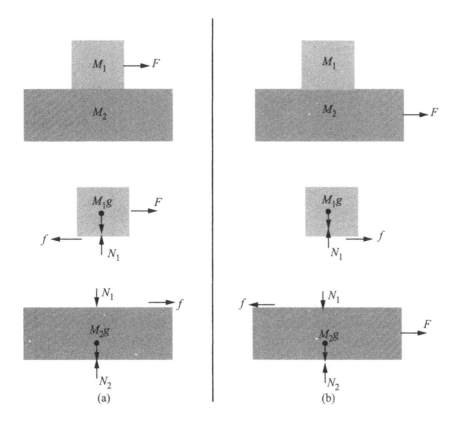

Figure 3.2   *Example 3.1: Free-body diagram of the blocks when (a) the external force acts on the top block, (b) external force acts on the bottom block.*

Clearly, $a = F/(M_1+M_2)$. However, this formula is valid only till $f = f_{max} = \mu N_1 = \mu M_1 g$, after which block M_2 will move with acceleration $a_2 = f_{max}/M_2 = \mu g M_1/M_2$, while the block M_1 will move ahead with $a_1 = (F - f_{max})/M_1$ until it gets off the lower block. A plot of acceleration vs. the applied force is shown in Fig. 3.3(a). The maximum common acceleration for both the blocks is

$$a_{max} = \frac{f_{max}}{M_2} = \frac{M_1}{M_2}\mu g,$$

and the maximum force for the common acceleration is

$$F_{max} = (M_1 + M_2)a_{max} = \frac{M_1}{M_2}(M_1 + M_2)\mu g.$$

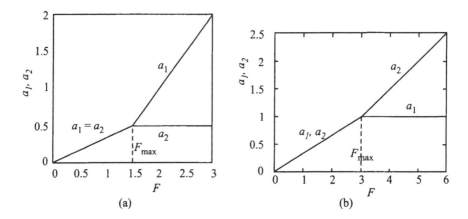

Figure 3.3   *Plot of accelerations of blocks M_1 and M_2 as a function of F (a) when the upper block is pulled with a force F; (b) when the lower block is pulled with a force F. We have used parameter values $M_1 = 1$ kg, $M_2 = 2$ kg, and $\mu = 0.1$. We take $g = 10 \, m/s^2$*

Case (b)   The force is applied to the lower block

The free-body diagrams of the blocks are shown in Fig. 3.2(b). The forces in the vertical direction again balance each other. For $f < f_{max} = \mu M_1 g$, both the blocks move together with a common acceleration $a = F/(M_1 + M_2)$. This formula is valid only till $f = f_{max} = \mu N_1 = \mu M_1 g$, after which block M_1 moves with acceleration $a_1 = f_{max}/M_1 = \mu g$, while block M_2 moves ahead with $a_2 = (F - f_{max})/M_2$.

Note that here $a_2 > a_1$, consequently the top block falls behind the bottom block. For this case, the maximum common acceleration for both the blocks is μg, and the maximum applied force during the maximum common acceleration is $F_{max} = (M_1 + M_2)\mu g$. See Fig. 3.3(b) for an illustration.

After our discussion on dry friction we will have a glance at the frictional force active during earthquakes. Since earthquakes affect our lives tremendously, the science of earthquakes is one of the most important areas of research in physics and engineering. It is a very difficult subject, and there are many unsolved questions. In our discussion, we will only give a qualitative picture of the frictional forces occurring during an earthquake.

3.3.2   Plate tectonics and earthquakes

The Earth's interior is divided into various parts: the inner core (0–1230 km), outer core (1230–3500 km), mantle (3500–6340 km), and crust (6340–6378 km). See Fig. 3.4 for an illustration. The topmost layer of the Earth consists of several plates (called tectonic plates) that are locked together. The thickness of the plates vary with an average of around 100 km. The mantle consists of molten metal that are convected due to temperature difference. These convective currents inside the Earth pushes these plates which generates major stresses at the boundaries of the plates. The plates withstand these pressures and are locked together for a long time, but sometimes they suddenly slip. The slippage causes earthquakes. A tremendous amount of stress and energy is released during this process that is carried away by mechanical waves on the Earth. The frictional force between the tectonic plates is poorly understood.

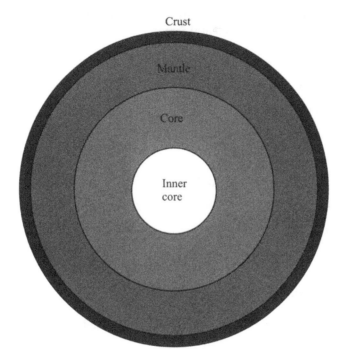

Crust

Mantle

Core

Inner core

Figure 3.4   *Structure of the Earth's interior*

Scientists are trying to understand the physics of earthquakes, and are hoping to make predictions regarding them. One of the major problems is modelling of the frictional forces between the plates. Note that the dry frictional force between two blocks has an origin in atomic forces, but the frictional force between tectonic plates has a macroscopic origin.

3.3.3   Wet friction

In Appendix B we discuss the frictional force acting on a ball of radius r that is moving in a fluid with velocity v. The viscosity of the fluid is η, and its density is ρ. We derive a phenomenological formula using experimental findings and dimensional analysis. The frictional force $F \propto \eta v r$ for small velocities (strictly speaking for small Reynolds number $\rho v r / \eta$); this is the viscous regime. For large velocities or Reynolds number, the frictional force is dominated by turbulence, and its magnitude is $F \propto \rho v^2 r$. These formulas are very useful in fluid dynamics and aerodynamics. Before closing this chapter we make a passing remark that turbulence is still an unsolved problem. Consequently, for large velocities, we do not yet have a theoretical model to compute the frictional force from first principles.

Exercises

1. Identify fundamental and derived forces in the following list. For derived forces, indicate the constituent fundamental forces.

   (a) Magnetic force between two magnets

   (b) Surface tension

   (c) Tension force in a string

   (d) Muscular force (for example when you hold a book)

   (e) van der Wall's force

   (f) The force due to pressure gradient

   (g) Viscous force

   (h) The force due to wind

2. Make a list of the derived forces around you.

3. Estimate the magnitude of the following forces

   (a) The electromagnetic force between the proton and an electron in the hydrogen atom.

   (b) The electromagnetic force between two protons inside a helium nucleus.

   (c) The force due to the air pressure on your body.

   (d) The force acting on a typical moving train.

   (e) The gravitational pull between the Earth and the Sun, and between the Earth and the moon.

   (f) The gravitational force with which the centre of the galaxy pulls our Solar System.

   (g) The reaction force at the foundation of a building due to its weight.

4. A block of mass $m = 0.2$ kg is placed on another block of mass $M = 2$ kg, and the system is placed on a horizontal table. There is no friction between the table and the lower block (M), whereas the coefficient of friction between the lower block and the upper block is 0.25. A horizontal force F is applied to the upper block. Find the accelerations of both the blocks, and the frictional force between them when (a) $F = 0.5$ N (b) $F = 2.0$ N. Take $g = 10$ m/sec^2.

5. A rope is wrapped around a fixed drum for n turns as shown in Fig. 3.5. The coefficient of friction between the rope and the drum is μ. Compute the ratio of tension at both ends of the rope when the rope covers an angle of $3\pi/2$ over the drum.

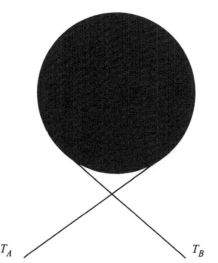

T_A T_B

Figure 3.5   *Exercise 5*

6. A beaker filled with water is placed on a table. The combined mass of the beaker and water is 5 kgs. Compute the reaction force exerted by the table on the beaker when

    (a) A one-kg iron block is completely immersed in water, and is touching the floor of the beaker.

    (b) A one-kg wooden block is floating in water.

    (c) A one-kg iron block is completely immersed in water, but is supported by a string. The block is not touching the floor of the beaker.

7. Estimate the frictional force acting on a jet plane while taking off and when it is in mid air.

8. Estimate the frictional force acting on a cyclist.

9. Water in an inverted glass can be supported by a wet paper sticking at the mouth of the glass. Why does it work? What could be the maximum size of the glass in this experiment? Do the experiment.

Projects

1. Perform an experiment to compute the coefficient of friction for glass–glass and wood–glass surface. Study the effect of lubricants.

4

Kinematics Vs. Dynamics

Newton's laws of motion provides us a prescription to solve for the motion of a mechanical system given its initial conditions and the forces acting on it. This method of studying the motion of material objects is called *dynamics*. However there are other ways to study mechanical systems. One such method is *kinematics in which we describe the motion of objects by prescribing the acceleration or velocity without worrying about the forces that produce the accelerations.*

Although we will focus on dynamics in this book, in the following we will briefly discuss kinematics. We will illustrate some ideas on kinematics using several examples.

4.1 Examples of kinematic motion

- For particles moving with constant acceleration a, we can derive an expressions for its position and velocity by integrating $dv/dt = a$ and $v = dx/dt$:

$$x = x_0 + v_0 t + \frac{1}{2}at^2,$$

$$v = v_0 + at,$$

  In this derivation, we did not have to think about the forces acting on the particle.

- In a uniform circular motion, the magnitudes of the velocity (v) and the acceleration (a) of the particle are constant with a relationship $a = v^2/R$, where R is the radius of the circle. The direction of velocity is tangential to the circle, and the direction of acceleration is radially inward.

- Recall that we discussed the motion of planets in Chapter 1. The ancient theories (by Eudoxus, Aristotle, and Ptolemy) as well as the medieval theories of Copernicus, Brahe and Kepler were all kinematical in nature. Thinkers before Newton did not know what caused the planets to move.

- For a ball rolling without slipping, we can express the velocity of the center of the ball (V_C) in terms of angular frequency (ω) and radius R:

$$V_C = \omega R.$$

It follows from the *kinematic constraint* that the velocity of the bottom-most point of the ball is zero in the laboratory frame of reference.

- Two wheels touch each other and are rotating without slipping as shown in Fig. 4.1. The radius of the wheels are r_1 and r_2 respectively. Their angular velocities are ω_1 and ω_2 respectively. Since the contact points of both the wheels have the same velocity,

$$\omega_1 r_1 = -\omega_2 r_2.$$

This is a constraint equation. Similar equations are used in the design of gears.

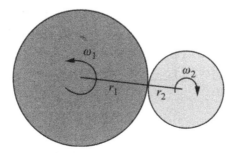

Figure 4.1 *Two wheels are touching each other and they rotate without slipping.*

The above examples do not require any application of Newton's laws. They are based on simple ideas of motion. More complex engineering problems involve complex calculations. One such problem is how to transfer power from the piston of an engine to the wheels of a vehicle. I would like to point out that though the forces behind a motion was not understood, pretty complex machines were designed before Newton discovered the laws of motion. Also complex planetary motion were studied, essentially using kinematics.

In this book, we will not deal with complex kinematic problems. Several simple kinematic examples are given below.

EXAMPLE 4.1   Plot the trajectories of particles whose displacements are given by:

1. $\mathbf{r}(t) = \mathbf{r}_0 + \mathbf{u}t$, where $\mathbf{u} = \hat{\mathbf{x}} + 2\hat{\mathbf{y}}$ and $\mathbf{r}_0 = \hat{\mathbf{x}} + 2\hat{\mathbf{y}}$ are constant vectors.

2. $\mathbf{r}(t) = \exp{(\alpha t)}\hat{\mathbf{x}} + \exp{(-\alpha t)}\hat{\mathbf{y}}$, where α is a constant.

3. $\mathbf{r}(t) = \cos{(\omega t)}\hat{\mathbf{x}} + \sin{(\omega t)}\hat{\mathbf{y}} + t\hat{\mathbf{z}}$ with $\omega = 1$.

SOLUTION   We plot the trajectories of the particle in Fig. 4.2.

1. $\mathbf{r}(t) = \mathbf{r}_0 + \mathbf{u}t$: The particle moves in a straight line in the direction of $\mathbf{u}$ as shown in Fig. 4.2(a). The velocity of the particle is constant.

2. $x = \exp{(\alpha t)}$ and $y = \exp{(-\alpha t)}$. Since $xy = 1 = $ const, the trajectory of the particle is a hyperbola as shown in Fig. 4.2(b). The directions of motion is to the right since x increase with time.

3. $x = \cos \omega t$; $y = \sin \omega t$; $z = t$. It follows that

$$x^2 + y^2 = 1,$$

hence the projection of the trajectory on the xy plane is a circle. Along z, the particle moves with uniform velocity. The direction of motion is counter-clockwise as seen by the evolution of x. The motion is depicted in Fig. 4.2(c).

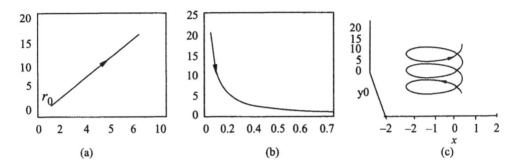

(a)   (b)   (c)

Figure 4.2   *Motion of a particle whose coordinate is (a)* $\mathbf{r}(t) = \mathbf{r_0} + \mathbf{u}t$; *(b)* $\mathbf{r}(t) = \exp(-\alpha t)\hat{\mathbf{x}} + \exp(\alpha t)\hat{\mathbf{y}}$ *with* $\alpha = 1$; *(c)* $\mathbf{r}(t) = \cos(\omega t)\hat{\mathbf{x}} + \sin(\omega t)\hat{\mathbf{y}} + t\hat{\mathbf{z}}$ *with* $\omega = 1$.

4.2 Dynamics with kinematic constraints

In engineering applications we encounter many examples of constrained motion. These problems are solved by combining Newton's laws with kinematic constraints. We illustrate these ideas using simple examples.

EXAMPLE 4.2   **Atwood machine–Two masses**
m_1 and m_2 are hanging on two sides of a massless frictionless pulley through a massless inextensible rope. Compute the accelerations of the masses.

SOLUTION   The forces acting on both the blocks are shown in Fig. 4.3. Let $x_{1,2}$ be the coordinates of the blocks $m_{1,2}$ according to the axis shown in the figure. The equations of motions of the blocks are

$$m_1 \frac{d^2 x_1}{dt^2} = T_1 - m_1 g$$

$$m_2 \frac{d^2 x_2}{dt^2} = T_2 - m_2 g.$$

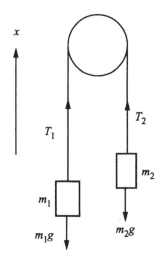

Figure 4.3   *Atwood machine*

Since the string is inextensible, $x_1 + x_2$ = constant. This is a kinematical constraint. By taking double derivative of the kinematic constraint we obtain,

$$a_1 = -a_2.$$

Also, the fact that the pulley and the string are massless yields another kinematic constraint,

$$T_1 = T_2.$$

Hence we have only two independent unknown variables: a_1 and T_1. The solutions of these variables are

$$a_1 = \frac{m_2 - m_1}{m_2 + m_1} g,$$

$$T = \frac{2m_1 m_2}{m_1 + m_2} g.$$

Let us check the limiting cases: (1) when $m_1 = m_2$, $a = 0$, which is reasonable; (2) when $m_2 \gg m_1$, $a_1 \approx g$.

EXAMPLE 4.3   A block of mass m slides without friction on a 45° wedge that itself is pushed along a table with constant acceleration A. Find the acceleration of the block.

SOLUTION   This is an example of a constrained motion. A block of mass m is constrained to move along the wedge as shown in Fig. 4.4. Therefore, in the moving frame, $dx' = -dy'$, that is, the distance moved along x is the same as the distance moved along $-y$ in the moving frame. Consequently $a'_x = -a'_y$. In the laboratory frame, the acceleration will be

$$\mathbf{a} = (A + a'_x)\hat{\mathbf{x}} - a'_x\hat{\mathbf{y}}$$

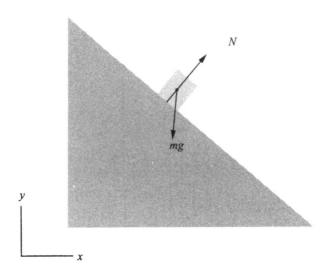

Figure 4.4   *A block of m sliding along a wedge that itself is being pushed with an acceleration A*

To solve for a'_x, we need some more inputs. Here we use Newton's laws (dynamics). The equation of motion of the block is

$$ma_x = m(A + a'_x) = \frac{N}{\sqrt{2}},$$

$$ma_y = -ma'_x = mg - \frac{N}{\sqrt{2}},$$

where N is the normal force on the block by the wedge. Solving the above equations yield

$$a'_x = \frac{g - A}{2}.$$

Therefore,

$$a_x = a'_x + A = \frac{g + A}{2},$$

$$a_y = -\frac{g - A}{2}.$$

When $A = 0$, we obtain the usual solution $a_x = g/2$ and $a_y = -g/2$. For $A < g$, $a_y < 0$ hence the block slides downward. However for $A > g$, $a_y > 0$, hence the block slides upward.

Exercises

1. The position and the velocity of a particle is denoted by **r** and **v** respectively. If **r** × **v** = **A** is a constant vector, then show that the acceleration of the particle is along

the radius vector. Conversely, if the acceleration of the particle is radial, show that
$\mathbf{r} \times \mathbf{v}$ is a constant of motion. Can you recognise the vector $\mathbf{r} \times \mathbf{v}$?

2. Atwood machine 2: Three masses m_1, m_2, and m_3 are hanging by three pulleys
   as shown in Fig. 4.5(a). Compute the acceleration of the masses. Make the same
   assumptions as in Example 4.2 for pulleys and rope (HINT: Write down the constraint
   equation that gives an additional equation).

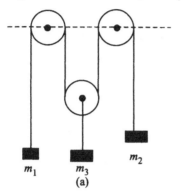 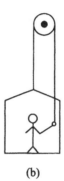

Figure 4.5   (a) Exercise 2; (b) Exercise 3.

3. A painter of mass 80 kg is hanging by a pulley as shown in Fig. 4.5(b). The mass of
   the supporting elevator is 20 kg. He pulls down the rope with such a force that he
   presses against the floor of the plank with $70g$ (g is the acceleration due to gravity).
   What is the acceleration of the painter? What is the tension on the rope?

4. Three gears A, B, and C have radius r_1, r_2, and r_3 respectively. Gear A is touching
   gear B, and gear B is touching gear C. What are the relationships between their
   angular velocities?

5. A swimmer is standing at point A on the bank of a river 200 metre wide. She wishes
   to reach point B that is directly opposite to her on the other side of the river. The
   river speed is 2 km/hour, and swimmer's average speed (in still water) is 2.5 km/hour.
   What should be the direction of swimmer's velocity with respect to (wrt) the water
   and wrt the ground? How long would the swimmer take to cross the river? What
   would be the trajectory of the swimmer as seen from the ground?

6. The swimmer of Exercise 5 is not interested in going to point B, but she moves with
   the following constraints.

   (a) Swimmer's velocity wrt water is always perpendicular to the river's velocity.

   (b) Swimmer's velocity wrt water is always pointing towards B.

   How long would she take to cross the river in both these cases? What is the
   trajectory of the swimmer? Where does she reach on the other side of the river?

7. What is the best way to cross the river in the shortest possible time for the parameters
   given in Exercise 5?

5

Motion in One Dimension

In Chapter 2, we discussed the statements of Newton's laws of motion. These laws are very powerful and they describe motion all around us. They are quantitative and mathematical, and can predict the future of mechanical systems. For some systems, the procedure to obtain the solution of a mechanical system is a simple mathematical equation, while for some others it is quite complex. The solutions for quite a few mechanical systems are not yet known. In this book we limit ourselves to simple problems. We begin with one-dimensional systems.

According to Newton's law, the equation of motion for a point particle of mass m experiencing a force $\mathbf{F}$ is

$$m\frac{d^2\mathbf{r}}{dt^2} = \mathbf{F},$$

where $\mathbf{r}$ is the position of the particle. If we assume that the particle moves in one dimension, then

$$m\frac{d^2x}{dt^2} = F(x, t), \tag{5.1}$$

where x is the coordinate of the particle along the direction of motion. The above equation is a second-order differential equation that requires two initial conditions and the force $F(x, t)$. We give the initial position and velocity as initial conditions and solve the differential equation. Finding the solution of a differential equation basically means determining the position of x at all times.

Let us study some simple examples:

1. **A particle falling under gravity**

   The equation of motion of a particle of mass m that is falling under gravity is

   $$m\frac{d^2x}{dt^2} = m_G g,$$

   where m_G is the gravitational mass of the particle, and g is the acceleration due to gravity. Since $m = m_G$ (refer to Section 3.1), we obtain

$$\frac{d^2x}{dt^2} = g, \tag{5.2}$$

or

$$\frac{dv}{dt} = g.$$

Integration of the above equation yields $v = v_0 + gt$. Using the definition $v = dx/dt$, we obtain

$$\frac{dx}{dt} = v_0 + gt,$$

integration of which yields

$$x(t) = x_0 + v_0 t + \frac{1}{2} gt^2.$$

Here x_0 and v_0 are the initial position and initial velocity respectively of the particle. This is the solution of differential equation (5.2). You can check the correctness of the solution by taking a double derivative of x and substituting it in the equation.

2. **A charged particle in a constant electric field**

   The equation of motion for a charged particle with charge q under the influence of electric field E is

$$\frac{d^2x}{dt^2} = \frac{qE}{m}.$$

Applying the same method as above, we obtain

$$x(t) = x_0 + v_0 t + \frac{1}{2} \frac{qE}{m} t^2.$$

3. **Spring–mass system**

   A block of mass m is connected to a spring of spring constant k. If x is the displacement of the block from its equilibrium position, then the equation of motion of the block is

$$m \frac{d^2x}{dt^2} = -kx.$$

We rewrite $k/m = \omega_0^2$, which yields

$$\frac{d^2x}{dt^2} = -\omega_0^2 x. \tag{5.3}$$

We cannot apply the procedure used in Examples 1 and 2 because we cannot directly integrate $dv/dt = -x$.

This simple example shows that Eq. (5.1) in general cannot be solved by a simple integration technique. Therefore we turn to a branch of mathematics that tells us how to solve various types of differential equations. Note however that not all types of differential equations can be solved analytically. You will learn about them in greater detail in advanced mathematics courses. Here, we will provide a working knowledge of simple differential equations so that you can solve some simple problems in mechanics. You will see that the mathematics used here is quite simple, yet very powerful.

5.1   Solving differential equations

5.1.1   Mathematical formalism

An equation involving an unknown variable and its derivatives is called a *differential equation* (DE). A *linear differential equation* is a special class of DE in which the maximum power of the unknown variable (here x) and its derivatives is one. If the highest derivative in the DE is of the nth order, then the DE is called an *nth order DE*. By this definition Eq. (5.3) is a second-order linear DE. It is customary to keep the unknown variable and its derivative in the left-hand side. So we rewrite Eq. (5.3) as

$$\frac{d^2x}{dt^2} + \omega_0^2 x = 0. \tag{5.4}$$

In the above equation, the coefficients of x and its derivatives are constants. These types of equations are called *linear differential equations with constant coefficients*. They are quite easy to solve. According to the theory of differential equations,

1. The solutions of a linear DE with constant coefficients are of the form $x = \exp(at)$, where a is a constant. A substitution of the above form of the solution in DE yields an algebraic equation.

2. For an nth order DE, the algebraic equation is of the nth order. Hence, we have n complex roots for a. These roots could be either distinct or repeated.

3. If the roots are distinct, then $x_i = \exp(a_i t)$ are n independent solutions of the equation.

4. The principle of superposition is valid for linear DEs. According to this principle if $y_1(t)$ and $y_2(t)$ are solutions of a DE, then $c_1 y_1(t) + c_2 y_2(t)$ (c_1 and c_2 are constants) is also a solution of DE (Exercise 1 in this chapter). When the roots of an algebraic equation are distinct, then according to the superposition principle, the general solution of DE is

    $$x(t) = \sum_{i=1}^{n} c_i x_i(t),$$

    where c_is are constants.

5. If the roots are repeated, then the solution for a general nth order DE is somewhat

involved. For a second order DE, they are $x_1 = \exp(at)$ and $x_2 = t\exp(at)$ (Exercise 5.2). Hence, the general solution of a second-order linear DE with repeated roots is

$$x(t) = c_1 \exp(at) + c_2 t \exp(at),$$

where c_1, c_2 are constants.

Thus the theory of DE helps us solve linear DE. We illustrate the above procedure by applying it to the DE of a linear oscillator (Eq. (5.4)).

5.1.2   Solution of a linear oscillator

Let us consider the differential equation of a linear oscillator (Eq. (5.4)). Before we proceed to solve this equation, we point out that the force on the particle is zero at $x = 0$, which is called an *equilibrium point*. Also, the force on the particle is always opposite to the direction of displacement. Hence the force always attempts to push the particle towards its equilibrium point. These types of systems are called *stable systems*.

Now let us return to the mathematical procedure described in Section 5.1.1. As prescribed, we substitute $x = \exp(at)$ in the DE which yields

$$a^2 + \omega_0^2 = 0, \tag{5.5}$$

or $a = \pm i\omega_0$. So $x_1 = \exp(i\omega_0 t)$ and $x_2 = \exp(-i\omega_0 t)$ are both solutions of DE (5.3). These are called independent functions, i.e., x_2 cannot be expressed in terms of x_1 and vice versa. The general solution of the DE is

$$x = c_1 x_1 + c_2 x_2 = c_1' \cos(\omega_0 t) + c_2' \sin(\omega_0 t) \tag{5.6}$$

with $c_1' = c_1 + c_2$ and $c_2' = i(c_1 - c_2)$. The above solution (Eq. (5.6)) can also be written as

$$x(t) = A\cos(\omega_0 t - \phi_0)$$

with

$$A = \sqrt{c_1'^2 + c_2'^2},$$

$$\phi_0 = \tan^{-1}\frac{c_2'}{c_1'}.$$

Note that sin and cosines are called harmonic functions in mathematics. Therefore, the above solution is called the *harmonic oscillation*.

The constants c_1 and c_2 (or $c_{1,2}'$) are computed using the initial conditions as follows. Suppose the initial position is $x(0)$, and the initial velocity is $v(0)$, then

$$x(0) = c_1'$$

$$v(0) = \omega_0 c_2'.$$

Therefore, the solution of the linear oscillator is

$$x(t) = x(0)\cos(\omega_0 t) + \frac{v(0)}{\omega_0}\sin(\omega_0 t).$$

Thus we solve the DE for the spring–mass system. Figure 5.1 shows a plot of $x(t)$ vs. t for $x(0) = 1$, $v(0) = 0$, and $\omega_0 = 1$ (dotted line).

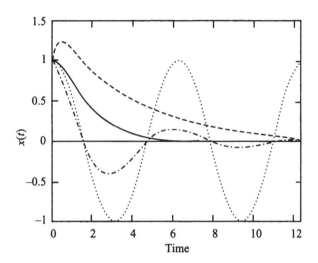

Figure 5.1   *Solution of an oscillator for the initial conditions $x(0) = 1$ and $v(0) = 0$: (a) undamped oscillator with $\omega_0 = 1$ (dotted line); (b) under-damped oscillator with $\omega_0 = 1$ and $\gamma = 0.3$ (chained line); (c) over-damped oscillator with $\omega_0 = 1$ and $\gamma = 2$ (solid line); (d) critically-damped oscillator with $\omega_0 = 1$ and $\gamma = 1$ (dashed line).*

When $\omega_0 = 0$, the roots of the Eq. (5.5) are $a = 0, 0$ (repeated). Hence the independent solutions are 1 and t (see rule 5 in Section 5.1.1). Therefore, the general solution of the DE (5.3) is

$$x(t) = c_1 + c_2 t.$$

Note that $\omega = 0$ corresponds to a free particle.

We can solve the DE for a frictional oscillator using a similar method. The DE for a frictional oscillator is

$$\frac{d^2 x}{dt^2} + 2\gamma\frac{dx}{dt} + \omega_0^2 x = 0, \tag{5.7}$$

where $2\gamma\dot{x}$ is the frictional term[1]. When we substitute $\exp(at)$ in the above equation, we obtain the following equation for a:

$$a^2 + 2\gamma a + \omega_0^2 = 0,$$

---

[1]The frictional force on a body is proportional to its velocity in a viscous regime (at low Reynolds number).

whose solution is

$$a_{1,2} = -\gamma \pm \sqrt{\gamma^2 - \omega_0^2}. \tag{5.8}$$

Hence the general solution of the DE is

$$x(t) = c_1 \exp(a_1 t) + c_2 \exp(a_2 t),$$

where $a_{1,2}$ of Eq. (5.8) come in the following three different forms:

1. $\omega_0 > \gamma$: For this case, $a_{1,2}$ are complex

$$a_{1,2} = -\gamma \pm i\sqrt{\omega_0^2 - \gamma^2},$$

and the general solution is

$$x(t) = \exp(-\gamma t)[c_1' \cos(\omega_1 t) + c_2' \sin(\omega_1 t)]$$

with $c_1' = c_1 + c_2$, $c_2' = i(c_1 - c_2)$, and $\omega_1 = \sqrt{\omega_0^2 - \gamma^2}$. The constants are determined using the initial conditions. The solution can also be written as

$$x(t) = \exp(-\gamma t)A \cos(\omega_1 t + \phi_0),$$

where A is a constant, and ϕ_0 is the initial phase. Figure 5.1 shows a plot of $x(t)$ vs. t for $\omega_0 = 1$, $\gamma = 0.3$, $x(0) = 1$, and $v(0) = 0$ (or $A = 1$ and $\phi_0 = 0$) as a chained line. This type of oscillator is called an *under-damped oscillator*.

2. $\omega_0 < \gamma$: Here $a_{1,2}$ are real and negative, and the general solution is

$$x(t) = c_1 \exp([-\gamma + \sqrt{\gamma^2 - \omega_0^2}]t) + c_2 \exp([-\gamma - \sqrt{\gamma^2 - \omega_0^2}]t)$$

Figure 5.1 shows a plot of $x(t)$ vs. t for $\omega_0 = 1$, $\gamma = 2$, $x(0) = 1$, and $v(0) = 0$ as a solid line. The above solution shows that for $t = T \sim \left[-\gamma + \sqrt{\gamma^2 - \omega_0^2}\right]^{-1}$ and beyond, $x(t) \to 0$. There is no oscillation. This type of oscillator is called an *over-damped oscillator*.

3. $\omega_0 = \gamma$: Here $a_1 = a_2 = -\gamma$ (repeated roots) and the solution is

$$x(t) = c_1 \exp(-\gamma t) + c_2 t \exp(-\gamma t).$$

For initial conditions $x(0) = A$ and $\dot{x}(0) = 0$, the solution is

$$x(t) = A \exp(-\gamma t)[1 + \gamma t].$$

Figure 5.1 shows a plot of $x(t)$ vs. t for $\omega_0 = 1$, $\gamma = 1$, $x(0) = 1$, and $v(0) = 0$ (or $A = 1$) as a dashed line. This kind of oscillator is called a *critically-damped oscillator*. This is an intermediate case between the under-damped and the over-damped oscillation.

Thus we are able to mathematically solve for the position and the velocity of an undamped and a damped oscillator.

5.1.3   Forced oscillator: An example of an inhomogeneous DE

The Eqs.(5.4) and (5.7) have terms involving an unknown variable and its derivatives (x, $\dot{x}$, and $\ddot{x}$) only. These kinds of equations are called *homogeneous differential equations*. When a DE has a term which is an explicit function of the independent variable (here t) only, then the equation is called an *inhomogeneous differential equation*. An example of an inhomogeneous equation is

$$\frac{d^2x}{dt^2} + \omega_0^2 x = F\cos(\omega_f t). \tag{5.9}$$

Here the right-hand side contains a term which is a function of time, not x or its derivatives. The procedure described in Section 5.1.1 cannot be used to solve the above equation. We will sketch a new scheme to solve an inhomogeneous DE.

The solution of an inhomogeneous equation is written as a sum of the homogeneous part x_{homog} and the particular part $x_{\text{particular}}$:

$$x = x_{\text{homog}} + x_{\text{particular}}. \tag{5.10}$$

x_{homog} satisfies the homogeneous equation

$$\frac{d^2 x_{\text{homog}}}{dt^2} + \omega_0^2 x_{\text{homog}} = 0$$

and is given by Eq. (5.6), while $x_{\text{particular}}$ satisfies

$$\frac{d^2 x_{\text{particular}}}{dt^2} + \omega_0^2 x_{\text{particular}} = F\cos(\omega_f t). \tag{5.11}$$

Clearly $x = x_{\text{homog}} + x_{\text{particular}}$ satisfies the Eq. (5.9). We can solve $x_{\text{particular}}$ by trial and error. There are other methods, but we will not discuss them here. We attempt

$$x_{\text{particular}} = A\cos(\omega_f t),$$

which yields

$$A = \frac{F}{\omega_0^2 - \omega_f^2}.$$

Hence the general solution is

$$x = c_1' \cos(\omega_0 t) + c_2' \sin(\omega_0 t) + \frac{F\cos(\omega_f t)}{\omega_0^2 - \omega_f^2},$$

For initial conditions $[x(0), v(0)]$, the above solution is

$$x(t) = x(0)\cos\omega_0 t + \frac{v(0)}{\omega_0}\sin\omega_0 t + \frac{F}{\omega_0^2 - \omega_f^2}(\cos\omega_f t - \cos\omega_0 t) \qquad (5.12)$$

The above solution is valid for $\omega_0 \neq \omega_f$. When the two frequencies are equal, the last term is of the 0/0 form. Application of L'Hospital's rule (applying derivative wrt ω_f and taking the limit $\omega_f \to \omega_0$) yields

$$x(t) = x(0)\cos\omega_0 t + \frac{v(0)}{\omega_0}\sin\omega_0 t - \frac{Ft\sin(\omega_0 t)}{2\omega_0}. \qquad (5.13)$$

This is called the *resonance* condition. We plot the solutions in Fig. 5.2 with initial conditions $x(0) = 1$ and $v(0) = 0$. The parameter values are $F = 2$, $\omega_0 = 1$, and $\omega_f = 5, 1.1, 1$. For $\omega_f = 5$, $x(0)\cos(t)$ of Eq. (5.12) dominates, and the solution is oscillatory. For $\omega_f = 1.1$, we obtain beats with periodicity of $T = 20\pi$. When $\omega_f = 1$, the term $Ft\sin(t)/2$ dominates, and the solution grows with time.

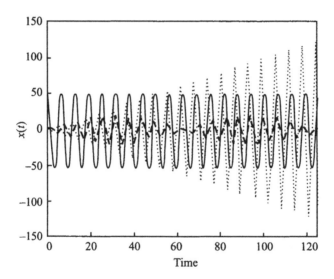

Figure 5.2 *The solution for a forced oscillator with initial conditions $x(0) = 1$ and $v(0) = 0$, and parameter values $F = 2$, $\omega_0 = 1$, and $\omega_f = 5$ (solid line; $50 \times x(t)$ for magnification along the y-axis), $\omega_f = 1.1$ (chained), $\omega_f = 1$ (dotted). We find oscillations for $\omega_f = 5$, beats for $\omega_f = 1.1$, and growth in amplitude for $\omega_f = 1$.*

The above example illustrates how to solve an inhomogeneous DE. The most critical part of the solution is $x_\text{particular}$, which is obtained here by trial and error.

In the next subsection, we will solve a differential equation that describes an unstable system.

5.1.4   Solution of an unstable system

Let us consider a DE

$$\frac{d^2x}{dt^2} = x. \tag{5.14}$$

For $x \neq 0$, the force is always in the direction of the displacement. Hence, the particle will run away either to $+\infty$ or to $-\infty$. The force on the particle is zero at $x = 0$. This system is an example of an *unstable system*. The point $x = 0$ is an *unstable equilibrium point*. Note that the particle has mass of one unit.

Let us obtain the solution of the above DE. We start with a trial solution $x(t) = \exp(at)$ which yields

$$a^2 = 1$$

or $a = \pm 1$. Hence $\exp(t)$ and $\exp(-t)$ are both solutions of the DE. Therefore, the general solution of the DE is

$$x(t) = c_1 \exp(t) + c_2 \exp(-t),$$

$$= c_1' \cosh(t) + c_2' \sinh(t),$$

where $\cosh(t) = [\exp(t) + \exp(-t)]/2$ and $\sinh(t) = [\exp(t) - \exp(-t)]/2$. The substitution of the initial conditions yields

$$x(t) = x(0)\cosh(t) + v(0)\sinh(t).$$

Let us understand the above solution for special cases. When $x(0) = v(0) = 0$, we obtain $x(t) = 0$, which is the equilibrium point. When $v(0) = 0$, then $x(t) = x(0)\cosh(t)$, implying that the particle goes to $\pm\infty$ depending on the sign of $x(0)$.

For a general situation

$$x(t) \sim \frac{[x(0) + v(0)]}{2} \exp(t),$$

because $\exp(-t) \rightarrow 0$ asymptotically for large t. Therefore $x(t) \rightarrow \infty$ or $x(t) \rightarrow -\infty$ depending on the sign of $x(0) + v(0)$. Note however that under a special situation when $v(0) = -x(0)$, $x(t) = x(0)\exp(-t) \rightarrow 0$.

We will discuss the above problem later in Section 5.2.2 with the energy method that provides us a better feel for the system.

EXAMPLE 5.1   Solve the following differential equations with the given initial conditions (IC):

1. $\ddot{x} + x = 0$ with IC: $x(0) = 0, \dot{x}(0) = 1$.

2. $\ddot{x} = x$ with IC: $x(0) = 0, \dot{x}(0) = 1$.

3. $\ddot{x} + \dot{x} + x = 0$ with IC: $x(0) = 2, \dot{x}(0) = 0$.

4. $\ddot{x} + x = \sin(2t)$ with IC: $x(0) = 1, \dot{x}(0) = 0$.

5. $\ddot{x} + x = \sin at$ with $a = 0.99$ and $a = 1$, and IC $x(0) = 0, \dot{x}(0) = 0$.

SOLUTION   We will solve the above equations using the prescription given in Section 5.1.1.

1. We substitute $x(t) = \exp(at)$ in the differential equation, which yields $a = \pm i$. Therefore,

$$x(t) = c_1 \exp(it) + c_2 \exp(-it)$$

The initial conditions yield

$$c_1 + c_2 = 0$$

$$i(c_1 - c_2) = 1$$

Hence $2c_1 i = 1$, or $c_1 = -i/2 = -c_2$. Therefore,

$$x(t) = -\frac{i}{2} \left[\exp(it) - \exp(-it)\right]$$

$$= \sin t.$$

We could also write the general solution as

$$x(t) = c_1' \cos t + c_2' \sin t,$$

by substituting $\exp(\pm it) = \cos(t) \pm i\sin(t)$. The substitution of the initial conditions in the DE implies that $c_1' = 0$ and $c_2' = 1$, which yields

$$x(t) = \sin t.$$

Clearly both the general solutions give us the same final solution. We can easily verify that the final solution satisfies the DE as well as the initial conditions.

2. The algebraic equation is $a^2 = 1$ whose solution is $a = \pm 1$. Therefore the general solution is

$$x(t) = c_1 \exp t + c_2 \exp(-t)$$

The initial conditions imply that $c_1 = 0.5$ and $c_2 = -0.5$. Hence,

$$x(t) = 0.5(\exp(t) - \exp(-t)) = \sinh(t).$$

3. The algebraic equation is

$$a^2 + a + 1 = 0.$$

The solution of the above algebraic equation is

$$a = \frac{-1 \pm i\sqrt{3}}{2}$$

Hence,

$$x(t) = c_1 \exp\left(-t/2\right) \exp\left(\frac{i\sqrt{3}}{2}t\right) + c_2 \exp\left(-t/2\right) \exp\left(-\frac{i\sqrt{3}}{2}t\right)$$

$$= \exp\left(-t/2\right) \left[c_1' \cos\left(\frac{\sqrt{3}}{2}t\right) + c_2' \sin\left(\frac{\sqrt{3}}{2}t\right) \right]$$

with $c_1 + c_2 = c_1'$ and $i(c_1 - c_2) = c_2'$. The initial conditions imply that $c_1' = 2$ and $c_2' = 2/\sqrt{3}$. Therefore the solution $x(t)$ is

$$x(t) = \exp\left(-t/2\right) \left[2\cos\left(\frac{\sqrt{3}}{2}t\right) + \frac{2}{\sqrt{3}} \sin\left(\frac{\sqrt{3}}{2}t\right) \right].$$

4. The procedure described earlier yields the homogeneous solution as

$$x_{\text{homog}} = c_1 \cos t + c_2 \sin t.$$

For the particular solution, we attempt $x_{\text{particular}}(t) = A \sin 2t$. The substitution of $x_{\text{particular}}$ in the equation yields $A = -1/3$. Therefore,

$$x(t) = x_{\text{homog}} + x_{\text{particular}} = c_1 \cos t + c_2 \sin t - \frac{1}{3} \sin 2t.$$

The initial conditions yield $c_1 = 1$ and $c_2 = 2/3$. Hence,

$$x(t) = \cos t + \frac{2}{3} \sin t - \frac{1}{3} \sin 2t.$$

Note that $x_{\text{particular}}(t)$ does not satisfy the initial conditions, but the complete solution does.

5. We have solved the above equation in Section 5.1.3. The solution is

$$x(t) = c_1 \cos t + c_2 \sin t + \frac{1}{1 - a^2} \sin at.$$

The initial conditions imply that $c_1 = 0$ and $c_2 = -a/(1 - a^2)$. Hence,

$$x(t) = \frac{\sin at - a \sin t}{1 - a^2},$$

Note that for $a = 0.99$, $x(t)$ is periodic with period $T = 200\pi$. Hence the solutions has beats as seen in Fig. 5.2. Denoting $a = 1 + \epsilon$ and taking the limit $\epsilon \to 0$ we obtain

$$x(t) = -\frac{t \cos t}{2} + \frac{\sin t}{2}.$$

This is the solution for $a = 1$. The reader is urged to plot all the solutions as a function of time.

EXAMPLE 5.2   A spherical ball of mass m falls under gravity through a viscous fluid. Find the position and velocity of the ball as a function of time. Assume that the mass starts at rest from a height h above the ground.

SOLUTION   The spherical ball experiences gravitational force mg and viscous force $\gamma\dot{x}$, where $\dot{x}$ is the velocity of the ball, and γ the frictional coefficient. We have assumed that the velocity (strictly speaking the Reynolds number) of the ball is small (see Appendix B). We choose the vertical direction as the positive x direction, with x and v as the particle's position and velocity respectively. The equation of motion of the ball is

$$m\ddot{x} = -mg - \gamma\dot{x}$$

which can be rewritten as $m\dot{v} + \gamma v = -mg$,

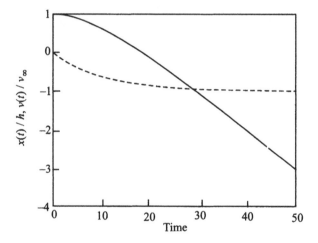

Figure 5.3   *The plot of velocity $v(t)/v_\infty$ (dashed line) and position $x(t)/h$ (solid line) of the spherical ball as a function of time. The parameter used are $g = 10$ m/s^2, $\gamma = 0.1$ Ns/m, $m = 1$ kg, $h = 10^3$ m.*

whose solution is

$$v(t) = c_1 \exp(-\gamma t/m) - \frac{mg}{\gamma}.$$

The first term is the homogeneous solution, and the second term is the particular solution. The initial condition $v(0) = 0$ yields

$$v(t) = \frac{mg}{\gamma}\left[\exp(-\gamma t/m) - 1\right].$$

We can integrate the above equation to obtain the position of the ball. For the initial condition $x(0) = h$, the solution is

$$x(t) = h - \frac{mg}{\gamma}t + \frac{m^2g}{\gamma^2}\left[1 - \exp(-\gamma t/m)\right].$$

In Fig. 5.3 we plot $v(t)$ and $x(t)/h$ vs. t for the parameter values of $g = 10$ m/s^2, $\gamma = 0.1$ N s/m, $m = 1$ kg, $h = 1$ km. For $t > m/\gamma = 10$ s, $v(t) \to v_\infty = -mg/\gamma = -100$ m/s. Asymptotically $x(t) \approx$ constant $- (mg/\gamma)t$.

In this section, we presented a method to solve linear DE. However, the equation of motion for many systems is not linear. In the example cited above, the linear equations had analytic solutions. However many nonlinear equations do not have analytic solutions. Therefore, we must look for other methods to analyse mechanical systems. One such method is based on energy. We describe this method below.

5.2   The energy method to analyse mechanical systems

The equation of motion of a mechanical system is a second-order DE. Sometimes these equations can be quite difficult to solve. For a class of systems for which a potential function $U(x)$ exists and force $F = -dU(x)/dx$, we can convert the equation of motion to a first-order DE that is easier to solve. Multiplying both sides of the equation of motion by v we obtain

$$mv\frac{dv}{dt} = -\frac{dU}{dx}\frac{dx}{dt}$$

or $$\frac{d}{dt}\left[\frac{1}{2}mv^2 + U\right] = 0,$$

or $$\frac{1}{2}mv^2 + U(x) = E, \tag{5.15}$$

where E is a constant called energy. The above equation yields a first-order DE

$$\frac{dx}{dt} = \pm\sqrt{2\left[E - U(x)\right]/m}. \tag{5.16}$$

The sign of the velocity of the particle is dictated by the initial condition of the particle. The solution of the above equation is

$$t = \int_a^x \frac{dx'}{\pm\sqrt{2\left[E - U(x)\right]/m}} \tag{5.17}$$

which can be solved either analytically or numerically, depending on the form of $U(x)$.

Let us visualise the motion of a particle under the potential $U(x)$ shown in Fig. 5.4. In the figure, we plot the potential $U(x)$ vs. x. The energy E of the particle is also shown in the plot. For x with $E > U(x)$, the particle has positive kinetic energy, and is in motion. At $E = U(x)$, the particle's velocity becomes zero. The region with $U(x) > E$ is a forbidden zone because kinetic energy would be negative in this region. Hence for the potential shown in Fig. 5.4, the particle is either confined in the region $a \leq x \leq b$ or it moves in the region

$x \geq c$. The sign of the velocity of the particle is chosen according to the initial condition. For example, if the particle starts from b, then its velocity will be negative.

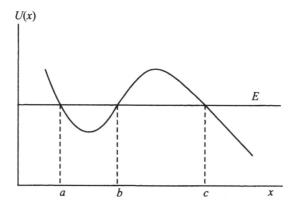

Figure 5.4   *A sketch of potential $U(x)$ vs. position x.*

In the region $[a, b]$, particle's motion is *periodic*. The turning points are $x = a$ and $x = b$ where the velocity becomes zero. This type of motion is called a *bounded motion*. If the particle starts from any point in the region of $[c, \infty)$, it asymptotically goes to $+\infty$, and the motion is *unbounded*.

As an example, let us consider spring–mass systems. We start the block from a fully stretched position $x = A$, where A is the amplitude of oscillation. At this position $v = 0$, therefore according to Eq. (5.15) $E = m\omega_0^2 A^2/2$. The velocity of the block is negative since it moves leftward immediately after its release. Hence,

$$\frac{dx}{dt} = -\sqrt{2\left[E - U(x)\right]/m}, \tag{5.18}$$

An integration of the above equation yields

$$\int_0^t dt' = -\int_A^x \frac{dx'}{\sqrt{2\left[E - \frac{1}{2}m\omega_0^2 x'^2\right]/m}}$$

$$t = -\frac{1}{\omega_0} \int_A^x \frac{dx'}{\sqrt{A^2 - x^2}}$$

$$t = \frac{1}{\omega_0}\left[-\sin^{-1}(x/A) - \phi_0 + \pi/2\right]$$

where ϕ_0 is a constant. Hence,

$$x(t) = A\cos(\omega_0 t + \phi_0), \tag{5.19}$$

which is what we had derived earlier by solving the equation of motion. In this example, we have only one constant ϕ_0 because Eq. (5.18) is a first-order equation. The velocity at

$t = 0$ can be obtained from the energy equation if E is given. Since $x(0) = A$, $\phi_0 = 0$.

We can describe the evolution of an oscillator in terms of its phase $\phi = \omega_0 t$ as shown in Fig. 5.5. The position coordinate x can be obtained by projecting point P along the the x-axis. At P, the particle is moving leftward, and its velocity is negative. The angular coordinates of the particle corresponding to points A,B,C,D are $\omega_0 t = 0, \pi/2, \pi, 3\pi/2$. The motion repeats after $t = 2\pi/\omega_0$.

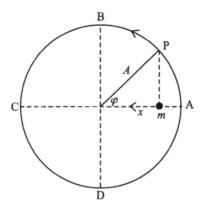

Figure 5.5   *Graphic description of an oscillator in terms of its phase $\phi = \omega_0 t$. The projection of point P on the x-axis describes the position of the particle.*

In the following sections we will discuss the motion of a particle in some generic potentials and illustrate the usefulness of the energy method.

5.2.1   Simple harmonic oscillator revisited ($U(x) = x^2/2$)

For simplicity, we will take $m = 1$. The energy equation for the oscillator is

$$\frac{1}{2}v^2 + \frac{1}{2}x^2 = E, \tag{5.20}$$

and its equation of motion is $\ddot{x} + x = 0$, which has been solved in Section 5.1.2.

Let us understand the motion of the particle for various values of E. At $x = 0$, the force $F = -dU/dx = 0$. Therefore, $x = 0$ is an equilibrium point. If the particle is kept there with $v = 0$, then it will stay there forever. This configuration corresponds to $E = 0$. For higher values of E (see Fig. 5.6(a)), the motion is confined between two points A and B. This is because the kinetic energy becomes negative for $x > x_A$ and $x < x_B$, which is impossible. We can find the positions of the extreme points from Eq. (5.20). At the extreme points, $E = U(x)$, hence $v = 0$ at these points. Therefore $x_A = \sqrt{2E}$ and $x_B = -\sqrt{2E}$. At A, the force $F = -dU/dx$ is negative, so the particle moves leftward. The opposite behaviour is seen at B. The points A and B are called *turning points*. We can also claim from Eq. (5.20) that the kinetic energy of the particle is maximum at $x = 0$, and decreases as the particle approaches the end points.

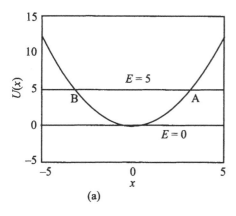

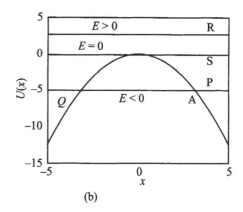

Figure 5.6   (a) Sketch of the potential and energy of a simple harmonic oscillator. (b) Sketch of the potential $U(x) = -x^2/2$ and energy.

5.2.2   Potential $U(x) = -x^2/2$

The energy equation for the system is

$$\frac{1}{2}v^2 - \frac{1}{2}x^2 = E, \tag{5.21}$$

and its equation of motion is $\ddot{x} = x$, which was solved earlier in Section 5.1.4.

The potential function is plotted in Fig. 5.6(b). Let us understand the dynamics of this system. At $x = 0$, $F = -dU/dx = 0$. Therefore, if we keep the particle at $x = 0$ with $v = 0$, it will stay there forever until it is disturbed. If disturbed, the particle will run away to infinity because the force $-dU/dx$ is in the direction of the displacement. Therefore, $x = 0$ is an unstable equilibrium point. This point with $v = 0$ yields $E = 0$.

Now consider the system when $E < 0$ (Fig.5.6(b)). If the particle starts from the point P with negative velocity, it will move leftward up to A where its velocity vanishes. Note that the kinetic energy of the particle decreases as it approaches A. At A, the particle stops but it experiences a force in the forward direction $(-dU/dx > 0)$, so it turns back and starts to move to the right. During its subsequent motion, its velocity keeps on increasing. As $t \to \infty$, $x \to \infty$.

If the starting velocity at P was positive, then the particle will continue to move to the right with increasing velocity. It is easy to check that a similar motion will take place if the particle starts from point Q. The only difference is that the direction is changed.

Now we will examine the motion of the particle for positive E. Suppose the particle starts from point R with negative velocity. The energy equation shows that the kinetic energy or speed of the particle decreases as it moves towards the origin $(x = 0)$. At the origin, the particle has negative velocity, so it continues to move leftward. To the left of $x = 0$, the kinetic energy of the particle starts to increase and the particle moves with increased speed. Asymptotically the particle goes to $x \to -\infty$. It is easy to verify that if the particle had positive velocity at point R at $t = 0$, it will move rightward and asymptotically go to $x = \infty$.

We have discussed earlier that the unstable equilibrium point with $v = 0$ has $E = 0$. However an interesting motion occurs if $v \neq 0$ for $E = 0$ case. If the particle starts from point S and moves leftward, its speed decreases as it climbs the potential hill. From the energy equation, we can deduce that $v(x) = -x$. A question arises: when will the particle reach the top of the potential hill? We can get the answer by integrating the energy equation

$$\dot{x} = -x$$

which yields

$$T(\epsilon) = - \int_{x_P}^{\epsilon} \frac{dx}{x} = \ln(x_P/\epsilon)$$

As $\epsilon \to 0$, $T(\epsilon) \to \infty$. Hence the particle cannot reach the top of the potential hill in a finite time. Note that this is the case only when $E = 0$. For $E > 0$, the particle crosses $x = 0$ in a finite time (verify!).

Before we end this discussion, we point out an important fact. For an unstable system, the particle goes to $\pm\infty$ with an exception; as discussed earlier, the particle would tend to the origin when the initial velocity of the particle, with $E = 0$, points towards the origin.

The solutions of DE for various initial conditions discussed earlier are consistent with the solution of the energy equation. After the discussion on the above potentials, we will discuss the motion of a pendulum which is a bit more complex.

5.2.3   Pendulum

A pendulum consists of a bob of mass m attached to a *rod* of length l. We assume the rod to be massless. If we choose the bottom-most point of the pendulum as a reference for the potential energy, then the potential energy of the pendulum is

$$U(\phi) = mgl(1 - \cos(\phi)).$$

We have taken the counter-clockwise displacement as positive.

The bob will perform a circular motion because of the rod. Therefore, the energy equation is

$$\frac{1}{2}ml^2\dot{\phi}^2 + mgl(1 - \cos(\phi)) = E \tag{5.22}$$

where E is the energy of the system. We plot the potential function in Fig. 5.7 and analyse the motion in terms of E. A word of caution is in order. The motion of the pendulum is really two-dimensional. However, since there is no motion in the radial direction, we can treat this as a one-dimensional motion.

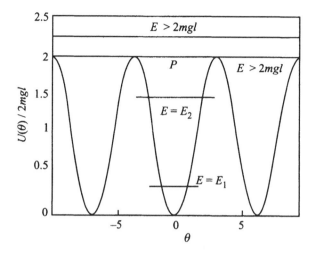

Figure 5.7   *The potential energy of a pendulum*

At $\phi = 0$, $F = -dU/d\phi = 0$, so $\phi = 0$ is an equilibrium point. If we take $\phi = \dot{\phi} = 0$, we obtain $E = 0$, which corresponds to the bob resting at the bottom-most position (Fig. 5.8(a)). The line $E = 0$ coincides with the x-axis in Fig. 5.7. We can expand the potential function near $\phi = 0$ as a Taylor series expansion, which yields

$$U(\phi) = mgl\left(1 - 1 + \frac{\phi^2}{2} + ...\right) \approx \frac{1}{2}mgl\phi^2.$$

Here we retain only the first two terms. The above potential matches the potential of a simple harmonic oscillator (Section 5.2.1). Therefore for small ϕ, the pendulum executes a simple harmonic motion (Fig. 5.8(b)). This kind of motion takes place for small E (e.g., $E = E_1$ in Fig. 5.7).

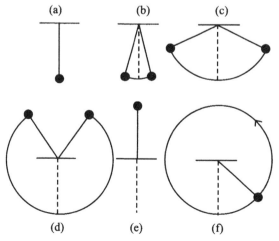

Figure 5.8   *Various configurations of a pendulum*

For a somewhat higher energy ($E = E_2$ of Fig. 5.7), the motion is still periodic. But the amplitude of oscillations is comparatively larger (e.g., Fig. 5.8(c), (d)). The solutions for large amplitudes cannot be expressed in terms of sin and cos function, hence they are not harmonic. If we increase the energy even further, we reach the maximum of the potential function.

The maximum of the potential occurs at $\phi = n\pi$ (n odd). Let us just focus on $\phi = \pi$. The force is zero at this point. The configuration ($\phi = \pi, \dot{\phi} = 0$) with $E = 2mgl$ corresponds to the pendulum standing up, which an unstable equilibrium (Fig. 5.8(e)). If we expand the potential function near $\theta = \pi$, we obtain

$$U(\phi) = mgl(1 - \cos(\pi - \Delta\phi)) = mgl(1 + \cos(\Delta\phi)) \approx 2mgl - mgl\frac{(\Delta\phi)^2}{2},$$

which is of the form discussed in Section 5.2.2. Therefore the motion near $\phi = \pi$ resembles the motion discussed for $U(x) = -x^2/2$.

If the bob starts from point P with $E = 2mgl$ and positive velocity, then it will move toward the potential maximum with slower and slower velocity. As discussed in Section 5.2.2, the bob will not be able to reach $\phi = \pi$ in a finite time. Similarly if the bob is given negative velocity at point P, then it will rotate clockwise and asymptotically approach $\phi = -\pi$.

After this case, we consider the motion of the pendulum for $E > 2mgl$. Under this situation, the pendulum rotates continuously clockwise or counter-clockwise (Fig. 5.8(f)). The velocity ($\dot{\phi}$) of the bob changes with ϕ (see the energy equation).

So far we have discussed the motion of a pendulum in a semi-qualitative manner. In the next section we will try to solve Eq. (5.22) analytically.

5.2.3.1   Analytic solution for the motion of a pendulum

In this section we will derive an analytic solution for the time period of the pendulum when it is executing a periodic motion (Fig. 5.8(b)–(d))

Suppose the maximum amplitude of a pendulum is ϕ_0, then its energy equation is

$$\frac{1}{2}l\dot{\phi}^2 = g(\cos\phi - \cos\phi_0)$$

Therefore,

$$t = \int \frac{d\phi}{\pm\sqrt{(2g/l)(\cos\phi - \cos\phi_0)}}.$$

We can obtain an expression for the time period of oscillation using the above equation:

$$T = 4\int_0^{\phi_0} \frac{d\phi}{\sqrt{(2g/l)(\cos\phi - \cos\phi_0)}}.$$

We make a change of variable

$$\frac{\sin(\phi/2)}{\sin(\phi_0/2)} = \sin\xi,$$

which yields

$$T = 4\sqrt{\frac{l}{g}} \int_0^{\pi/2} \frac{d\xi}{\sqrt{1 - k^2 \sin^2 \xi}} = 4\sqrt{\frac{l}{g}} K(k),$$

where $k = \sin(\phi_0/2)$, and the function $K(k)$ is called the *elliptic integral of the first kind*.

Since $K(0) = \pi/2$, we obtain $T = 2\pi\sqrt{l/g}$ as the time period for small oscillations. For ϕ_0 small,

$$T \approx 4\sqrt{\frac{l}{g}} \int_0^{\pi/2} d\xi [1 + \frac{k^2}{2} \sin^2 \xi + ...]$$

$$\approx 2\pi\sqrt{\frac{l}{g}} (1 + \frac{1}{16}\phi_0^2 + ...).$$

Therefore, the leading order correction to the time period is the second term in the above expression. For $\phi_0 = \pi/2$, the correction is only about 15%. That is the reason why $2\pi\sqrt{l/g}$ is a reasonably robust formula for the time period of a pendulum.

Due to the above approximation, Galileo could observe that the time period of a pendulum is independent of its amplitude. This property is also the key to mechanical clocks.

EXAMPLE 5.3   A particle of mass 2 units is moving in a potential of $x^2/2$. At $t = 0$, its position and velocity are $x(0) = 1, \dot{x}(0) = 1$. At what time will the particle reach its maximum position?

SOLUTION   We compute the energy E of the particle using the initial conditions: $E = m\dot{x}^2/2 + x^2/2 = 3/2$. The maximum position occurs at the turning point at which $\dot{x} = 0$. (see Fig. 5.9(a)) Hence $E = x_{\max}^2/2 = 3/2$, implying that $x_{\max} = \sqrt{3}$. Therefore, the range of the particle's position is $(-\sqrt{3}, \sqrt{3})$. The equation for the energy is

$$\frac{1}{2}m\dot{x}^2 = E - \frac{x^2}{2},$$

which implies that

$$\dot{x} = \pm\sqrt{\frac{2E - x^2}{m}} = \pm\sqrt{\frac{3 - x^2}{2}}$$

We take the positive sign because v is positive at $t = 0$. The particle is travelling from $x = 1$ to $x = \sqrt{3}$. Now the integration of the above equation yields

$$\int_0^T dt = \int_1^{\sqrt{3}} dx \frac{\sqrt{2}}{\sqrt{3 - x^2}}$$

$$T = \sqrt{2} \sin^{-1} \frac{x}{\sqrt{3}} \Big|_1^{\sqrt{3}}$$

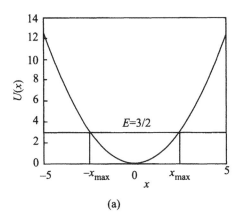

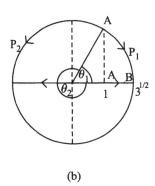

(a)   (b)

Figure 5.9   (a) Plot of the potential $U(x) = x^2/2$ as a function of x. The energy of the particle is shown as a horizontal line. (b) Particle's position in terms of phases.

$$T = \sqrt{2}\left[\frac{\pi}{2} - \sin^{-1}\frac{1}{\sqrt{3}}\right] = \sqrt{2}\left[\frac{\pi}{2} - \frac{\pi}{6}\right] = \sqrt{2}\pi/3.$$

NOTE   The equation of motion is

$$2\ddot{x} = -x,$$

whose solution is

$$x(t) = c_1 \cos t/\sqrt{2} + c_2 \sin t/\sqrt{2}.$$

The substitution of the initial conditions yield $c_1 = 1$, and $c_2 = \sqrt{2}$. We want to determine time T when $x(T) = \sqrt{3}$. The solution can be obtained from

$$\sqrt{3} = \cos T/\sqrt{2} + \sqrt{2}\sin T/\sqrt{2},$$

which yields

$$T = \sqrt{2}\cos^{-1}(1/\sqrt{[3]}) = \sqrt{2}\pi/3,$$

which is consistent with the previous solution.

We could also solve this problem graphically. The motion of the particle in terms of angular coordinates is shown in Fig. 5.9(b). The positions $x = 1$ and $x = \sqrt{3}$ correspond to points A and B respectively. If we take the velocity to be positive, then the particle takes path (P1), and the angular distance between the two points is $\pi/2 - \pi/6 = \pi/3$. The angular frequency of the oscillator is $\omega = \sqrt{k/m} = 1/\sqrt{2}$, and the time period of the oscillator is $2\pi/\omega = 2\sqrt{2}\pi$, which is the time required to cover angle 2π. Therefore, the time required to go from point A to B is $\sqrt{2}\pi/3$ (same as the earlier answer).

If the particle had negative velocity at $t = 0$, i.e., if $v(0) = -1$, then the particle would take path (P2) (see Fig. 5.9(b)) to reach B covering an angle of $5\pi/3$. Hence the time required for path (P2) will be $\sqrt{2}(5\pi/3)$. Hence the particle could reach $x = \sqrt{3}$ at two different times depending on the path taken.

In this chapter we discussed how to analyse mechanical systems by solving the equation of motion (second-order differential equations) and by solving the energy equation. The energy method provides us a very good qualitative idea about the system. There is another powerful method to analyse mechanical systems called *phase space visualisation*; here we plot both the position and momentum of a particle. The phase space description of mechanical systems will be discussed in the next chapter.

Exercises

1. Prove the superposition principle for linear differential equations.

2. Suppose a is the repeated root of the algebraic equation corresponding to a second-order differential equation

$$\ddot{y} + A\dot{y} + By = 0.$$

Show that $\exp(at)$ and $t\exp(at)$ would be solutions of this equation.

3. Solve the following differential equations with initial conditions (IC):

   (a) $\ddot{x} + 2x = 0$ with IC: $x(0) = 5, \dot{x}(0) = 1$.
   (b) $\ddot{x} = 2x$ with IC: $x(0) = 1, \dot{x}(0) = 1$.
   (c) $\ddot{x} + \dot{x} + 4x = 0$ with IC: $x(0) = 1, \dot{x}(0) = 0$.
   (d) $\ddot{x} + x = \sin(2t)$ with IC: $x(0) = 1, \dot{x}(0) = 1$.
   (e) $\ddot{x} + x = t$ with IC $x(0) = 0, \dot{x} = 0$.

4. The parameters for a linear oscillator are (a) $\omega_0 = 5$, $\gamma = 0.5$, (b) $\omega_0 = 5$, $\gamma = 15$. Solve for $x(t)$ when the initial conditions are $[x(0) = A, v(0) = 0]$.

5. Write down the equation of motion and general solution for a forced spring–mass system with forcing (a)$A\sin(\alpha t)$; (b) $k\exp(-\beta t)$; (c) At. Assume the mass and the spring constant to be 1 each. Obtain the solution for all the three cases with initial condition $[x(0) = A, v(0) = 0]$.

6. Consider a potential $U(x) = -x^2/2$.

   (a) At $t = 0$, position and velocity of a particle of mass 1 unit are $x(0) = 2, \dot{x}(0) = 1$. Write down the equation of motion and the solution for the particle. Find the time required for the particle to reach the point $x = 3$?
   (b) Three particles of unit mass each start from $x = -2$ with velocities 1, 2, and 3 units. How long would these particles take to reach $x = 0$? What would be the speed of the particles at the origin?

7. A particle of mass $m = 2$ units is moving under the influence of potential $U(x) = x^2/2$. The initial condition of the particle is $x(0) = 2, \dot{x}(0) = 1$. How long will the particle take to reach the origin? What would be the speed of the particle at the origin?

8. A particle is in motion under the influence of a potential $U(x) = 4/x^3 - 3/x^2$. Describe the motion of the particle for energy $E < 0$, $E = 0$, and $E > 0$.

9. What is the period of a one-dimensional motion of a particle of mass m and energy E in a potential field

    (a) $U = x^4/4$
    (b) $U = -\dfrac{U_0}{\cosh^2 \alpha x}$ with $-U_0 < E < 0$
    (c) $U = U_0 \tan^2(\alpha x)$

10. A particle of mass m is moving under the influence of potential $U(x) = A|x|^n$, where $n > 0$. Argue that the particle will execute a periodic motion. Derive an expression for the time period of the particle with total energy E. What is the dependence of time period with n? Is your result consistent with the oscillator problem?

11. A bug of mass 2 gm rests on a speaker diaphragm. The diaphragm performs a simple harmonic motion in the vertical direction with amplitude of 2 cm and frequency 100 Hz. Find the maximum and minimum apparent weight experienced by the bug. Also find the range of frequencies at which the bug will remain in contact with the diaphragm throughout the motion.

Projects

1. Design a parachute.

6

Phase Space Description of Mechanical Systems

In the last chapter we discussed how to solve the equation of motion and the energy equation for a particle moving under the influence of a force. In this chapter we will introduce the phase space analysis of a mechanical system.

In dynamics, a particle traces a trajectory in real space (also called configuration space). The trajectory of a particle provides very useful information about its motion, yet this description has several serious limitations. As an example, consider a particle moving with a constant velocity along the x-direction. The trajectory, which is a straight line, does not provide us any information regarding the particle's velocity. A slow moving particle as well as a fast moving one follow the same trajectory.

To overcome this difficulty, we can plot the trajectories of a system in an abstract space called *phase space* whose axes are the position and the velocity (x, v).[1] For a free particle

$$\dot{x} = v_0,$$

$$x = v_0 t + x_0,$$

which is depicted in Fig. 6.1. Faster particles have larger y coordinates than the slower ones. Hence the phase space plot provides a unique description of the mechanical system, and it contains more information than the real space trajectories.

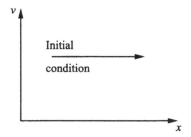

Figure 6.1 *Phase space plot of free particle*

---

[1]Strictly speaking the variables of the phase space are the position, x, and the momentum, p. But for many mechanical systems $p = mv$, so we can also choose velocity instead of momentum as the axis.

We know from Newton's laws that given its initial position and velocity, the future of a particle is uniquely determined. Hence, a system follows a unique path from a point in the phase space. For a particle moving in three dimensions, the phase space is six-dimensional with axes (x, y, z, v_x, v_y, v_z). This six-dimensional space is quite difficult to visualise. Therefore, scientists have devised clever techniques to study phase space in higher dimensions. These ideas are beyond the scope of this course; here we will only study the basics of the two-dimensional phase space structure using several illustrative examples.

6.1   Simple harmonic oscillator revisited

Consider a simple harmonic oscillator whose block has unit mass. Hence the linear momentum $p = v$. We can rewrite the equation of motion $\ddot{x} = -x$ as two first-order differential equations:

$$\dot{x} = v, \tag{6.1}$$

$$\dot{v} = -\omega_0^2 x. \tag{6.2}$$

So $$\frac{dv}{dx} = \frac{-\omega_0^2 x}{v},$$

or $$\frac{v^2}{2} + \omega_0^2 \frac{x^2}{2} = \text{const} = E.$$

Note that the constant is the total energy of the spring–mass system. The phase space trajectories for different values of E are ellipses as shown in the Fig. 6.2. The larger ellipses in the figure have proportionally larger energies.

The physical interpretation of the phase space trajectories is as follows. Point A has zero velocity, hence it corresponds to the maximum stretched configuration of the oscillator. When the mass is released from this position, it moves to the left with negative velocity. Hence the motion of the system along the phase space trajectory is clockwise, and it goes toward point B. The speed of the particle increases during this motion. At B, the speed is maximum. Subsequently the speed of the particle decreases. At C, which corresponds to the maximum stretched configuration to the left, the velocity of the particle is zero. After C, the particle changes direction of motion, and moves towards D. During this process, the velocity of the particle is positive. At D, the velocity of the particle is maximum. Subsequently the velocity of the particle decreases and it reaches A, which is the original configuration of the oscillator. The above motion repeats after that.

Phase space trajectories for an oscillator are concentric ellipses except at the origin. The elliptical trajectories correspond to different amplitudes of the oscillator depending on the values of energy E.

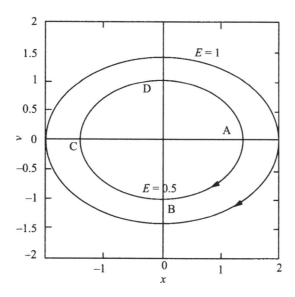

Figure 6.2   *Phase space picture of an oscillator whose frequency* $\omega_0 = 1/\sqrt{2}$.

A single non-elliptic phase space trajectory is the origin itself. If the system starts from the origin, it stays there forever because $\dot{x} = 0$ and $\dot{v} = 0$ at the origin. Hence, for the initial conditions $[x(0) = 0, v(0) = 0]$, we have $[x(t) = 0, v(t) = 0]$ at all times. These types of points are called fixed points for obvious reasons.

DEFINITION   *The points at which $\dot{x} = 0$ and $\dot{v} = 0$ are called fixed points* (FP).

If a system starts from a fixed point, it will remain there forever. It turns out that fixed points play a major role in characterising mechanical systems. We cannot discuss these issues in detail. The phase space trajectories for the simple harmonic oscillator are ellipses in the neighbourhood of the FP. These types of FPs are called *centre*.

In the next example, we will consider the potential $U(x) = -x^2/2$, which is the potential of a generic unstable systems near the equilibrium points.

6.2   Potential function $U(x) = -x^2/2$

In the following discussion, we will construct the phase space trajectories for a system having the potential function $U(x) = -x^2/2$. The equation of motion is

$$\ddot{x} = -\frac{dU}{dx} = x,$$

and $x = 0$ is the unstable equilibrium point. We can rewrite the above equation in terms of two first-order DEs:

$$\dot{x} = v,$$

$$\dot{v} = x.$$

So

$$\frac{dv}{dx} = \frac{x}{v},$$

or

$$\frac{v^2}{2} - \frac{x^2}{2} = \text{const} = E,$$

where E is the energy of the system. The phase space trajectories for various energies are shown in Fig. 6.3.

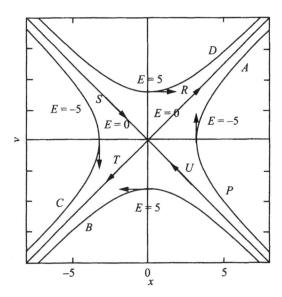

Fig. 6.3 *Phase space structure of an unstable system near its equilibrium point.*

The phase space trajectories have interesting interpretations. The curves B and D have $E > 0$. Suppose we start from some point on the curve D with $x_0 < 0$, then the particle moves right, and its velocity decreases till $x = 0$. After $x = 0$, the particle moves to the right with an increasing velocity as it continues to move to ∞. The trajectory B represents the motion of the particle from the right to left with similar features.

The curves A and C have $E < 0$. Suppose the particle starts from point P on curve A. The particle moves left and its velocity decreases, and at some point it becomes zero. Then the particle starts to move to the right. The particle's velocity increases as the particle moves to the right. Curve C has similar interpretation.

The condition $\dot{x} = \dot{y} = 0$ yields the fixed point $(0,0)$. This point has $E = 0$. It is curious to note that the two straight lines $v = \pm x$ also correspond to $E = 0$. In Section 5.2.2 we studied the motion of a particle when $E = 0$. The lines R and T correspond to the particle moving away from the origin, while the lines S and U are the phase space trajectories of the particle when it is moving towards the origin. Note that along S and U the particle cannot reach $x = 0$ in finite time. It asymptotically reaches there at $t = \infty$.

Do the lines TR and SU intersect each other? The answer is no! A particle placed on curves S or U will never reach the origin, hence S and U never join. Similarly the lines R and T do not start from the origin; if the particle starts from $(0,0)$, it would stay at the origin forever. Hence, SU and RT never intersect at the origin. In the limiting case, S and U approach the origin as $t \to \infty$, and R and U approach the origin as $t \to -\infty$. Hence, the future of systems at $x = 0$ and on the curves R, S, T, U never collide with each other, which is consistent with the statement of Newton's laws that the initial conditions of a system uniquely determine the future of the system.

The behaviour of the unstable system near this FP is very different from the corresponding behaviour of the FP of a spring–mass system. Technically this type of fixed point is called a *saddle*. All unstable systems with potential $-x^2$ near their equilibrium point have such behaviour near the fixed points.

General mechanical systems are much more complex than the simple harmonic oscillator or the unstable system discussed above. To analyse these systems, one typically starts with the analysis of the fixed points. The local behaviour of the fixed points provide us some clue as to how the mechanical system behaves at other points. We illustrate the procedure using an example.

EXAMPLE 6.1   Consider the potential $x^2/2 - x^4/4$. Assume $m = 1$. Write down the equation of motion, and sketch the phase space plots.

SOLUTION   The potential of the particle is $U(x) = x^2/2 - x^4/4$ which is plotted in Fig. 6.4(a). The equation of motion for the particle is

$$\ddot{x} = -\frac{dU}{dx} = -x + x^3.$$

that can be rewritten as

$$\dot{x} = p$$

$$\dot{p} = -x + x^3.$$

The condition $\dot{x} = \dot{p} = 0$ yields $(0,0)$ and $(\pm 1, 0)$ as fixed points. The potential plot is given in Fig. 6.4(a). Clearly $(0,0)$ is a stable FP, and $(\pm 1, 0)$ are unstable FPs. Near $(0,0)$, the equations are

$$\dot{x} = p,$$

$$\dot{p} = -x,$$

which is the equation of the oscillator. Hence, near $(0,0)$ the behaviour will be the same as that near the oscillator. Now near $(1,0)$, we will make a change of variable $x' = x - 1$. In terms of (x', p), the equations are

$$\dot{x}' = p',$$

$$\dot{p}' = 2x',$$

which are the equations for the unstable potential discussed in Section 6.2. The only difference is the factor of 2, which only changes the slope of the trajectory near the FP. Hence, the phase space around $(1,0)$ should look like that in Fig. 6.3. The same thing holds for $(-1,0)$ as well. From the potential plot we can see that $(\pm 1, 0)$ are unstable points. When we put together all the ingredients, the resulting phase space plot is as seen in Fig. 6.4(b). Interpret all the phase space trajectories.

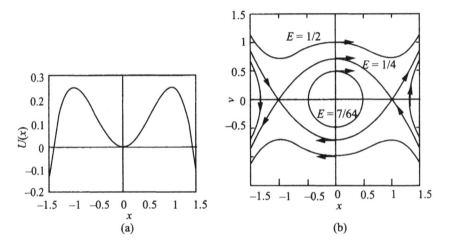

Fig. 6.4   (a) Plot of potential $U(x) = x^2/2 - x^4/4$ as a function of x. (b) Phase space trajectory for a particle moving under the potential $U(x) = x^2/2 - x^4/4$.

6.3   No intersection theorem

The future of a mechanical system is uniquely determined given its initial condition. From this property follows the *no intersection theorem* stated as

The phase space trajectories cannot intersect except at the saddle points.

The proof of the no intersection theorem is as follows.

PROOF   If two phase space trajectories intersect at a point say P, then at P the system has two futures as shown in the Fig. 6.5. This is impossible because a system's future is uniquely determined given its initial conditions. Therefore, our assumption that phase space trajectories can intersect is incorrect. Hence, the no intersection theorem. The above argument however does not rule out asymptotic intersection of the phase space trajectories at the fixed points. If the system is at the fixed point, it remains there forever. The trajectories that appear to intersect the fixed point only reach the fixed point asymptotically.

Hence, there is no contradiction with Newton's laws that the future of a mechanical system is uniquely determined given its initial conditions.

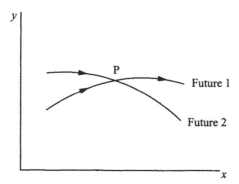

Figure 6.5   *We reach a contradiction if two phase space trajectories intersect at a point P. The system would have two different futures from P onwards.*

In this section, we provided a brief introduction to the phase space description of mechanical systems. We only performed the phase space analysis for one-dimensional systems. Two-dimensional and three-dimensional mechanical systems have four-dimensional and six-dimensional phase space respectively, which are harder to visualise. These analyses are beyond the scope of this book.

The phase space description of mechanical systems is very useful for the analysis of complex nonlinear systems. In Chapter 15, we will briefly discuss this topic using several examples. Here we will end our brief discussion on phase space with a discussion on some common misunderstandings students have regarding potential and energy.

6.4   Some tricky points regarding potential and energy

1. **Does a particle always move from a higher potential to a lower potential?**

   No! If a particle starts from rest, then it moves from a higher potential to a lower potential. However, if a particle has a finite kinetic energy, then it could go from a lower potential to a higher potential, at least for a while. As an example, for potential $U = -x^2/2$, when the particle starts with positive velocity and negative position, it moves from lower potential to higher potential. Note that the direction of force is locally from a higher to a lower potential (negative of the direction of potential slope), which is not necessarily the direction of motion.

2. **Can the total energy of a system be negative?**

   Yes! The total kinetic energy cannot be negative, however the total energy of a system, which is the sum of the kinetic and potential energy, can be negative. We saw this for the potential $U(x) = -x^2/2$. Also note that we can always add a constant to the potential, hence to the total energy.

3. Are equilibrium points and fixed points the same?

No! Equilibrium points are the points in configuration (or real) space where the force is zero, or the potential is extremum. If a particle has zero velocity at this point, then it will stay there forever. However, if a particle has a finite velocity at the equilibrium point then it will move either away from the equilibrium point, or toward the equilibrium point. For the oscillator and the unstable system discussed in this chapter, $x = 0$ was a equilibrium point.

Fixed points are points in phase space where $\dot{x}=0$ and $\dot{v} = 0$. If a system starts from a fixed point, it will stay there forever. Note that the condition $\dot{v} = 0$ is an equilibrium condition, but $\dot{x} = 0$ is an additional condition for the fixed point. For the simple harmonic oscillator and unstable system $(x = 0, v = 0)$ is the fixed point, not $(x = 0)$ which is a line in the phase space.

So far we have discussed the solution of equations of motion, the energy method, and phase space. We have illustrated the ideas by solving some simple examples analytically. However many mechanical systems are not amenable to analytic treatment. To understand these systems, we use computers. In the following section, we will briefly discuss a numerical procedure to solve the equations of motion.

6.5   Solution of an equation of motion on a computer

Newton's equation can be solved analytically only for a small set of physical systems. For most systems, we either use approximate methods, or numerical techniques. In this section, we will provide a numerical prescription for solving Newton's equation of motion for relatively simple systems.

Newton's equation for a particle of mass m moving in one dimension is

$$\ddot{x} = f(x, t).$$

We rewrite the above second-order differential equations (DE) as two first-order DEs:

$$\dot{x} = v,$$

$$\dot{v} = f(x.t).$$

We could solve the above equation using a small computer program in FORTRAN, C, C++ or Java. However we use a more comfortable approach. We use a computer software named *Matlab* to solve the above equation. Octave is another software that is very similar to Matlab; most Matlab programs run in Octave without any modification and vice versa. Simple usage of Matlab and Octave is briefly discussed in Appendices D.

We can solve a DE using the following Matlab function and commands. For explanations refer to Appendix D.

1. Save the function definition in a m-file $f.m$

```
function xdot = f (x, t)
xdot= [x(2); f];
```

where f has to be replaced by the function definition. Here v is $x[2]$.

2. Use the following command to solve DE.

```
tspan=[0 2*pi];
initcond=[1 0];
[t,x]=ode45(@f, tspan, initcond);
```

When the commands have been executed, array x would contain the position of the particle at various values of t in the range $[0, 2\pi]$ for the initial condition $[x(0) = 1, v[0] = 0]$. We illustrate the usage of the above procedure using several examples.

EXAMPLE 6.2   For an oscillator, $f(x, t)/m = -\omega^2 x$. We take $\omega = 2$. So we replace the second line of $f.m$ by

```
omega=2.0;
xdot = [x(2); -omega^2*x(1)];
```

and run the commands of item 2. We illustrate the plot of x vs. t and v vs. t in Fig. 6.6. The phase space plot of x vs. v can also be plotted. The phase space would look similar to Fig. 6.2

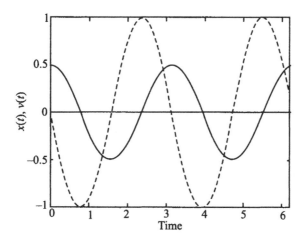

Figure 6.6   *Plot of $x(t)$ vs. t (solid line) and $v(t)$ vs. t (dashed line) for the equation of motion $\ddot{x} = -4x$. We take the initial condition $(x, v) = (0.5, 0)$.*

EXAMPLE 6.3   To solve $\ddot{x} = x - x^3$ using Matlab, we replace the second line of $f.m$ by

```
xdot= [x(2); x(1)*(1-x(1)^2)];
```

and run the *ode45* commands again for various initial conditions. The phase space plot of x vs. v is plotted in Fig. 6.7.

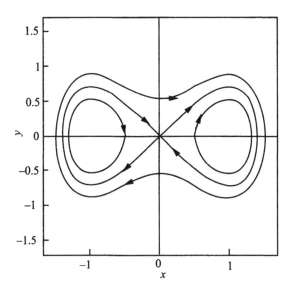

Figure 6.7   *Phase space trajectory of the particle whose equation of motion is $\ddot{x} = x - x^3$.*

Exercises

1. Sketch the phase space trajectories for a charged particle moving in a constant electric field. Assume one-dimensional motion.

2. Consider a particle of mass m moving under the influence of the following potentials. Write down the equation of motion, and sketch the phase space plots.

   (a) $U(x) = -x^2/2 + x^4/4$
   (b) $U(x) = x^2/2 - x^4/4$
   (c) $U(x) = x^2/2 - x^3/3$

3. The equation of motion of a particle is given by

   $$\ddot{x} = 1 - x.$$

   Consider the mass of the particle to be 1. The particle starts from rest at $x = 0$.

   (a) Identify x at which the velocity will again become zero. How long would it take to reach there?

   (b) Consider the phase space for the particle. Identify the fixed point(s) of the system.

(c) Write down the equation for the phase space trajectories.

(d) Draw the phase space trajectories of the particle.

4. Consider a particle of mass 1 unit moving under the influence of the following potentials. Write down the equation of motion, and solve it using a computer for various initial conditions. Plot $x(t)$, $v(t)$, and phase space trajectories.

(a) $U(x) = -x^2/2 + x^4/4$

(b) $U(x) = x^2/2 - x^4/4$

(c) $U(x) = x^2/2 - x^3/3$

(d) $U(x) = 1 - \cos(x)$.

Projects

1. Consider the following nonlinear oscillator

$$\ddot{x} + \dot{x} + x + \alpha x^3 = F_0 \cos t.$$

This equation is called the *Duffing oscillator*. Numerically solve the above equation for various values of α and F_0.

7

Symmetry Properties of
Newton's Equation

In this chapter we will discuss the *symmetry* properties of Newton's equations of motion. These properties are quite simple to state, yet they are very powerful. As an example, Newton's law of motion has the same form in all inertial frames—this is a symmetry property. Because of this property we can choose any convenient reference frame to study a mechanical system. There are other deep connections, e.g., conservation laws (conservation of linear momentum, angular momentum, and energy) are related to certain symmetry properties. We will discuss some of these issues in this chapter.

Experiments show that at microscopic scales or at large speeds, Newton's laws are not valid, yet some of the symmetry properties exhibited by the laws are valid even in those regimes. Hence, symmetries are more fundamental than Newton's laws.

In this chapter we will discuss basic symmetries: space translation, space rotation, time translation, time reversal, and mirror. We will also discuss symmetry under Galilean transformation. Let us start the chapter with a definition of symmetry.

7.1 Definition of symmetry

When we rotate a cylinder about its axis, the configuration or the view is the same as before. However, a cube when rotated by 10° about the vertical axis does not have the original configuration or view from a fixed angle. The rotation about an axis is an example of an operation, and objects that remain unchanged (or have the same appearance as before) under the operation are said to be *symmetric under the operation*. Formally,

A body is symmetric or invariant under some operation when it has the same appearance before and after the operation.

As discussed before, a cylinder is symmetric under an arbitrary rotation about its axis. A cube is not symmetric under arbitrary rotation, but it is symmetric under a 90° rotation about its symmetry axis. In a similar manner we can also define the symmetry of physical laws. Symmetry for physical objects could be visualised comparatively easily, but the symmetry of physical laws are more abstract and mathematical. The symmetry of a phys-ical law is stated as follows:

A physical law is symmetric or invariant under some transformation when the mathematical expression of the law looks the same before and after the transformation.

In the following sections we will discuss the operations given below and test whether New-ton's laws are invariant under them:

1. Shift of coordinate system

2. Rotation of coordinate system

3. Time translation

4. Mirror reflection

5. Time reversal

6. Galilean transformation

7.2   Invariance under the shift of coordinate system

In this section we will explore whether Newton's equation of motion is invariant if the coordinate system is shifted. Let us consider two inertial frames I, and I' that is shifted by a vector $-\mathbf{S} = -S_x\hat{\mathbf{x}}$ relative to $I(x, y, z, t)$ (see Fig. 7.1). The transformation between the two coordinate systems is $\mathbf{x}' = \mathbf{x} + \mathbf{S}$, or

$$x' = x + S_x; \ \ y' = y; \ \ z' = z, \tag{7.1}$$

and $t' = t,$ \hfill (7.2)

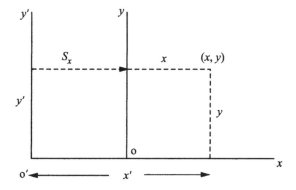

Figure 7.1   *Two inertial frames I and I'. I' is shifted to the left by S_x along the x-axis.*

i.e., time remains unchanged. As an example, consider an experiment in which we investigate the motion of a test particle of charge q and mass m in the field of a large stationary charge Q placed at the origin of reference frame $I(x, y, z, t)$. The equation of motion for the test particle in the reference frame I is

$$m\frac{d^2\mathbf{x}}{dt^2} = \mathbf{F} \tag{7.3}$$

Now let us derive the equation of motion in the other reference frame $I(x', y', z', t')$. Simple algebra shows that the acceleration of the test particle is the same in both frames:

$$\frac{d^2\mathbf{x}}{dt^2} = \frac{d^2\mathbf{x}'}{dt^2} \tag{7.4}$$

The mass of the test charge is also the same in both the frames of reference, i.e., $m' = m$. Since the force between the charges q and Q depend on the distance between the charged particles (Coulomb's law), the electric force acting on the charge particle is the same in both inertial frames, i.e., $\mathbf{F}' = \mathbf{F}$.

Equations (7.3) and (7.4), and the equality of the force immediately yields

$$m\frac{d^2\mathbf{x}'}{dt^2} = \mathbf{F}' \tag{7.5}$$

Hence *Newton's equation of motion is invariant under the coordinate shift if the forces acting on the particle are independent of the absolute position of particles (force depends on the relative separation between particles).* It is easy to check that the above property holds for any arbitrary shift of the coordinate system while keeping the orientation unchanged.

This symmetry property has important consequences. Newton's equation of motion is applicable everywhere irrespective of where an experiment is performed. We do not need to worry about where we are going to perform the experiment. This property of space is referred to as *homogeneity of space*. Note that shifting of the setup is equivalent to a shift of the coordinate system in the opposite direction.

An important ingredient of the above symmetry argument is the invariance of forces under coordinate shift. Experiments show that all the fundamental forces of nature are invariant under this symmetry operation, hence the laws of motion involving fundamental interactions are invariant under coordinate shift. Note however, that the condition $\mathbf{F} = \mathbf{F}'$ does not hold under every situation. For example, the gravitational force acting on a stone near the surface of the Earth varies with distance. To preserve the symmetry of coordinate shift, we need to translate the whole system, Earth+stone. Hence care must be taken while applying symmetry arguments.

7.3   Invariance under the rotation of coordinate system

In this section we will consider the symmetry operation of space rotation and test if Newton's equation of motion is invariant under this operation. We take two coordinate systems I and I' that make an angle θ with respect to each other as shown in Fig. 7.2. We wish to write down the equation of motion for a test particle located at point P. The coordinates of the test particle are different in the two reference frames, but are related to each other by the following transformation:

$$x' = x\cos\theta + y\sin\theta \tag{7.6}$$

$$y' = -x\sin\theta + y\cos\theta. \tag{7.7}$$

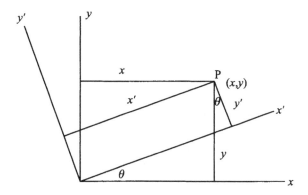

Figure 7.2 *The reference frame I' makes an angle θ wrt I. The coordinates of point P in the two frames are (x, y, z) and (x', y', z') respectively.*

The force acting on the particle at P is the same in both the reference frames, hence the components of force will also transform in the same way as $\mathbf{r}$, i.e.,

$$F'_x = F_x\cos\theta + F_y\sin\theta$$

$$F'_y = -F_x\sin\theta + F_y\cos\theta.$$

Time and mass remain unchanged under rotation. Hence,

$$m\ddot{x}' = m\ddot{x}\cos\theta + m\ddot{y}\sin\theta$$

$$= F_x\cos\theta + F_y\sin\theta$$

$$= F'_x$$

$$m\ddot{y}' = -m\ddot{x}\sin\theta + m\ddot{y}\cos\theta$$

$$= -F_x\sin\theta + F_y\cos\theta$$

$$= F'_y$$

Therefore, the form of Newton's equation of motion is the same in both frames of reference. We can choose any one of them for our analysis. Hence, *Newton's equation of motion is invariant under the rotation of coordinate system if the force is independent of the absolute orientation of the coordinate system.* Here we have shown the invariance of Newton's laws under the rotation about the z-axis. This symmetry property holds for an arbitrary rotation, however its derivation is beyond the scope of this book.

Because of the above symmetry property, we can orient our setup in any direction we wish, and still apply the same equation of motion. Note that the rotation of a setup by an angle is equivalent to a rotation of the coordinate axis by the same angle in the opposite direction.

The fundamental forces between two particles do not depend on the absolute orientation of the coordinate system. This property of space is called *isotropy of space. There is no preferred direction in the universe.* However this property does not hold in the presence of an external force. The solution of Newton's equation is not symmetric under rotation in an external field. For example, the projectile motion in a gravitational field $g\hat{x}$ depends on the initial angle of the velocity vector with respect to $\hat{x}$. However, the result is unchanged if we rotate both the initial velocity vector and the external field. Thus, if we take all the influencing agencies in a system and turn all of them by an angle, the result remains the same. Thus the isotropy of space is again preserved.

Experiments show that all the fundamental forces of physics are invariant under the rotation of the coordinate axis, or the rotation of the whole setup. There is no preferred direction in space.

7.4 Modern definition of vectors

The laws of motion are unchanged in whatever coordinate system we measure the components. Therefore, it is sensible to treat $\mathbf{r}$, $\mathbf{v}$, $\mathbf{F}$ etc. as objects (vectors) and write the physical laws in terms of these objects, rather than writing the laws in terms of components. When we need to measure physical quantities, we choose a coordinate system and measure their components.

In school textbooks, vectors are defined as quantities having magnitudes and directions. This definition is not very useful for mathematical formulation of physical laws. The modern definition of vectors is based on symmetry operation.

Any quantity whose components transform in the same way as the components (x, y, z) is called a vector.

Some examples of vectors are velocity, force, acceleration etc. Using the definition of vectors, it is clear that Newton's law $\mathbf{F} = m\mathbf{a}$ is invariant under the rotation of coordinate systems.

In addition to vectors in physics, we come across *quantities that do not change under rotation; these quantities are called scalars.* Some examples of scalars are mass of a particle, kinetic energy $mv^2/2$, potential energy etc.

If $\mathbf{A}$ and $\mathbf{B}$ are two vectors then we can show that $\mathbf{A} \cdot \mathbf{B}$ is a scalar. The components of vector $\mathbf{A}$ transform as following under the rotation of the coordinate system about the z-axis by an angle θ:

$$A'_x = A_x \cos\theta + A_y \sin\theta$$

$$A'_y = -A_x \sin\theta + A_y \cos\theta.$$

$$A'_z = A_z.$$

The components of vector $\mathbf{B}$ transform in a similar manner. It is easy to check that

$$A'_x B'_x + A'_y B'_y + A'_z B'_z = A_x B_x + A_y B_y + A_z B_z$$

or $\mathbf{A'} \cdot \mathbf{B'} = \mathbf{A} \cdot \mathbf{B}$. Hence $\mathbf{A} \cdot \mathbf{B}$ is a scalar. It follows from this result that $v^2 = \mathbf{v} \cdot \mathbf{v}$ is also a scalar.

There are other physical quantities like the dielectric constant which are neither scalar or vector. They are called *tensors*. For details on tensors, please refer to Appendix E.

Invariance under rotation implies that the quantities appearing in the equations of physics must transform as a scalar, vector, or tensor. We can rule out many expressions, e.g., addition of a scalar and component of a vector, $m\ddot{x} = F_y$, etc. These features make symmetry properties very useful in physics.

7.5   Invariance under time translation

The symmetry operation for time translation is $t = t' + s$ and $\mathbf{x'} = \mathbf{x}$. Clearly, $d^2\mathbf{x'}/dt'^2 = d^2\mathbf{x}/dt^2$. If we assume that the force is invariant under time translation, i.e. $\mathbf{F} = \mathbf{F'}$, then we can deduce that

$$m\frac{d^2\mathbf{x'}}{dt^2} = \mathbf{F'},$$

given $md^2\mathbf{x}/dt^2 = \mathbf{F}$. These arguments prove the invariance of Newton's equation of motion under time translation if the forces are symmetric under time translation. Experiments till date show that all the fundamental forces of nature are symmetric under time translation, i.e., they do not depend on time explicitly. External time-dependent forces do not obey this symmetry property.

We illustrate time translation symmetry using an experiment. If I shoot a bullet at 7 AM every morning, I expect the range of the bullet to be the same every morning assuming that the gravitational force every morning is the same ($\mathbf{F'} = \mathbf{F}$). If the Earth were to lose a lot of its matter for some reason, then the force experienced by the bullet on two different mornings would not be the same, and Newton's law would give different predictions for the range of the bullets. Hence, the conditions on the Earth has to be the same for the symmetry properties to hold. Note that fundamental forces, like the electric force between two electrons separated by a fixed distance are found to the same at all times.

Consider a kind of force $\mathbf{F}$ which can be written as a derivative of a potential $U(\mathbf{r})$, where $\mathbf{r}$ is the position of the particle. The electrostatic force on a test charge due to a fixed charge is an example of such force. These types of force or potentials do not depend explicitly on time. In one dimension, the equation of motion of the charged particle is

$$m_e \frac{d^2x}{dt^2} = m_e \frac{dv}{dt} = -\frac{dU(x)}{dx}.$$

By multiplying both sides by v, and using $v = dx/dt$, we obtain

$$\frac{d}{dt}\left(\frac{1}{2}mv^2 + U(x)\right) = 0$$

Hence $\frac{1}{2}mv^2 + U(x)$, which is energy, is a constant in time. The above calculation suggests that the conservation of energy may be a consequence of the invariance of Newton's law under time translation. We will show later that it is indeed the case.

The above simple calculation shows that the symmetry properties are connected to conservation laws.

7.6   Invariance under time reversal

This is an important symmetry property. Imagine a set of particles that are moving due to internal forces. After a certain time T, we revert the velocity of each particle. What will happen to the particles? It can be shown that for gravitational and electromagnetic forces, the system will revert to its initial state. This is called time-reversal symmetry. Let us define the above symmetry in more detail.

Suppose a particle starts to move from point A at time $t = 0$, and at $t = t_f$ the particle reaches point B whose coordinate is x_f. At $t = t_f$, we reverse the direction of velocity and let the particle go as shown in Fig. 7.3. This transformation is called time reversal. Mathematically, the time reversal transformation is $t' = t_f - t$ and $\mathbf{x}' = \mathbf{x}$.

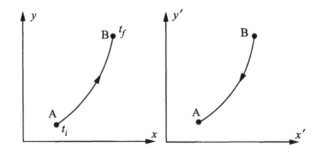

Figure 7.3   (a) Particle starts from point A at $t = 0$, and reaches point B at time $t = t_f$. (b) The velocity of the particle is reversed at B. The particle will reach point A if the force remains the same in the reversed path.

A simple mathematical analysis shows that under time reversal

$$\frac{d^2\mathbf{x}'}{dt'^2} = \frac{d^2\mathbf{x}}{dt^2}$$

If the force does not change with time ($\mathbf{F}' = \mathbf{F}$), then

$$m\frac{d^2\mathbf{x}'}{dt'^2} = \mathbf{F}'$$

Hence, the form of Newton's equation remains unchanged under time reversal. Therefore, *Newton's equation of motion is symmetric under time reversal if the force is invariant under time reversal.*

In the second experiment, the initial position is $\mathbf{x}_f$, and the initial velocity is opposite to that in the first experiment. If the force acting on the particle remains the same at each point, then the trajectories traced will be the same in both the experiments. This result can be shown from the equation of motion.

The above symmetry hinges on the invariance of forces under time reversal. Experiments show that gravity and electromagnetic forces are invariant under time translation. Surprisingly there is a small violation of the time-reversal symmetry in some nuclear interactions. Time-reversal symmetry is also broken for some macroscopic forces like friction. Consider a block moving on a rough surface. When we perform the time-reversal operation, the frictional force points in a direction opposite to the original direction. Consequently, the block does not return to its original position. In other words, the original trajectory and the time-reversed trajectory are not the same in the presence of friction.

7.7   Invariance under mirror reflection

Symmetry under mirror reflection is one of the most curious symmetries of physics. Suppose we perform two experiments A and B. The experiment B uses instruments that are exact replicas of the mirror image of the instruments used in experiment A. Parts of the instruments having no handedness will be identical in both A and B. However, parts having handedness (like coil, screws etc.) will be different in the two setups. If some coil is right-handed in experiment A, then its corresponding coil in B will be left-handed, and vice versa. Now the question is: Will the outcome of the experiments be different in the two setups?

Mathematically the symmetry operation (called mirror transformation) is

$$z' = z, \quad y' = -y, \quad x' = x; \quad t' = t.$$

We have used a right-handed coordinate system for both coordinate systems. See Fig. 7.4 for an illustration.

By taking the time derivative of the coordinates we obtain

$$a'_z = a_z, \quad a'_y = -a_y, \quad a'_x = a_x.$$

If the force transforms as

$$F'_z = F_z, \quad F'_y = -F_y, \quad F'_x = F_x \tag{7.8}$$

then $m\mathbf{a}' = \mathbf{F}'$.

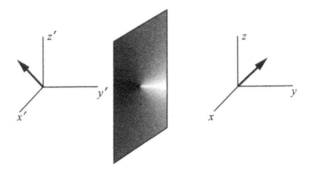

Figure 7.4   *Transformation under mirror reflection.*

Hence, Newton's law of motion is invariant under mirror transformation if the force obeys Eq. (7.8).

The transformation law for the force Eq. (7.8) appears obviously correct. Before 1957 every one thought so, and expected the mirror symmetry to be an exact symmetry of nature. But in an experiment, in 1957, Wu and her collaborators showed a violation of this symmetry. They found that the electrons from beta decay are emitted preferentially in the direction opposite to the angular momentum of the nucleus. Mirror symmetry is violated in this experiment. The arguments are as follows.

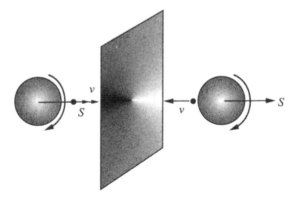

Figure 7.5   *In a beta decay experiment, the emitted electrons have preferential velocity opposite to the spin direction of the nucleus (right side). In the mirror image, the directions of velocity and spin are preferentially aligned, thus breaking the mirror symmetry.*

In the experiment of Wu and coworkers (A), the emitted electrons (beta particles) have preferential velocity opposite to the spin direction of the Cobalt-60 nucleus. Note that the angular momentum or spin along the mirror (y)-axis does not change sign under mirror reflection. If an object is turning clockwise with the rotation axis along the y-axis, then in the mirror too, it will be turning clockwise along the y-axis. Hence in the mirror image, the electrons would preferentially move in the direction of the spin as shown in the left side of

Fig. 7.5. Thus the law that the beta particles have preferential velocity opposite to the spin direction is violated in the mirror experiment. This is how mirror symmetry is violated in a beta decay experiment.

Careful investigations show that the above kind of anomaly is seen only in phenomena involving weak nuclear forces. Hence, the weak nuclear force is not mirror-symmetric. The other fundamental forces—gravity, electromagnetism, and strong nuclear forces—obey mirror symmetry.

In the macroscopic world we find many instances of mirror asymmetry. The DNA helix, shells of many biological species, and many molecules have definite handedness. These asymmetries are believed to have arisen due to various processes (sometime chance) happening in nature. The reader could find more details in the Internet or books.

EXAMPLE 7.1   Discuss symmetries of electric field due to a point charge Q.

SOLUTION   If the position of the charged particle and the test charge are $\mathbf{R}$ and $\mathbf{r}$ respectively, then the electric field at the test particle is

$$\mathbf{E}(\mathbf{r}) = \frac{Q}{4\pi\epsilon_0 |\mathbf{R} - \mathbf{r}|^3}(\mathbf{r} - \mathbf{R})$$

The above field is unchanged (a) if we shift both charge Q and the test charge by $\mathbf{s}$ (space translation); (b) if we shift the time by any amount (time translation); (c) under time reversal. (d) Also if we fix the origin at the source charge and rotate the axis by any angle, the electric field $\mathbf{E}$ is radial with the same magnitude (space rotation); (e) if we keep a mirror passing through the point charge Q. Hence Newton's equation of motion $m\ddot{\mathbf{r}} = q\mathbf{E}$ is symmetric under the above operations.

EXAMPLE 7.2   Discuss symmetries of electric field due to an infinite charged sheet.

SOLUTION   If the charged sheet is in the yz-plane, then the electric field $\mathbf{E}$ is along the x-axis. Hence $\mathbf{E}$ is symmetric under (a) translation along the x-axis; (b) rotation about any axis perpendicular to the yz-plane; (c) time translation; (d) time reversal; (e) mirror operation when the mirror coincides with the charged sheet. Note that if we rotate or translate the whole assembly (source and test charge), the equation remains symmetric. In the above examples, we kept the source fixed and studied the symmetry of the generated electric field.

7.8   Invariance under Galilean transformation

We discussed in Chapter 2 that Newton's laws of motion are applicable in all inertial frames. It immediately implies that Newton's laws of motion are invariant under transformations that take us from one inertial frame to another. Let us prove this result.

We take reference frames I and I', with I' moving with velocity $U\hat{\mathbf{i}}$ relative to reference frame I. The transformation rules connecting the coordinates of the two reference frames are

$$x = x' + Ut$$

$$t = t'$$

$$y = y'$$

$$z = z'$$

Note that absolute space and absolute time are inherently assumed in the above transformations.

We assume that the mass and the force remain the same in both the frames of reference, i.e., $m = m'$ and $\mathbf{F} = \mathbf{F}'$. Clearly

$$\mathbf{v} = \mathbf{v}' + U\hat{\mathbf{i}},$$

$$\mathbf{a} = \mathbf{a}'$$

Hence

$$m'\mathbf{a}' = \mathbf{F}'.$$

Therefore, Newton's laws are invariant under the above transformation, which is called Galilean transformation in honour of the man who thought for the first time that the law of inertia is valid in all reference frames.

Both the reference frames I and I' are equivalent, and one can apply Newton's law in both the reference frames. Let us state some simple observations consistent with Galilean invariance.

1. A ball is thrown upward with velocity v inside a closed tram (reference frame I') that is moving with velocity U. The ball returns to the original position in time $T = 2v/g$. On the platform frame (reference frame I), the ball covers a distance $UT = 2Uv/g$ along the x-direction, yet the time taken by the ball to return to the bottom surface is the same. *The time interval between two events is always the same in the Newtonian framework of mechanics.*

2. A metre stick of one metre length is placed horizontally in the train. We can measure the length very easily in the train using a ruler. To measure the length of the stick from the platform, we need to take a photograph of the stick. We can do this effectively by shining many laser beams *released simultaneously* and recording the reflections. If the absolute space hypothesis is correct, then the length of the stick would be the same in both the reference frames.

Galilean transformation is consistent with the assumption of absolute space and time. Einstein later showed that the time difference between two events are not the same in I and I'. Also, the length measurement of the metre stick is different in the two reference frames. Hence, Galilean symmetry is not obeyed in nature. Einstein showed that Lorentzian symmetry rather than Galilean symmetry is an exact symmetry of nature. Lorentzian symmetry will be discussed in Chapter 19 when we study relativity.

We will show below that Newton's law are not applicable in reference frames that are accelerating relative to inertial frames.

7.9   Accelerating frames of reference

Consider an inertial reference frame I, and a noninertial frame I' that is accelerating with respect to the inertial frame I with acceleration $A\hat{x}$. Newton's law $\mathbf{F} = m\mathbf{a}$ is valid in the inertial frame I. We assume that the origin of I and I' coincide at $t = 0$, and the initial velocity of I' is zero. The transformation rules to go from I to I' are

$$x = x' + \frac{1}{2}At^2$$

$$t = t'$$

$$y = y'$$

$$z = z'$$

$$m = m'$$

$$\mathbf{F} = \mathbf{F}'$$

Application of derivatives on the above equations yield

$$\mathbf{v} = \mathbf{v}' + At\hat{x}$$

$$\mathbf{a} = \mathbf{a}' + A\hat{x}$$

The substitution of the above relationships in Newton's equation $m\mathbf{a} = \mathbf{F}$ yields

$$m\mathbf{a}' + mA\hat{x} = \mathbf{F}'$$

or

$$m\mathbf{a}' = \mathbf{F}' - mA\hat{x} \qquad (7.9)$$

The form of equation is not the same in I and I' . Hence, Newton's laws are not invariant under the above transformation. Since $m\mathbf{a}' \neq \mathbf{F}'$, Newton's law of motion is not applicable in the accelerating frame I'. However if we postulate a *pseudo force* mA in the direction opposite to that of the acceleration, then the equation of motion in the form of Eq. (7.9) can be applied.

Let us summarise various symmetry principles discussed in this chapter. The first three symmetry operations discussed are coordinate shift, rotation of the coordinate system, and time translation. The four fundamental forces known so far are invariant under these transformations, hence Newton's laws are symmetric under these operations. These symmetry properties yield three conservation laws: Conservation of linear momentum due to the invariance of Newton's laws under coordinate shift, conservation of angular momentum due to the invariance of Newton's laws under coordinate rotation, and conservation of energy due to the invariance under time translation. We gave a crude proof of the conservation of energy in Section 7.5. In later chapters we will discuss conservation laws in more detail.

Mirror symmetry and time-reversal symmetry are approximate symmetries of nature. They do not hold for interactions involving nuclear forces. Galilean symmetry is a key symmetry of Newton's laws of motion. Experiments show that they work very well for speeds much less than the speed of light. However, when the particles move with speeds close to that of light (relativistic limit), Galilean symmetry does not hold. Einstein discovered that Lorentz symmetry rather than the Galilean symmetry is valid for these systems.

With these comments we close our discussion on the symmetries of Newton's laws.

Exercises

1. We discussed many symmetry transformations in the present chapter. Which of the symmetry properties are obeyed for the following forces? Keep the source fixed.

    (a) Gravitational force between a source point particle and a test point particle.

    (b) Force on a charged particle due an infinite line charge.

    (c) Gravitational field due to an infinite material plate.

    (d) Gravitational forces acting on the surface of the Earth.

    (e) Frictional force.

    (f) Force between two current-carrying conductors.

    (g) Weak nuclear force.

2. Show that $\mathbf{A} \times \mathbf{B}$ transforms as a vector given that $\mathbf{A}$ and $\mathbf{B}$ are vectors.

3. Which of the following quantities transform as vector $\mathbf{r}$ under rotation about the z-axis?

    (a) (x^2, y^2, z^2)

    (b) $(-x, y, z)$

    (c) (xy, yz, zx)

    (d) (xz, yz, z^2)

    (e) $(-y, x, z)$

    (f) $(x/r, y/r, z/r)$, where $r = \sqrt{x^2 + y^2 + z^2}$.

4. Consider a ball falling under gravitational field. Prove that this system is invariant under time reversal by solving the equation of motion under appropriate initial conditions.

5. Classify the following objects as scalar, vector, or none of these:

   (a) $(\mathbf{A} \times \mathbf{B}) \cdot \mathbf{C}$

   (b) $A_x^2 - A_y^2 + A_z^2$

   (c) $(\mathbf{A} \cdot \mathbf{B})\mathbf{C}$

   (d) $(\mathbf{A} \cdot \mathbf{B})(C_x + C_y + C_z)$

8

Two-dimensional Motion;
Central Force Problem

In Chapter 8, we applied Newton's law to study motion in one-dimensional systems. In this chapter we will do the same for two-dimensional systems. First we will solve the projectile motion in the Cartesian system. Then we will introduce an important coordinate system called the radial–polar coordinate system. This system is very useful for solving the central force problem. We will solve Kepler's problem and discuss the solution.

In the following section, we will solve projectile motion with and without air drag.

8.1 Two-dimensional motion in the Cartesian coordinate system

In the Cartesian coordinate system, Newton's equation of motion in the x and y direction is

$$m\ddot{x} = F_x,$$

$$m\ddot{y} = F_y.$$

Once the force and initial conditions are specified, we can solve for $x(t)$ and $y(t)$. We will illustrate the application of Newton's equation of motion using several examples.

Projectile motion without air drag

We choose a coordinate system in which gravity acts along the $-y$ direction (Fig. 8.1). The projectile starts with velocity $\mathbf{v}(0) = v_x(0)\hat{\mathbf{x}} + v_y(0)\hat{\mathbf{y}}$ from the origin. The equation of motion is

$$m\dot{v}_x = 0,$$

$$m\dot{v}_y = -mg,$$

whose solution is

$$v_x(t) = \text{const} = v_x(0) \tag{8.1}$$

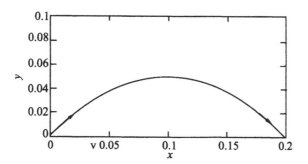

Figure 8.1 *Projectile motion with initial velocity* $\mathbf{v}(0) = (\hat{\mathbf{x}} + \hat{\mathbf{y}})$ *m/s and* $g = 10$ *m/s*^{2}.

$$v_y(t) = v_y(0) - gt. \tag{8.2}$$

Therefore

$$x(t) = v_x(0)t \tag{8.3}$$

$$y(t) = v_y(0)t - \tfrac{1}{2}gt^2. \tag{8.4}$$

The vertical velocity of the projectile starts with $v_y(0)$ and decreases until it becomes zero at $t^* = v_y(0)/g$. After $t = t^*$, the projectile starts a downward descent and hits the ground at $t_{\text{final}} = 2v_y(0)/g$. Therefore, the horizontal range of the projective is

$$L = v_x(0)t_{\text{final}} = \frac{2v_x(0)v_y(0)}{g},$$

and the maximum height the projectile can reach is

$$y_{\text{max}} = v_y(0)t^* - \frac{1}{2}g(t^*)^2$$

$$= \frac{v_y(0)^2}{2g}.$$

The trajectory is plotted in Fig. 8.1 for parameters $v_x(0) = v_y(0) = 1$ m/s and $g = 10$ m/s^2. Projectile motion is a trivial problem. We can make the problem more complex by introducing air drag.

Projectile motion with air drag

We assume that the air drag on the projectile is $-\gamma\mathbf{v}$, where γ is the frictional coefficient. This formula is valid for small velocities (strictly speaking for small Reynolds number; see Appendix B). In addition to the above, we assume that the frictional force is much smaller than the gravitational force, i.e.,

$$\gamma v \ll mg.$$

The equation of motion for the projectile is

$$m\dot{v}_x = -\gamma v_x,$$

$$m\dot{v}_y = -mg - \gamma v_y,$$

whose solution is

$$v_x(t) = v_x(0) \exp\left(-\frac{\gamma}{m}t\right), \tag{8.5}$$

$$v_y(t) = \left(v_y(0) + \frac{mg}{\gamma}\right) \exp\left(-\frac{\gamma}{m}t\right) - \frac{mg}{\gamma}, \tag{8.6}$$

with $[v_x(0), v_y(0)]$ as the initial velocity. Integration of the above equations yield

$$x(t) = \frac{mv_x(0)}{\gamma} \left[1 - \exp\left(-\frac{\gamma}{m}t\right)\right], \tag{8.7}$$

$$y(t) = -\frac{m}{\gamma}\left[v_y(0) + \frac{mg}{\gamma}\right]\exp\left(-\frac{\gamma}{m}t\right) - \frac{mgt}{\gamma} + \frac{m}{\gamma}\left[v_y(0) + \frac{mg}{\gamma}\right] \tag{8.8}$$

We have chosen the initial condition for the position as $x(0) = y(0) = 0$.

   The above solution appears a bit complex. We can plot these functions for various values of γ using a computer. It is also important to understand them by taking various limits. We use $\alpha = \gamma v_y(0)/(mg)$ as a small parameter for our analysis. Since $t \leq 2v_y(0)/g$, $\gamma t/m$ is small at all times, and we can expand $\exp(-\gamma t/m)$ as a Taylor series and keep only the leading order terms. Therefore

$$\exp\left(-\frac{\gamma}{m}t\right) = 1 - \frac{\gamma}{m}t + ...,$$

The time required for the projectile to reach the topmost point is obtained by setting $v_y(T/2) = 0$ in Eq. (8.6) and using the expansion of $\exp(-\gamma t/m)$. These operations yield

$$T \approx \frac{2v_y(0)}{g},$$

to the leading order. We can get an approximate value of the range for $\alpha \ll 1$ by substituting the above value of T in Eq. (8.7). The result is

$$x(T) = v_x(0)T - v_x(0)\frac{\gamma T^2}{2m} + ...,$$

The second term $\Delta L \approx -v_x(0)\gamma T^2/(2m)$ is a correction in the range of the projectile to a leading order. The relative decrease in the range of the projectile is approximately

$$\frac{\Delta L}{L} \approx \frac{\gamma T}{m} \approx \frac{2\gamma v_y(0)}{mg} \ll 1.$$

We could also rewrite the solutions $x(t)$ and $y(t)$ in dimensionless form. The time-scale of the projectile is $v_y(0)/g$. The length scales of the projectile along x and y directions are $v_x(0)T = v_x(0)v_y(0)/g$ and $v_y(0)T = [v_y(0)]^2/g$. Therefore the non-dimensional coordinates are

$$x' = x/\left(v_x(0)v_y(0)/g\right),$$

$$y' = y/\left([v_y(0)]^2/g\right),$$

$$t' = t/(v_y(0)/g).$$

In terms of these non-dimensional variables, the position of the projectile is

$$x' = \frac{1}{\alpha}\left[1 - \exp(-\alpha t')\right],$$

$$y' = -\frac{1}{\alpha}\left[1 + \frac{1}{\alpha}\right]\exp(-\alpha t') - \frac{t'}{\alpha} + \frac{1}{\alpha}\left[1 + \frac{1}{\alpha}\right]$$

In the limit $\alpha \to 0$, we obtain $x' = t'$ and $y' = t' - t'^2/2$ which is consistent with the results of the earlier example without any drag. In Fig. 8.2, we plot the trajectory (x', y') for $\alpha = 0$ (dotted line), $\alpha = 0.1$ (dashed line), $\alpha = 0.5$ (chained line), and $\alpha = 1$ (solid line),

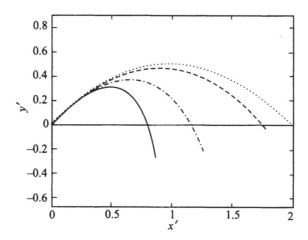

Figure 8.2   *Projectile motion with air drag for various values of* $\alpha = \gamma v_y(0)/(mg)$: $\alpha = 0$ *(dotted line),* $\alpha = 0.1$ *(dashed line),* $\alpha = 0.5$ *(chained line), and* $\alpha = 1$ *(solid line).*

In the above two examples, we used Cartesian coordinates for the analysis. In the next section we will introduce the radial–polar coordinate system.

8.2   Radial–polar coordinate system

The radial–polar coordinate system is very useful for analysing two-dimensional systems. In the radial–polar coordinate system, the position of a point particle is expressed using r and ϕ, where r is the distance of the particle from the origin, and ϕ is the angle of the radius vector with the x-axis. (See Fig. 8.3(a) for an illustration.) A common convention is to take counter-clockwise rotation as positive. We can easily derive the following transformation rules between (r, ϕ) and (x, y).

$$x = r \cos \phi,$$

$$y = r \sin \phi,$$

$$r = \sqrt{x^2 + y^2},$$

$$\phi = \tan^{-1} \frac{y}{x}.$$

In Cartesian coordinates, the position vector $\mathbf{r}$ is

$$\mathbf{r} = x\hat{\mathbf{x}} + y\hat{\mathbf{y}},$$

while in radial–polar coordinates $\mathbf{r}$ is

$$\mathbf{r} = r\hat{\mathbf{r}}.$$

The unit vectors in the radial–polar coordinate system are $\hat{\mathbf{r}}$ and $\hat{\phi}$. Note that these unit vectors change with the position of the particle. Hence, $d\hat{\mathbf{r}}/dt$ and $d\hat{\phi}/dt$ are non-zero unlike $d\hat{\mathbf{x}}/dt, d\hat{\mathbf{y}}/dt$ which are always zero (since $\hat{\mathbf{x}}$ and $\hat{\mathbf{y}}$ are constant unit vectors). Using simple geometry we can derive the following transformation rules between the unit vectors $(\hat{\mathbf{r}}, \hat{\phi})$ and $\hat{\mathbf{x}}, \hat{\mathbf{y}}$:

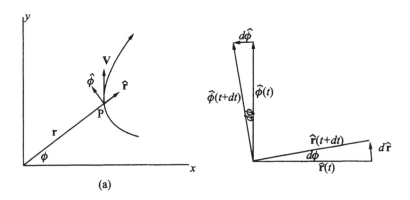

(a)

Figure 8.3   *(a) Radial–polar coordinate system. (b) Variation of unit vectors $\hat{\mathbf{r}}$ and $\hat{\phi}$ as a function of time.*

$$\hat{\mathbf{r}} = \hat{\mathbf{x}} \cos\phi + \hat{\mathbf{y}} \sin\phi, \tag{8.9}$$

$$\hat{\phi} = -\hat{\mathbf{x}} \sin\phi + \hat{\mathbf{y}} \cos\phi. \tag{8.10}$$

As the particle moves, both r and ϕ change as a function of time. The variation of the unit vectors after a short time Δt is shown in Fig. 8.3(b). We can easily compute $d\hat{\mathbf{r}}/dt$ and $d\hat{\phi}/dt$ by taking the limit $\Delta t \to 0$ as

$$\frac{d\hat{\mathbf{r}}}{dt} = \lim_{\Delta t \to 0} \frac{\hat{\mathbf{r}}(t + \Delta t) - \hat{\mathbf{r}}(t)}{\Delta t} = \lim_{\Delta t \to 0} \frac{\Delta\phi}{\Delta t}\hat{\phi} = \dot{\phi}\hat{\phi},$$

$$\frac{d\hat{\phi}}{dt} = \lim_{\Delta t \to 0} \frac{\hat{\phi}(t + \Delta t) - \hat{\phi}(t)}{\Delta t} = \lim_{\Delta t \to 0} -\frac{\Delta\phi}{\Delta t}\hat{\mathbf{r}} = -\dot{\phi}\hat{\mathbf{r}}.$$

We can also derive the above equations by taking time derivatives of eqs. (8.9) and (8.10) as follows:

$$\frac{d\hat{\mathbf{r}}}{dt} = -\hat{\mathbf{x}}\dot{\phi}\sin\phi + \hat{\mathbf{y}}\dot{\phi}\cos\phi = \dot{\phi}\hat{\phi}, \tag{8.11}$$

$$\frac{d\hat{\phi}}{dt} = -\hat{\mathbf{x}}\dot{\phi}\cos\phi - \hat{\mathbf{y}}\dot{\phi}\sin\phi = -\dot{\phi}\hat{\mathbf{r}}. \tag{8.12}$$

The velocity of the particle in the radial–polar coordinate is

$$\dot{\mathbf{r}} = \frac{d}{dt}(r\hat{\mathbf{r}})$$

$$= \frac{dr}{dt}\hat{\mathbf{r}} + r\frac{d\hat{\mathbf{r}}}{dt}$$

$$= \dot{r}\hat{\mathbf{r}} + r\dot{\phi}\hat{\phi}.$$

A particle going around the circle has velocity only along $\hat{\phi}$, while a particle going outward along a straight line has velocity only along $\hat{\mathbf{r}}$. A particle performing an arbitrary motion has velocity components along both these directions as shown in Fig. 8.4.

We can derive the acceleration of the particle in a similar manner:

$$\dot{\mathbf{v}} = \ddot{\mathbf{r}} = \frac{d}{dt}(\dot{r}\hat{\mathbf{r}} + r\dot{\phi}\hat{\phi})$$

$$= \ddot{r}\hat{\mathbf{r}} + \dot{r}\frac{d\hat{\mathbf{r}}}{dt} + (\dot{r}\dot{\phi} + r\ddot{\phi})\hat{\phi} + r\dot{\phi}\frac{d\hat{\phi}}{dt},$$

$$= (\ddot{r} - r\dot{\phi}^2)\hat{\mathbf{r}} + (2\dot{r}\dot{\phi} + r\ddot{\phi})\hat{\phi}.$$

The terms $(\ddot{r} - r\dot{\phi}^2)$ and $(2\dot{r}\dot{\phi} + r\ddot{\phi})$ are the components of acceleration along $\hat{\mathbf{r}}$ and $\hat{\phi}$ respectively. The following examples illustrate the nature of these terms.

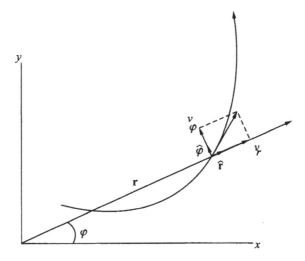

Figure 8.4   *A particle moving in two dimension has in general both tangential and radial components of velocity.*

1. **A particle going around a circle with constant angular velocity**
   Here $\dot{r} = \ddot{r} = 0$, $\dot{\phi} = \omega$, and $\ddot{\phi} = 0$. Substitution of these values in the expressions for the velocity and acceleration yields

   $$\mathbf{v} = \omega r \hat{\phi},$$

   $$\mathbf{a} = -r\omega^2 \hat{\mathbf{r}}.$$

   Newton's law $\mathbf{F} = m\mathbf{a}$ implies that a radial force is acting on the particle. This force is called the *centripetal force*.

2. **A particle going around a circle with angular acceleration**
   Here $\dot{r} = \ddot{r} = 0$, but $\ddot{\phi} \neq 0$. For this case,

   $$\mathbf{v} = \omega r \hat{\phi},$$

   $$\mathbf{a} = -r\omega^2 \hat{\mathbf{r}} + r\ddot{\phi}\hat{\phi}.$$

   The second term of $\mathbf{a}$ is due to the angular acceleration.

3. **A particle moving radially outward**
   Here $\dot{\phi} = \ddot{\phi} = 0$. Hence,

   $$\mathbf{a} = \ddot{r}\hat{\mathbf{r}}.$$

   This is a one-dimensional motion of the particle.

4. A particle moving radially outward with constant velocity ($\dot{r}$) and rotating with constant angular velocity ($\dot{\phi}$)

The velocity of the particle is

$$\mathbf{v} = \dot{r}\hat{\mathbf{r}} + r\dot{\phi}\hat{\phi},$$

and its acceleration is

$$\mathbf{a} = \frac{d}{dt}\dot{r}\hat{\mathbf{r}} + \frac{d}{dt}[r\dot{\phi}\hat{\phi}]$$

$$= [\dot{r}\frac{d\hat{\mathbf{r}}}{dt}] + [\dot{r}\dot{\phi}\hat{\phi} - r\dot{\phi}^2\hat{\mathbf{r}}]$$

$$= \dot{r}\dot{\phi}\hat{\phi} + [\dot{r}\dot{\phi}\hat{\phi} - r\dot{\phi}^2\hat{\mathbf{r}}]$$

$$= 2\dot{r}\dot{\phi}\hat{\phi} - r\dot{\phi}^2\hat{\mathbf{r}}.$$

The first term $\dot{r}\dot{\phi}\hat{\phi}$ arises because the radial velocity $\dot{r}\hat{\mathbf{r}}$ is rotating with angular velocity $\dot{\phi}$. The second term $\dot{r}\dot{\phi}\hat{\phi}$ results due to the increase in tangential velocity; in time dt, the tangential velocity $r\dot{\phi}$ increases to $(r + \dot{r}dt)\dot{\phi}$. These two terms put together yield acceleration $2\dot{r}\dot{\phi}\hat{\phi}$. The last term is the usual centripetal (item 1).

EXAMPLE 8.1   A wheel of radius R is rolling in a straight line without slipping. The centre of the wheel moves with constant speed V. A small pebble is glued at the end of a spoke of the wheel. The initial position of the bead is $(x = 0, y = 0)$. Find the pebble's position, velocity, and acceleration as functions of time.

SOLUTION   We assume that the wheel rolls along the x-axis as shown in Fig. 8.5(a).

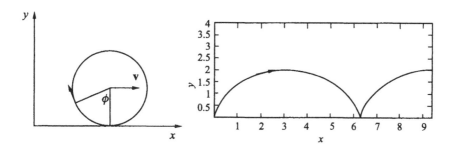

Figure 8.5   *(a) Motion of a pebble that is glued at the end of a spoke of the wheel. The wheel is rolling without slipping. (b) The trajectory of the pebble. The trajectory is called a cycloid.*

The initial position of the pebble is taken to be the origin $(0, 0)$. In the reference frame attached to the wheel, the velocity of the pebble is $-R\omega\hat{\phi}$, where $\hat{\phi}$ is the unit vector along the tangential direction in the rotating frame. We have chosen the counter-clockwise direction to be the positive direction (standard convention). We can compute the velocity of the pebble in the laboratory frame using the Galilean transformation, which yields

$$\mathbf{v} = V\hat{\mathbf{x}} - R\omega\hat{\phi}.$$

The no-slip condition at the bottom of the wheel implies that $V = \omega R$, hence the velocity of the pebble is

$$\mathbf{v} = R\omega(\hat{\mathbf{x}} - \hat{\phi}).$$

We could have also derived the above equation by computing $\dot{\mathbf{r}} = \dot{\mathbf{R}}_C + \mathbf{r}'$, where $\mathbf{r}'$ is the coordinate of the pebble in the reference frame moving horizontally with the wheel, and $\mathbf{R}_C$ is the coordinate of the centre of the wheel.

The components of the velocity vector in the xy coordinate system is

$$\dot{x} = \omega R - \omega R \cos \omega t,$$

$$\dot{y} = \omega R \sin \omega t.$$

Integrating both the equations in time and denoting $\omega t = \phi$ yields

$$x(t) = R(\phi - \sin \phi),$$

$$y(t) = R(1 - \cos \phi),$$

Note that we have chosen the initial conditions $(x(0), y(0)) = (0, 0)$. The above solution is the equation of a cycloid, and is depicted in Fig. 8.5(b). We obtain the acceleration of the pebble by differentiating velocity with respect to time, that is

$$\mathbf{a} = -\omega^2 R\hat{\mathbf{r}},$$

where $\hat{\mathbf{r}}$ is the unit vector along the radial direction in the wheel frame.

EXAMPLE 8.2   A particle moves outward along a spiral with the equation of its trajectory given by $r = \exp(\alpha\phi)$. ϕ increases in time according to $\phi = t^2/2$. Sketch the trajectory of the particle.

SOLUTION   The velocity and acceleration of the particle are

$$\mathbf{v} = \dot{r}\hat{\mathbf{r}} + r\dot{\phi}\hat{\phi}$$

$$= \alpha \exp(\alpha\phi)\dot{\phi}\hat{\mathbf{r}} + r\dot{\phi}\hat{\phi}$$

$$= rt(\alpha\hat{\mathbf{r}} + \hat{\phi})$$

$$\mathbf{a} = \dot{\mathbf{v}}$$

$$= (\alpha rt\dot{\phi} + r)(\alpha\dot{\phi}\hat{\phi} - \dot{\phi}\hat{\mathbf{r}})$$

$$= rt(\alpha t^2 + 1)(\alpha\hat{\phi} - \hat{\mathbf{r}}).$$

Note that the velocity and acceleration of the particle are perpendicular to each other. For $\alpha = 1$, the trajectory of the particle is shown in Fig. 8.6.

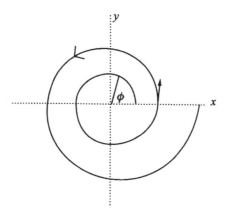

Figure 8.6   *Trajectory $r = \exp(\alpha\phi)$ with $\alpha = 0.1$.*

EXAMPLE 8.3   A planet of mass m is orbiting a star of mass M. The planet experiences a small drag force $\mathbf{F} = \eta\mathbf{v}$ due to its motion through the dense atmosphere of the star. Here η is the coefficient of friction. Assuming an essentially circular orbit with radius $r = r_0$ at $t = 0$, calculate the time dependence of the radius.

SOLUTION   The net force acting on the planet is the sum of the gravitational and frictional forces. Therefore the equations of motion for the planet is

$$m\ddot{\mathbf{r}} = -\frac{\alpha}{r^2}\hat{\mathbf{r}} - \eta\mathbf{v}.$$

Since the orbit is approximately circular, we take $\mathbf{v} = r\dot{\phi}\hat{\phi}$. In radial–polar coordinates, the equations of motion are

$$m(\ddot{r} + r\dot{\phi}^2) = -\frac{\alpha}{r^2}, \tag{8.13}$$

$$m(2\dot{r}\dot{\phi} + r\ddot{\phi}) = -\eta r\dot{\phi} \tag{8.14}$$

$\ddot{r}$ can be ignored due to the approximate circular nature of the orbit. As a consequence, Eq. (8.13) approximates as

$$mr\dot{\phi}^2 = -\frac{\alpha}{r^2},$$

The time derivative of the above equation yields (after a bit of algebra)

$$r\ddot{\phi} = -\frac{3}{2}\dot{\phi}\dot{r}.$$

When we substitute the above in Eq. (8.14) we obtain

$$\frac{m}{2}\dot{r}\dot{\phi} = -\eta r\dot{\phi},$$

or $\dot{r} = -\dfrac{2\eta}{m}r.$ (8.15)

The solution of the above equation is

$$r(t) = r(0)\exp(-2\eta t/m),$$

which is a spiral. The planet is falling toward the centre of gravity. In this problem η is small, so the rate of fall to the centre is gradual.

We can check from the solution whether our approximation is reasonable or not. By differentiating $r(t)$ twice, we obtain

$$\ddot{r} = \frac{4\eta^2}{m^2}r.$$

Therefore the ratio of the two terms in the radial acceleration is

$$\frac{\ddot{r}}{r\dot{\phi}^2} \approx \frac{\eta^2 r^3}{m\alpha}.$$

Here we have used the approximation of circular orbit $r\dot{\phi}^2 \approx \alpha/(mr^2)$. Let us compare the above ratio with the ratio between frictional and gravitational force

$$\frac{F_{\text{fric}}}{F_{\text{gravity}}} \approx \frac{\eta r^2 \dot{\phi}}{\alpha} \sim \frac{\eta r^{3/2}}{\sqrt{m\alpha}} \ll 1.$$

Clearly,

$$\frac{\ddot{r}}{r\dot{\phi}^2} \sim \left(\frac{F_{\text{fric}}}{F_{\text{gravity}}}\right)^2 \ll 1.$$

Hence our approximation that $\ddot{r} \ll r\dot{\phi}^2$ is consistent.

In this section, we have introduced the radial–polar coordinate system. With this background we are now ready to solve Kepler's problem of planetary motion.

8.3   Kepler's problem of planetary motion and its solution

As we discussed in Chapter 1, Kepler, using Brahe's observational data, showed that

1. The orbit of a planet around the Sun is in a plane, and the trajectory of the planet is an ellipse in this plane.

2. The radius vector from the Sun to the planet sweeps equal areas in equal intervals of time.

3. $T^2 = (1/C)R^3$, where T is the period of revolution, R is the semi-major axis of the ellipse, and C is a constant.

Using these empirical laws, Newton could show that a force called gravitation ($F \propto r^{-2}$) is responsible for the above motion of the planet. Newton worked out the force acting on the planet given the properties of the planetary motion. Here we do the converse; we derive Kepler's three laws assuming that the gravitational force is acting on the planet. To simplify the problem we assume that

1. The central mass (Sun) is much heavier than the mass of the planet, and that the Sun does not move.

2. The planet is a point particle. This approximation is good if the radius of the planet is much smaller than the distance between the Sun and the planet.

The gravitational force acting on the planet is

$$\mathbf{F} = -\frac{GMm}{r^2}\hat{\mathbf{r}} = -\frac{\alpha}{r^2}\hat{\mathbf{r}},$$

where M is the mass of the Sun, m is the mass of the planet, and $\alpha = GMm$. We assume that $M \gg m$ because of which the Sun does not move appreciably by the force $-\mathbf{F}$. The gravitational potential is given by

$$U(r) = -\frac{\alpha}{r}.$$

Before we proceed to solve the equations, recall the definition of angular momentum as

$$\mathbf{L} = \mathbf{r} \times \mathbf{p}$$

$$= (mr^2\dot{\phi})(\hat{\mathbf{r}} \times \hat{\phi})$$

which is a constant for a central force problem (Exercise 4.1). Since $\mathbf{L} \cdot \mathbf{r} = (\mathbf{r} \times \mathbf{p}) \cdot \mathbf{r} = 0$, the motion of the planet will be in a plane perpendicular to $\mathbf{L}$. Conservation of angular momentum implies that the rate of area swept by the planet $r^2\dot{\phi}/2 = L/(2m)$ is a constant; this is Kepler's second law for planetary motion.

Since the planetary motion is in a plane, and the force is radial, it is best to use radial–polar coordinates. Using $\mathbf{F} = m\mathbf{a}$, we obtain

$$\ddot{r} - r\dot{\phi}^2 = -\frac{\alpha}{mr^2}, \tag{8.16}$$

$$2\dot{r}\dot{\phi} + r\ddot{\phi} = 0. \tag{8.17}$$

Equation (8.17) implies that

$$\frac{d}{dt}(r^2\dot{\phi}) = 0,$$

or $r^2\dot\phi = \text{const} = L/m$, the same result as stated above (conservation of angular momentum).

The substitution of $r^2\dot\phi = L/m$ in Eq. (8.16) yields

$$\ddot{r} = \frac{L^2}{m^2 r^3} - \frac{\alpha}{mr^2} \tag{8,18}$$

The above force equation will be solved in Section 8.3.3. Here we solve Keplér's problem using the energy method:

$$\frac{1}{2}mv^2 + U(r) = E,$$

$$\frac{1}{2}m(\dot r^2 + r^2 \dot\phi^2) - \frac{\alpha}{r} = E$$

$$\frac{1}{2}m\dot r^2 + U_{\text{eff}}(r) = E \tag{8.19}$$

with

$$U_{\text{eff}}(r) = -\frac{\alpha}{r} + \frac{L^2}{2mr^2}.$$

Equation (8.19) is an equation of a particle that is moving in an effective potential $U_{\text{eff}}(r)$. Please note that the particle is really moving in two dimensions, but we are able to map it on to a one-dimensional problem. A plot of $U_{\text{eff}}(r)$ vs. r is shown in Fig. 8.7. The potential $U_{eff}(r)$ has a minimum $U_0 = -\frac{\alpha^2 m}{2L^2}$, at $r_0 = \frac{L^2}{m\alpha}$. Note that $U(r)$ depends on angular momentum L, and not on E.

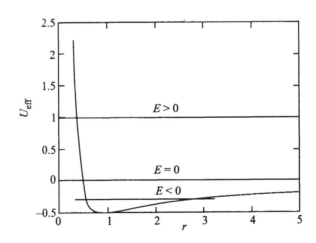

Figure 8.7   *Plot of effective potential $U_{eff}(r)$ vs. r*

It is possible to deduce several important conclusions from the effective potential itself. We will observe the behaviour of the system for $E < 0, E = 0$, and $E > 0$.

1. $E < 0$: For $E < 0$, we have two values of r where $\dot{r} = 0$ ($E = U_{\text{eff}}(r)$). Let us call them $r_{\min}$ and $r_{\max}$. The condition that $m\dot{r}^2/2$ must be positive implies that the motion is confined between $r_{\min}$ and $r_{\max}$. These are the closest and farthest points of the orbit from the centre of the potential. A system with $E < 0$ is called a *bound system* because we would need to give energy to free the planet from the Sun. A planet bound to a star has $E < 0$.

   For $E = U_{\min}$, $r_{\min} = r_{\max} = r_0$. Hence the trajectory is a circle of radius r_0.

2. $E \geq 0$: There is only one point where $\dot{r} = 0$. The particle comes closest to this point, and then goes off to infinity. We will now discuss these orbits.

The equation (8.19) can be rewritten as

$$\frac{1}{2}m\left(\frac{dr}{dt}\right)^2 + \frac{L^2}{2mr^2} - \frac{\alpha}{r} = E, \tag{8.20}$$

or $$\frac{1}{2}m\left(\frac{dr}{d\phi}\dot{\phi}\right)^2 + \frac{L^2}{2mr^2} - \frac{\alpha}{r} = E.$$

The substitution of $r = 1/u$ yields

$$\frac{du}{d\phi} = \left(\frac{2mE}{L^2} + \frac{2m\alpha}{L^2}u - u^2\right)^{1/2}.$$

Hence,

$$\int \frac{du}{\sqrt{A - (u - B)^2}} = \int d\phi, \tag{8.21}$$

with $B = \dfrac{m\alpha}{L^2}$,

$$A = \frac{m^2\alpha^2}{L^4} + \frac{2mE}{L^2}.$$

Note that $A > 0$ even for $E < 0$ (show this!). The solution of Eq. (8.21) is

$$\sin^{-1}\frac{u - B}{\sqrt{A}} = \phi + C,$$

or $$\frac{u}{B} = 1 + \frac{\sqrt{A}}{B}\sin(\phi + C),$$

or $\dfrac{p}{r} = 1 + e\cos\phi,$ (8.22)

with $C = \pi/2$ (we make this choice so that the nearest point corresponds to $\phi = 0$), and

$$p = \frac{1}{B} = \frac{L^2}{m\alpha},$$

$$e = \frac{\sqrt{A}}{B} = \sqrt{1 + \frac{2EL^2}{m\alpha^2}}.$$

Equation (8.22) is an equation of conic sections with one focus at the origin; $2p$ is called the *latus rectum* of the orbit, and e the *eccentricity*. There are four possible solutions:

1. $e = 0$ $(E < 0)$; Circle.

2. $0 < e < 1$ $(E < 0)$; Ellipse.

3. $e = 1$ $(E = 0)$; Parabola.

4. $e > 1$ $(E > 0)$; Hyperbola.

Solutions 1,2 give bound orbits, and 3,4 are unbound orbits. Planetary orbits are either of type 1 or type 2. Thus Kepler's first law that planetary orbits are ellipses is verified. Parabolic and hyperbolic trajectories can be observed in the Coulomb scattering of charged particles.

8.3.1   Classification of Kepler orbits

8.3.1.1   Circle ($e = 0$)

The planetary orbit is a circle when

$$E = -\frac{m\alpha^2}{2L^2},$$

and

$$r = p = \frac{L^2}{m\alpha}.$$

This orbit corresponds to the bottom of the effective potential $U_{eff}(r)$ and minimum energy.

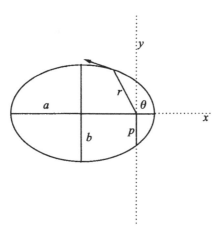

Figure 8.8   *Elliptical orbit of a planet with e = 0.75*

8.3.1.2   Ellipse $(0 < e < 1)$

When $e < 1$, the trajectories are ellipses (Fig. 8.8) with r bound between (r_{min}, r_{max}).

$$r_{min} = \frac{p}{1+e} > \frac{p}{2},$$

$$r_{max} = \frac{p}{1-e},$$

$$a = (r_{min} + r_{max}) = \frac{p}{1-e^2} = \frac{\alpha}{2|E|}, \tag{8.23}$$

$$y = r\sin\phi,$$

$$x = r\cos\phi,$$

$$b = y_{max} = \frac{p}{\sqrt{1-e^2}} = \frac{L}{\sqrt{2m|E|}}. \tag{8.24}$$

Let us shift the origin of the coordinate system to the centre of the ellipse. The amount of shift is $a - r_{min} = pe/(1-e^2) = ae$. The coordinates in the shifted system are denoted by (x', y'):

$$x' = x + ae = \frac{p\cos\phi}{1+e\cos\phi} + ae = a\frac{e+\cos\phi}{1+e\cos\phi},$$

$$y' = y = r\sin\phi = \frac{p\sin\phi}{1+e\cos\phi} = b\frac{\sqrt{1-e^2}\sin\phi}{1+e\cos\phi}.$$

A simple manipulation shows that

$$\frac{x'^2}{a^2} + \frac{y'^2}{b^2} = 1,$$

which is the equation of an ellipse with a and b as major and minor axes respectively.

We can compute the time period of the orbit as follows. Using

$$L = mr^2\dot{\phi},$$

we obtain

$$\int dt = \frac{2m}{L} \int \frac{1}{2}r^2 d\phi,$$

$$T = \frac{2m}{L}A = \frac{2m}{L}\pi ab,$$

$$T = \pi\alpha\sqrt{\frac{m}{2|E|^3}}. \tag{8.25}$$

Since $a \propto |E|^{-1}$ [Eq. (8.23)], we obtain $T^2 \propto a^3$, which is Kepler's third law for planetary orbits.

8.3.1.3   Parabola $(e = 1)$

For $E = 0$ we have $e = 1$. For this case

$$r = \frac{p}{1 + \cos\phi}.$$

The plot is shown in Fig. 8.9 (dashed line).

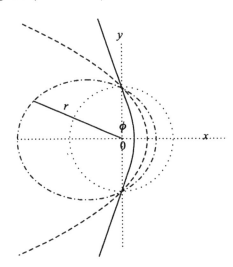

Figure 8.9   *Trajectories of Kepler's problem for various eccentricities: $e = 0$ (circle, dotted line), $e = 0.5$ (ellipse, chained line), $e = 1$ (parabola, dashed line), and $e = 3$ (hyperbola, solid line).*

Clearly $r_{\min} = p/2$. We can easily rewrite the above equation in Cartesian coordinates as

$$r + r \cos \phi = p,$$

or $r = p - x.$

By squaring the above equation, we obtain

$$2px = p^2 - y^2,$$

which is an equation of a parabola. A coordinate shift $x' = x + p/2$ transforms the above equation to

$$x' = -\frac{1}{2p} y^2,$$

which is the usual form for a parabola.

8.3.1.4   Hyperbola $(e > 1)$

When $E > 0$, we have $e > 1$. For this case, $r \to \infty$ when $\cos \phi_0 = -1/e$. The range of allowed ϕ is $(-\phi_0, \phi_0)$. By several algebraic manipulations we obtain

$$\frac{x'^2}{a^2} - \frac{y'^2}{b^2} = 1,$$

which is the equation of a hyperbola (see Fig. 8.9, solid line).

Thus, for a given angular momentum L, we have discussed trajectories for various energies. The two parameters (L, E) determine the nature of the trajectory of the particle.

8.3.2   Non-dimensionalised Kepler's problem

We can non-dimensionalise Kepler's problem using appropriate length- and time-scales. The length-scale is chosen to be the radius of the circle, which is $r_0 = L^2/m\alpha$. The velocity-scale is the speed of the circular path: $v_0 = L/(mr_0)$. Therefore, time-scale will be $t_0 = r_0/v_0$. The energy-scale is chosen to be $E_0 = L^2/(mr_0^2)$. We make a change of variable

$$r = r_0 \tilde{r},$$

$$t = t_0 \tilde{t},$$

$$v = v_0 \tilde{v},$$

$$E = E_0 \tilde{E}.$$

When we substitute the above in the energy equation Eq. (8.20) we obtain

$$\frac{1}{2}\left(\frac{d\tilde{r}}{d\tilde{t}}\right)^2 + \frac{1}{2\tilde{r}^2} - \frac{1}{\tilde{r}} = \tilde{E}. \tag{8.26}$$

The angular momentum equation in terms of the new variables is

$$\tilde{r}^2\frac{d\phi}{d\tilde{t}} = 1,$$

substitution of which in the new energy equation yields (with $\tilde{r} = \tilde{u}^{-1}$)

$$\frac{1}{2}\left(\frac{d\tilde{u}}{d\phi}\right)^2 + \frac{\tilde{u}^2}{2} - \tilde{u} = \tilde{E}. \tag{8.27}$$

The solution of the above equation is

$$\tilde{u} = \frac{1}{\tilde{r}} = 1 + \sqrt{1 + 2\tilde{E}}\cos\phi. \tag{8.28}$$

Note that in the dimensionless form, $p = 1$ and $e = \sqrt{1 + 2\tilde{E}}$. In Fig. 8.9, we plot $(\tilde{r}, \phi)$ for various values of $\tilde{E}$.

Clearly $\tilde{E} = -1/2$ yields a circular orbit. For the circular orbit, the normalised potential energy and kinetic energy ($\tilde{T}$) are given by

$$\tilde{U} = -\frac{\alpha/r_0}{E_0} = -1,$$

$$\tilde{T} = \frac{\frac{1}{2}mv_0^2}{E_0} = \frac{1}{2}.$$

Hence $\tilde{E} = \tilde{T} + \tilde{U} = -1/2$, consistent with our earlier conclusion. For $-1/2 < \tilde{E} < 0$, we get elliptical orbits. For $\tilde{E} = 0$ and $E > 0$, we obtain parabolic and hyperbolic orbits respectively. You can easily check that the non-dimensional results are consistent with our earlier results with dimensions.

8.3.3   Solving the force equation of a planet

We can also derive the solution of the planetary orbit (Eq. (8.28)) from the force equation Eq. (8.16). The non-dimensionalised force equation in terms of u is

$$\frac{d^2\tilde{u}}{d\phi^2} = \tilde{u} - 1,$$

whose solution is

$$\tilde{u} = 1 + c_1\cos(\phi + c_2).$$

By choosing $r_{\min}$ at $\phi = 0$, we obtain

$\tilde{u} = 1 + c_1 \cos \phi.$

At $r_{\min}$ or $u_{\max}$, $du/d\phi = 0$. Hence from Eq. (8.27)

$$\frac{\tilde{u}_{\max}^2}{2} - \tilde{u}_{\max} = \tilde{E},$$

that yields

$$\tilde{u}_{\max} = 1 + c_1 = 1 + \sqrt{2\tilde{E} + 1},$$

or $c_1 = \sqrt{2\tilde{E} + 1}$. Therefore,

$$\tilde{u} = 1 + \sqrt{2\tilde{E} + 1} \cos \phi,$$

which is the same as Eq. (8.28). This is how we solve for the equation of the orbit using force equation.

In the next sub-section, we will discuss the similarity solution for Kepler's problem.

8.3.4   Similarity solution (another symmetry)

The Newton's equation of motion for a central potential is

$$m\ddot{\mathbf{r}} = -\frac{d}{dr}U(r)\hat{\mathbf{r}} = -\frac{\alpha}{r^2}\hat{\mathbf{r}}. \tag{8.29}$$

We have solved the above equation and obtained the trajectories for $U(r) = -\alpha/r$. Suppose for a given energy E and angular momentum L, the trajectory is an ellipse. Now we make the following transformation:

$r = \beta r',$

$t = \gamma t'.$

Substitution of the above in Eq. (8.29) yields

$$\frac{\beta}{\gamma^2}m\ddot{\mathbf{r}}' = -\frac{\alpha}{\beta^2}\frac{1}{r'^2}\hat{\mathbf{r}}'. \tag{8.30}$$

If $\beta^3 = \gamma^2$, then we obtain the same equation as before. What does it mean? A major consequence of the above result is

$$\left(\frac{r}{r'}\right)^3 = \left(\frac{t}{t'}\right)^2,$$

which is Kepler's third law. In fact, the dependence of the time period on powers of r is because of the symmetry under scaling.

The above result also implies that a trajectory scaled by a factor β is allowed if the time required is also scaled by $\beta^{3/2}$. If planet A with major radius a and minor radius b takes time Δt to go from point S to point T, then another planet B having major and minor radii $2a$ and $2b$ respectively will take $2^{3/2}\Delta t$ time to cover the corresponding S′−T′ distance.

Scaling is one of the most powerful symmetry principle employed in physics. You can derive similar relationships for the oscillator and other problems. While discussing dimensional analysis (Appendix B) we hypothesise that the time period of a pendulum, atomic time scales etc. are functions of certain powers of the physical parameters. This hypothesis is valid essentially due to the symmetry under scaling or similarity transformation.

In the next section we will compare the potential energy and the kinetic energy of a planetary system.

8.3.5   Virial Theorem

We define the time average of a quantity Q as

$$\langle Q \rangle = \frac{1}{T} \int_0^T Q(t)dt,$$

where T is the time period of the orbit. Let us take $Q = \mathbf{p} \cdot \mathbf{v}$, where $\mathbf{p}$ is the momentum of the planet. Using the product rule

$$\mathbf{p} \cdot \mathbf{v} = \frac{d}{dt}(\mathbf{p} \cdot \mathbf{r}) - \mathbf{a} \cdot \mathbf{r},$$

where $\mathbf{a} = -\alpha \mathbf{r}/r^3$ is the acceleration of the particle. Temporal averaging of each of the terms of the above equation yields $\langle \mathbf{p} \cdot \mathbf{v} \rangle = 2 \langle T \rangle$, $- \langle \mathbf{a} \cdot \mathbf{r} \rangle = \langle \alpha/r \rangle = - \langle U \rangle$, and

$$\left\langle \frac{d}{dt}(\mathbf{p} \cdot \mathbf{r}) \right\rangle = \lim_{T \to \infty} \frac{1}{T} \int dt \frac{d}{dt}(\mathbf{p} \cdot \mathbf{r}) = \lim_{T \to \infty} \frac{1}{T} [\mathbf{p} \cdot \mathbf{r}]_{t=0}^{t=T} = 0$$

because $(\mathbf{p} \cdot \mathbf{r})$ is finite for a bounded orbit. Therefore,

$$2 \langle T \rangle = - \langle U \rangle .$$

This is called the *virial theorem*. Clearly $\langle E \rangle = \langle T \rangle + \langle U \rangle = - \langle T \rangle = \langle U \rangle /2$. This procedure is quite general, and it is used to derive the relationship between kinetic and potential energy for other forms of central potentials (Exercise 7.23).

It is possible to derive the above relation by an explicit computation of the integral. That is left as an exercise.

EXAMPLE 8.4   Halley's comet is in an elliptical orbit around the Sun whose mass is 2×10^{30} kg. The eccentricity of the comet's orbit is 0.967, and the period is 76 years. Using these data, determine the closest and the farthest distance of Halley's comet from the Sun. Determine the speed of the comet when it is closest to the Sun?

SOLUTION   Recall that for an elliptical orbit

$$T = 2\pi a^{3/2}\sqrt{m/\alpha},$$

$$= \pi\alpha\sqrt{m/2|E|^3},$$

with $\alpha = GM_s m$, where m is the mass of the comet and $M_s = 2 \times 10^{30}$ kg is the mass of the Sun. Using the above equations, we obtain

$$a = \left(\frac{\alpha T^2}{4m\pi^2}\right)^{1/3} = 2.69 \times 10^{12} \text{ m},$$

$$E = -\left(\frac{m\pi^2\alpha^2}{2T^2}\right)^{1/3},$$

$$= -m\left(\frac{\pi GM_s}{T\sqrt{2}}\right)^{2/3} = -m \times 2.48 \times 10^7 \text{J}.$$

Therefore,

$$r_{\min} = a(1 - e) = 8.9 \times 10^{10}\text{m} \sim 0.59 \text{ AU},$$

$$r_{\max} = a(1 + e) = 5.3 \times 10^{12}\text{m} \sim 3.53 \text{ AU}.$$

At the closest point to the Sun

$$E = -\frac{GM_s m}{r_{\min}} + \frac{1}{2}mv^2,$$

which yields the speed $v = 54.3$ km/s, that is approximately 1.8 times the average orbital speed of the Earth.

8.4 A bit of astronomy: Planets

Solar planets follow elliptical orbits around the Sun. The moons of planets follow elliptical orbits around the corresponding planets. The following lists out several important points regarding the orbits of these astronomical objects:

1. The point of closest approach in an elliptical orbit is called the *periapsis* or *pericentre,* and the furthest point is called the *apoapsis, apocentre* or *apapsis.* These are Greek words that are used to describe planetary and lunar orbits. Various related terms are used for different celestial objects. For example, *perigee* and *apogee* refer to orbits around the Earth, and *perihelion* and *aphelion* refer to orbits around the Sun. Please refer to Wikipedia for details.

2. Planetary distances are measured in the units of the Sun–Earth distance that is called an Astronomical Unit (AU).

3. A planet going around the Sun is affected by other planets in the Solar System, and by its own moon. Several centuries of work have shown that these effects could have

a major influence on the trajectories of the planet. Theoretical work show that the trajectories of the planet in the three-body problem (e.g., Sun, Earth and Moon) could be regular or irregular depending on the values of parameters (like energy, angular momentum etc.) and the initial conditions. Irregular trajectories are called *chaotic motion*, which will be discussed in Chapter 15. A critical question to be answered is whether a planet under the influence of the other two masses is stable or not. Many scientists starting from Poincaré have studied this problem for more than a century, yet many issues remain unresolved. We quote a recent work by Malhotra, Holman, and Ito (2001) where they argue that the planetary orbit is unstable, but the time-scale of instability of the planets is over 5–10 million years, and the escape time-scale of the planets is beyond the age of the Sun. These studies show that even though a two-body problem is exactly solvable, three-body and many-body problems are much more complex.

4. Planets are not point particles—typically, they are spinning rigid bodies. These aspects bring in additional complexity, some of which will be discussed in Chapters 17 and 18.

5. Every planet has its unique features in terms of its atmosphere, geological structure, magnetic field etc. These aspects are major areas of research for space scientists and geo-scientists.

8.5   General central force problem

A general attractive central force potential is of the form

$$U = -\frac{\alpha}{r^n},$$

with $\alpha, n > 0$ or $U = \beta r^m$,

with $\beta, m > 0$. The effective potentials $U_{eff} = U + L^2/(2mr^2)$ for these potentials have been plotted in Fig. 8.10.

For $0 < n < 2$, bounded orbits are possible when the energy is negative. For $n > 2$, and $n = 2$ with $\alpha > 1/2$, the orbit spirals into the centre. Centrifugal force wins for $n = 2$ with $\alpha < 1/2$, and the orbit is unbounded. For βr^m potentials, all the orbits are bounded as shown in the figure. A detailed study of the trajectories for general potential is quite complex, and is beyond the scope of this book. However we list some major results here.

Exact solution exists only for $n = 1$ and $m = 2$. The case $m = 2$ is called an isotropic oscillator for obvious reasons (Exercise 17). It has also been shown that the orbit is closed only for these two cases. For the general case, we can deduce some results using effective potentials, however, a closed form solution is impossible to obtain. For details, refer to Landau (1976) and Goldstein, Poole, and Safko (2002).

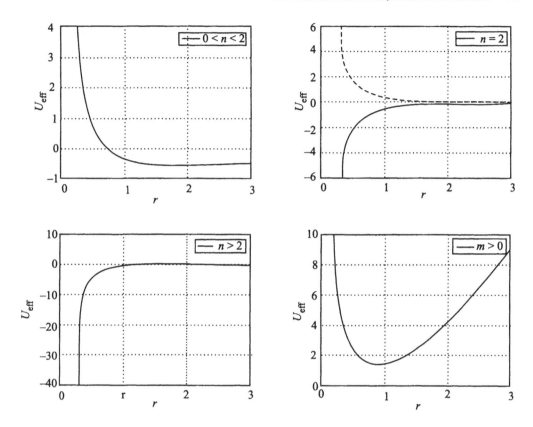

Figure 8.10   *Effective potential for central force potentials with centrifugal potential as $1/2r^2$: (a) $0 < n < 2$ with $\alpha = 1$, (b) $n = 2$ with $\alpha = 1$ (solid line) and $\alpha = 0.1$ (dashed line), (c) $n > 2$ with $\alpha = 1$, (d) $m > 0$ with $\beta = 1$.*

EXAMPLE 8.5   An object of unit mass orbits in a central potential $U(r)$. Its orbit is $r = a \exp(-b\phi)$ with $b > 0$.

1. Sketch the trajectory.

2. What are the constants of motion? Find the potential of the object.

SOLUTION   The plot of the trajectory is sketched in Fig. 8.11. At $\phi = 0$, $r = a$. As ϕ increases, r decreases. The velocity of the particle is

$$\mathbf{v} = \dot{r}\hat{\mathbf{r}} + r\dot{\phi}\hat{\phi}$$

$$= a\exp(-b\phi)(-b\dot{\phi})\hat{\mathbf{r}} + r\dot{\phi}\hat{\phi}$$

$$= r\dot{\phi}(-b\hat{\mathbf{r}} + \hat{\phi}).$$

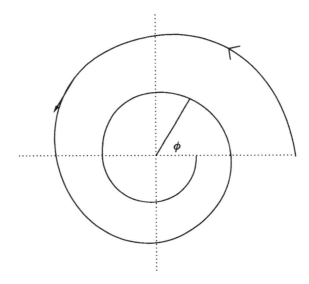

Figure 8.11   *Trajectory $r = a\exp(-b\phi)$ with $b = 0.1$.*

(2) The constants of motion are angular momentum L and energy E:

$$L = r^2\dot{\phi}$$

$$= a^2 \exp\left(-2b\phi\right)\dot{\phi},$$

$$E = \frac{1}{2}(\dot{r}^2 + r^2\dot{\phi}^2) + U(r)$$

$$= \frac{1}{2}r^2\dot{\phi}^2(b^2 + 1) + U(r).$$

$$= \frac{L^2}{2r^2}(b^2 + 1) + U(r)$$

Since E is a constant, $U(r) = \alpha r^{-2}$ with

$$\alpha = -L^2(b^2 + 1)/2,$$

and $E = 0$. The force is $\sim r^{-3}$ whose direction is radially inward.

In the next chapter we will apply Newton's laws to three-dimensional mechanical systems.

Exercises

1. Find the angle for which a projectile on the surface of the Earth has maximum range. Ignore air friction.

2. Find the angle for which the projectile with air drag ($\gamma/m = 0.5\,\mathrm{s}^{-1}$) has maximum range. You can use computers to calculate the result.

3. Estimate the maximum velocity for which the $\gamma v \ll mg$ approximation holds for a projectile.

4. In Example 8.1 we studied the trajectory of a pebble that is stuck at the outer edge of a rolling wheel. Compute the trajectory of the pebble if

   (a) it is fixed at a distance a inside the rim.

   (b) it is fixed at a distance a outside the rim.

   Plot the trajectories of the pebble.

5. Describe the motion of a charged particle with charge q and mass m, whose initial condition is $\mathbf{r}(0) = 0$ and $\mathbf{v}(0) = v_0\hat{\mathbf{x}}$, and is subjected to the following external field. Give both a mathematical and physical reasoning.

   (a) Electric field $\mathbf{E} = E_0\hat{\mathbf{y}}$

   (b) Electric field $\mathbf{E} = E_0\hat{\mathbf{x}}$

   (c) Magnetic field $\mathbf{B} = B_0\hat{\mathbf{y}}$

   where E_0 and B_0 are constants.

6. A point mass moves in the xy plane such that its velocity at any point is given by

   $$\mathbf{v} = \hat{\mathbf{i}} + x\hat{\mathbf{j}}.$$

   At $t = 0$, the particle is at the origin. Find the equation of its path. Write a similar velocity form as above for motion along a circular path, centred at the origin, with a constant speed in the anti-clockwise direction.

7. A particle moves in a plane with constant radial velocity $\dot{r} = 1$ m/s, and constant angular velocity $\dot{\phi} = 2$ rad/s. The initial position of the particle is $\phi = 0$ and $r = 1$ m.

   (a) Derive an expression for the trajectory of the particle.

   (b) When the particle is 5 m from the origin, find its velocity and acceleration.

   (c) Sketch the trajectory of the particle.

8. A particle has constant angular velocity, but its radial velocity is proportional to its distance from the origin. Derive an equation for the trajectory of the particle?

9. Evaluate approximately the ratio of the mass of the Sun to that of the Earth, using only the length of the year and of the lunar month (27.3 days), and the mean radii of the Earth's orbit 1.49×10^8 km and of the moon's orbit 3.8×10^5 km.

10. A shell is fired from a Bofors gun with velocity $100 \, \text{m/s}$. The angle of the projectile is $45°$ with respect to the horizontal. Compute the trajectory using (a) $\mathbf{g} = \text{const}$; (b) $\mathbf{g} \propto 1/r^2$(Kepler's problem). Show that both the results are consistent.

11. Perilion and aphelion of Pluto's orbit are 29.658 AU and 49.305 AU respectively. The corresponding quantities for the Earth are 0.999860 AU and 1.0000 AU. Compute the orbital time period of Pluto given the fact that Earth's time period is 365.256 solar days.

12. The eccentricity of Earth's orbit is 0.0167. Compute the ratio of its maximum and minimum speed. Do the same exercise for Pluto whose eccentricity is 0.24.

13. Devise schemes to measure the mass of the moon.

14. Compute mass of the Sun using the orbital properties of the Earth.

15. r_{min} and r_{max} for a Earth's satellite are 10000 km and 6000 km respectively. The mass of the satellite is 2000 kg. Compute the eccentricity, energy, angular momentum, and minimum and maximum speed of the satellite.

16. A space vehicle of mass 2000 kg is orbiting around the Earth in a circular orbit with the radius of orbit as $2R_e$. We wish to transfer the vehicle to a circular orbit of radius $4R_e$ (see Fig. 8.12). One of the schemes to transfer the vehicle is to use a semi-elliptical orbit as shown in the figure. What velocity changes are required at the points of intersection, A and B? What is the change in energy of the system in the two configurations?

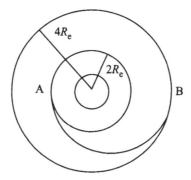

Figure 8.12   *Exercise 16*

17. A particle is moving under the influence of a potential $U(r) = kr^2$ with $k > 0$. Derive the trajectory of the particle. Do this problem in both Cartesian and polar coordinates. The Cartesian coordinate system is easier compared to the polar coordinate system.

18. [HARD PROBLEM]   A particle of mass m is moving under the influence of central potential $U(r) = \alpha/r^2$. Describe the motion of the particle in terms of α, angular momentum, and energy of the particle.

19. An object of unit mass orbits in a *central potential* $U(r)$. Its orbit is $r = 1/(1 + \cos\phi)$, and its angular momentum is L.

   (a) Sketch the trajectory.

   (b) Express object's velocity in terms of ϕ and L.

   (c) What is the potential?

20. The trajectory of a particle is described by $x = a\cos(\omega t)$, $y = b\sin(\omega t)$. Determine the force that acts on the particle at every point of the path.

21. The trajectory of a particle is given by a curve

$$\frac{x^2}{a^2} + \frac{y^2}{b^2} = 1.$$

The acceleration of the particle is always along the y-axis. At $t = 0$, the particle was at $(x = 0, y = b)$ and had velocity v_0. Determine the force acting on the particle.

22. A particle of mass one unit is moving with a constant speed v on a curve $y = A\sin x$. Compute the velocity, acceleration, and force of the particle at $x = \pi/2, \pi$ and $3\pi/2$.

23. A particle is moving under the influence of a $-\alpha/r$ potential, and it has negative energy. Compute the average kinetic and potential energy for a circular and elliptical orbit by directly computing the integral using the solution $r = r(\phi)$. Verify Virial theorem.

24. A particle is moving under the influence of potential $U = -\alpha/r^n$ with $\alpha, n > 0$. For $E < 0$, the orbit of the particle is bounded. We could define time average of Q as

$$\langle Q \rangle = \lim_{T \to \infty} \frac{1}{T} \int_0^T Q(t)dt.$$

Using function $Q = \mathbf{p} \cdot \mathbf{v}$, show that $2\langle T \rangle = -n\langle U \rangle$. Similarly, prove for $U = \beta r^m$ with $\beta, m > 0$ that $2\langle T \rangle = m\langle U \rangle$.

25. Derive similarity solutions for potentials given in Exercise 24.

26. For Kepler's problem, prove that $\mathbf{v} \times \mathbf{L} - \alpha\mathbf{r}/r$ is a constant. This vector is called the Runge–Lenz vector.

Projects

1. Verify Kepler's third law for planetary orbit using the data given in Appendix G. Obtain the proportionality constant between T^2 and R^3. Does the computed proportionality constant match with the one derived in this book (Eq. (8.25))?

2. Study the orbits of comets of the Sun.

9

Three-dimensional Motion

In Chapters 5 and 8 we studied one-dimensional and two-dimensional motion respectively. In this chapter we will study how to setup and solve the equation of motion in three dimensions. Several coordinate systems, e.g., Cartesian, cylindrical, spherical–polar etc. are employed for this purpose. As we will see, a given problem can be solved more conveniently in one coordinate system than the other. Hence, the choice of a coordinate system depends on the problem at hand, and we have to learn how to identify the appropriate coordinate system for a given problem.

First, we will solve some problems in the Cartesian coordinate system.

9.1   Three-dimensional motion in the Cartesian coordinate system

In the Cartesian coordinate system, Newton's equations of motion can be written as

$$m\ddot{x} = F_x,$$

$$m\ddot{y} = F_y,$$

$$m\ddot{z} = F_z.$$

Once the force and initial conditions are specified, we can solve for $x(t), y(t),$ and $z(t)$. We will illustrate the method using some examples.

1. **Motion of a charged particle in a constant magnetic field**
   The force experienced by a charged particle of mass m and charge q in a constant magnetic field $\mathbf{B}$ is $q\mathbf{v} \times \mathbf{B}$, where $\mathbf{v}$ is the velocity of the particle. Therefore, the equation of motion of the particle is

$$m\frac{d\mathbf{v}}{dt} = q\mathbf{v} \times \mathbf{B}$$

We assume the magnetic field to be along the z-direction. In terms of components, the equations of motion are

$$m\frac{dv_x}{dt} = qv_y B, \tag{9.1}$$

$$m\frac{dv_y}{dt} = -qv_x B, \tag{9.2}$$

$$m\frac{dv_z}{dt} = 0. \tag{9.3}$$

Equation (9.3) implies that $v_z = v_{z0} = \text{const}$. By multiplying Eq. (9.2) with i and adding it to Eq. (9.1), we obtain

$$m\frac{d\hat{v}}{dt} = -iqB\hat{v},$$

where $\hat{v} = v_x + iv_y$ is a complex velocity.[1] The solution of the above equation is

$$\hat{v}(t) = \hat{v}_0 \exp(-i\omega t)$$

with $\omega = qB/m$. We take the initial velocity as $\mathbf{v}(0) = -v_0\hat{\mathbf{y}} + v_{z0}\hat{\mathbf{z}}$. Note that the velocity should be such that the net force on the particle is towards the centre. The initial complex velocity is $\hat{v}_0 = -iv_o$ and

$$\hat{v}(t) = -iv_o \exp(-i\omega t).$$

By separating the real and imaginary parts, we obtain

$$v_x = -v_0 \sin \omega t,$$

$$v_y = -v_0 \cos \omega t.$$

Note that the charged particle is moving clockwise. Using $\mathbf{v} = d\mathbf{r}/dt$, we obtain

$$x(t) = (v_0/\omega) \cos \omega t + x_0$$

$$y(t) = -(v_0/\omega) \sin \omega t + y_0$$

$$z(t) = v_{z0}t + z_0,$$

where (x_0, y_0) is the centre of the circle. We choose $(x_0, y_0) = (0, 0)$. Clearly the trajectory of the particle is a helix as shown in Fig. 9.1. The projection of the trajectory on the xy plane is a circle. The direction of the arrow specifies the direction of motion.

---

[1] A complex velocity here is purely a mathematical construct. It helps to solve the above problem.

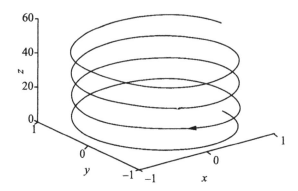

Figure 9.1   *Motion of a charged particle in a constant magnetic field $B\hat{z}$. We take $\omega = 1, V_0 = 1$ and $V_{z0} = 2$. The trajectory is a helix.*

2. **Motion of a charged particle in $\mathbf{E} \times \mathbf{B}$ field**

A charged particle of mass m and charge q moves in a constant uniform magnetic field $\mathbf{B}\hat{z}$ and a constant uniform electric field $\mathbf{E}\hat{x}$. The force on the charge particle is $q\mathbf{v} \times \mathbf{B} + q\mathbf{E}$. Therefore, the equation of motion of the particle is

$$m\frac{d\mathbf{v}}{dt} = q\mathbf{v} \times \mathbf{B} + q\mathbf{E}.$$

In terms of components, they are

$$m\frac{dv_x}{dt} = qv_y B + qE, \tag{9.4}$$

$$m\frac{dv_y}{dt} = -qv_x B, \tag{9.5}$$

$$m\frac{dv_z}{dt} = 0. \tag{9.6}$$

Equation (9.6) implies that $v_z = v_{0z} = 0$. By multiplying Eq. (9.5) with i and adding it to Eq. (9.4), we obtain

$$m\frac{d\hat{v}}{dt} + iqB\hat{v} = qE,$$

where $\hat{v} = v_x + iv_y$. The solution of the above equation is

$$\hat{v} = A \exp\left(-i\omega t\right) - i\frac{E}{B},$$

with $\omega = qB/m$. We take the initial condition as in the previous problem. Substitution of the initial condition yields $A = i(E/B - v_0)$. Therefore,

$$\hat{v} = i(\frac{E}{B} - v_0) \exp(-i\omega t) - i\frac{E}{B}.$$

By separating the real and imaginary parts, we obtain

$$v_x = \left(\frac{E}{B} - v_0\right) \sin \omega t,$$

$$v_y = \left(\frac{E}{B} - v_0\right) \cos \omega t - \frac{E}{B}.$$

Using $\mathbf{r} = \int \mathbf{v} dt$, we obtain

$$x(t) = -\frac{\left(\dfrac{E}{B} - v_0\right)}{\omega} \cos \omega t,$$

$$y(t) = \frac{\left(\dfrac{E}{B} - v_0\right)}{\omega} \sin \omega t - \frac{E}{B}t.$$

In a reference frame moving with velocity $-(E/B)\hat{\mathbf{y}}$, the motion is a circle. The above drift motion is called $\mathbf{E} \times \mathbf{B}$ *drift*. The trajectory of the particle is depicted in Fig. 9.2.

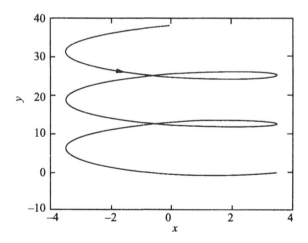

Figure 9.2   *Motion of a charged particle in a constant electric and magnetic field $E\,\hat{\mathbf{x}} + B\hat{\mathbf{z}}$. We take $\omega = 1$, $v_0 = 1$, and $E/B = 2$.*

In the above two examples we used Cartesian coordinates for analysing the motion of the particle. In the next two sections, we will introduce cylindrical and spherical–polar coordinate systems, which are useful for studying three-dimensional motion.

9.2   The cylindrical coordinate system

The cylindrical coordinate system is useful when axial symmetry (rotation symmetry about an axis) is present. Recall the helical motion of the charged particle in a constant magnetic field. This trajectory can be conveniently described in the cylindrical coordinate system.

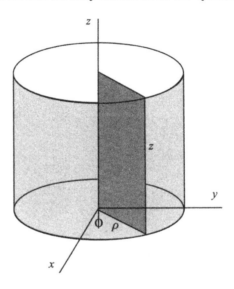

Figure 9.3   *The cylindrical coordinate system. The position of a particle is specified by* (ρ, ϕ, z).

In cylindrical coordinates, a particle's position is specified by (ρ, ϕ, z) as shown in Fig. 9.3. The angle ϕ is measured along the counter-clockwise direction with the x-axis as a reference. The transformation rules to shift from the cylindrical to the Cartesian system are

$$x = \rho \cos \phi,$$

$$y = \rho \sin \phi,$$

$$z = z.$$

The inverse transformation rules are

$$\rho = \sqrt{x^2 + y^2},$$

$$\phi = \tan^{-1} \frac{y}{x},$$

$$z = z.$$

The unit vectors in cylindrical coordinates are $\hat{\rho}, \hat{\phi}, \hat{z}$. The vectors $\hat{\rho}$ and $\hat{\phi}$ remain parallel to the xy plane as in the radial–polar coordinate system discussed in Chapter 8, while $\hat{z}$ is parallel to the z-axis. Recall the transformation rules between the unit vectors $(\hat{\rho}, \hat{\phi})$ and $(\hat{x}, \hat{y})$:

$$\hat{\rho} = \hat{\mathbf{x}} \cos\phi + \hat{\mathbf{y}} \sin\phi,$$

$$\hat{\phi} = -\hat{\mathbf{x}} \sin\phi + \hat{\mathbf{y}} \cos\phi.$$

The unit vectors $\hat{\rho}$ and $\hat{\phi}$ vary as the particle moves. The rate of change of these vectors have been derived earlier in Section 8.2.

EXAMPLE 9.1   Write down the coordinates of a charged particle that is moving in a constant magnetic field.

SOLUTION   In Section 9.1 we showed that the position of the charged particle is

$$x(t) = (v_0/\omega) \cos\omega t,$$

$$y(t) = -(v_0/\omega) \sin\omega t,$$

$$z(t) = v_{z0} t.$$

In cylindrical coordinates, the position of the particle is

$$\rho = \frac{v_0}{\omega},$$

$$\phi = -\omega t,$$

$$z = v_{z0} t.$$

$\phi = -\omega t$ implies that the charged particle is rotating clockwise.

9.3   The spherical–polar coordinate system

The spherical–polar coordinate system is useful for problems having spherical symmetry, e.g., for solving Schrödinger's equation for the hydrogen atom. This coordinate system is also used for describing dynamics on a surface of a sphere, e.g., for the motion of aeroplanes, or wind etc. on the surface of the Earth. In this system, a particle's position is specified by (r, θ, ϕ) as shown in Fig. 9.4. The *polar angle* θ is the angle which the radial vector $\mathbf{r}$ makes with the z-axis. The *azimuth* ϕ is the angle made by the projection of $\mathbf{r}$ with the x-axis in the xy plane.

The transformation rules for converting spherical–polar coordinates to Cartesian coordinates are

$$z = r \cos\theta,$$

$$x = r \sin\theta \cos\phi,$$

$$y = r \sin\theta \sin\phi.$$

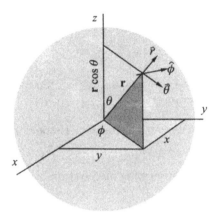

Figure 9.4   *Spherical–polar coordinate system. The position of a particle is specified by* (r, θ, ϕ).

The inverse transformation rules are

$$r = \sqrt{x^2 + y^2 + z^2},$$

$$\theta = \cos^{-1} \frac{z}{r},$$

$$\phi = \tan^{-1} \frac{y}{x}.$$

The range of θ and ϕ are $(0, \pi)$ and $(0, 2\pi)$ respectively. In this coordinate system, the unit vectors are $\hat{\mathbf{r}}, \hat{\theta}, \hat{\phi}$ as shown in the figure. Since the unit vectors vary with the trajectory of the particle, the time derivatives of these unit vectors are non-zero. The derivation of these derivatives for general situations is quite involved, so we will not discuss them here. The reader is referred to *mathworld's website*[2] for details.

The position and velocity vectors of a particle are

$$\mathbf{r} = r\hat{\mathbf{r}},$$

$$\mathbf{v} = \dot{r}\hat{\mathbf{r}} + r\frac{d\hat{\mathbf{r}}}{dt}.$$

The computation of $\frac{d\hat{\mathbf{r}}}{dt}$ is cumbersome for a general situation. However, it can be quite simple for some specific situations.

As mentioned before, the spherical–polar coordinate system is useful for describing the motion on the surface of a sphere. We will illustrate the motion using some examples given below.

1. An aeroplane travels from the North Pole to the South Pole in 12 hours passing through Kanpur as shown in Fig. 9.5 (path A). We take the origin of the coordinate system to coincide with the centre of the Earth, z-axis to pass through the North Pole,

---

[2]http://mathworld.wolfram.com

and x-axis to pass through the zero degree longitude. Hence the azimuth of Kanpur will be $\phi = 80°$ (longitude of Kanpur) and $r = 6400$ km. The polar angular velocity $\Omega = \pi/(12 \text{ hours})$, and the motion is along a circular path A shown in Fig. 9.5. The polar angle $\theta = \Omega t$. Along path A

$$\frac{d\hat{\mathbf{r}}}{dt} = \Omega\hat{\theta}.$$

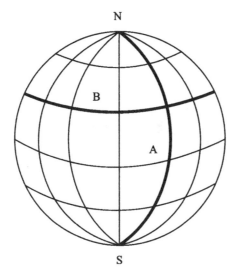

Figure 9.5   *Motion of the aeroplanes on the surface of the Earth.*

Therefore the velocity of the aeroplane is

$$\mathbf{v} = \frac{d\mathbf{r}}{dt} = r\frac{d\hat{\mathbf{r}}}{dt} = r\Omega\hat{\theta}.$$

2. An aeroplane travels around the Earth from west to east at the latitude of 45° in 24 hours. The orbit of the aeroplane is as shown in Fig. 9.5 (path B). Here $\theta = 45° = (\pi/8)$ rad and $r = 6400$ km. The azimuthal angular velocity is $\Omega = 2\pi/T = (2\pi/86400)$ rad/s, and $\phi = \Omega t$. Let us find $\frac{d\hat{\mathbf{r}}}{dt}$. The component of $\hat{\mathbf{r}}$ along the z-axis does not change, so the component of $\hat{\mathbf{r}}$ parallel to the xy plane will contribute to $\frac{d\hat{\mathbf{r}}}{dt}$. Using simple geometry we find that

$$\frac{d\hat{\mathbf{r}}}{dt} = \sin\theta\frac{d\hat{\rho}}{dt}$$

$$= (\sin\theta)\Omega\hat{\phi}.$$

Here $\hat{\rho}$ is the unit vector in the plane of motion of the aeroplane away from the axis and perpendicular to $\hat{\phi}$. Consequently

$$\mathbf{v} = r(\sin\theta)\Omega\hat{\phi}.$$

3. A rocket is sent vertically upward with a constant velocity v_0 from Kanpur. The coordinate of the rocket is $\phi = 80°$, $\theta = (90 - 26) = 64°$, and $r = v_0 t$.

EXAMPLE 9.2   On June 22, for an observer at latitude λ (a) what is the trajectory of the Sun in Earth's frame?; (b) what is the length of the day. Ignore refractive effects. Assume the Earth to be a spherical body.

SOLUTION   In the Northern Hemisphere of the Earth, the day is longest on June 22. In Fig. 9.6(a) we illustrate the Sun rays at noon. In the following, we will study the motion of the Sun rays for an observer at a latitude λ. We use a local coordinate system xyz as shown in the figure.

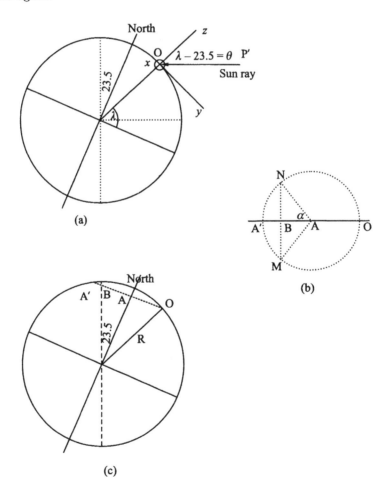

Figure 9.6   *The trajectory of Sun rays on June 22. (a) The Sun rays fall vertically at noon at point O; (b) The point makes a circle OMA'NO in 24 hours. When O moves in the circle, it is lighted in the sector MON, but dark in the sector NA'M; (c) The trajectory of O on the Earth.*

The angle the Sun rays make with the local vertical is $\theta = \lambda - 23.5$. At a given time the Sun rays make an angle ϕ with the x-axis (180 in the morning, 90 at noon, and 0 in the evening). The projection of the Sun's rays along the x-axis is $OP' = R_{ES} \sin \phi$, where R_{ES} is the distance of the Earth from the Sun. Hence,

$$x = OA = R_{ES} \cos \phi,$$

$$z = R_{ES} \sin \phi \cos \theta,$$

$$y = R_{ES} \sin \phi \sin \theta.$$

In Fig. 9.6(b) we illustrate the trajectory of a point O on Earth's surface as the Earth spins about the North–South axis. The trajectory of O as seen from the North Pole is shown in Fig. 9.6(c). The point O gets sunlight in sector MON, and it does not get sunlight in the sector NA'M. A bit of algebra shows that

$$OA = R \cos \lambda,$$

$$AB = R \sin \lambda \tan 23.5,$$

where R is the radius of the Earth. Hence, the length of the day at latitude λ is

$$\text{day} = 24 \text{ hours} \times \frac{\pi - \alpha}{\pi}$$

where

$$\cos \alpha = \frac{AB}{NA} = \frac{AB}{OA}$$

$$= \tan \lambda \tan 23.5$$

The plot of day-length vs. latitude is given in Fig. 9.7.

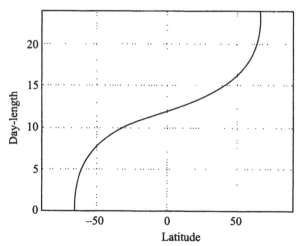

Figure 9.7 *Plot of day-length as a function of latitude.*

Note that the length of the day is 24 hours for latitude greater than 66.5 °, and 0 hours for latitude south of −66.5°.

These examples illustrate the use of the cylindrical and the spherical–polar coordinate system. Before we close our discussion, we remark that the gravitation potential $U(r) = -\alpha/r$ is a three-dimensional potential. However, conservation of angular momentum reduces the problem to a two-dimensional one where the radial–polar coordinate system can be applied successfully. Quantum treatment of the hydrogen atom however requires a three-dimensional system.

After this discussion on two- and there-dimensional motion we will move onto noninertial frames.

Exercises

1. Describe the motion of a charged particle with charge q and mass m whose initial condition is $\mathbf{r}(0) = 0$ and $\mathbf{v}(0) = v_{0x}\hat{\mathbf{x}} + V_{0z}\hat{\mathbf{z}}$, and which is subjected to the following external fields. Give both a mathematical and physical reasoning.

   (a) Electric field $\mathbf{E} = E_0\hat{\mathbf{x}}$.

   (b) Magnetic field $\mathbf{B} = B_0\hat{\mathbf{z}}$ and electric field $\mathbf{E} = (E_x\hat{\mathbf{x}} + E_y\hat{\mathbf{y}} + E_z\hat{\mathbf{z}})$.

   (c) Electric field $\mathbf{E} = E_0\hat{\mathbf{z}}$ and magnetic field $\mathbf{B} = (B_x\hat{\mathbf{x}} + B_y\hat{\mathbf{y}} + B_z\hat{\mathbf{z}})$.

   (d) Electric field $\mathbf{E} = E_0\hat{\mathbf{z}}$ and magnetic field $\mathbf{B} = B_0(\cos\alpha\hat{\mathbf{x}} + \sin\alpha\hat{\mathbf{y}})$.

2. A particle slides down a helical groove of pitch h and radius R under gravity. Assume initial velocity of the particle to be zero, and the surface of the groove to be frictionless. What would be the velocity of the particle after the nth turn?

3. Express $d\hat{r}/dt$, $d\hat{\theta}/dt$, and $d\hat{\phi}/dt$ in terms of Cartesian coordinates.

4. A man in a circus is riding a motorcycle within a spherical dome of radius 10 metres.

   (a) He wants to move in a horizontal circle of radius 5 metres in the lower hemisphere. What should be his speed?

   (b) Is it possible to move in a horizontal circle of radius 5 metres in the upper hemisphere? If yes, what should be his speed?

   (c) He wants to move in a vertical circle. What should be his minimum speed?

5. A conical pendulum makes a circular orbit in the horizontal plane. The mass of the bob is m, the length of the string is l, and the half-angle of the cone is α. Compute the time period of the orbit.

6. Describe the motion of (a) a geostationary satellite; (b) a distant star; and (c) the moon in the Earth's frame of reference.

7. [HARD PROBLEM] In Example 9.2 we computed the length of the day on June 22 as a function of latitude. Extend the solution to any arbitrary day of the year.

8. A particle of mass m moves on the inner side of a cone whose half-angle is α. The axis of the cone is vertical. At $t = 0$, the particle starts with velocity $\mathbf{v} = r_0\omega\hat{\phi}$, where ω is a constant, and $\hat{\phi}$ is along the horizontal direction as in the cylindrical coordinate system.

   (a) Write down the equation of motion.

   (b) For a certain value of r_0, the particle moves in a circle. What is that particular value of r_0?

   (c) What are the conserved quantities of the system?

   (d) Reduce the above problem to a one-dimensional problem using effective potential. What is the range of r within which the particle moves?

   (e) Solve the above problem numerically.

9. Solve Exercise 8 for a hemisphere.

10. A long hollow cylinder is rotating about its axis with angular velocity ω. What should be the coefficient of friction of the wall so that a particle of mass m at the wall does not fall down?

11. A pole of one metre length is placed vertically on a horizontal plane at the surface of the Earth at the latitude of λ. Describe the motion of the tip of the shadow of the pole as a function of time.

Projects

1. There are many computer programs that track stars and planets, e.g., Google-Earth, Kstar etc. Study the motion of several stars and planets in your sky using these programs, and verify the trajectories from observations.

2. Design a Sun dial for your city.

3. Write a computer program to date solar and lunar eclipses from the year 1500 to year 2100. Verify your results with earlier results.

10

Motion in a Noninertial Reference Frame

Newton's laws of motion are valid in any inertial reference frame. These laws can also be extended to noninertial frames by introducing some new terms in the equation. In this chapter we will derive the equations of motion for bodies in reference frames which are accelerating along a straight line, or which are rotating about an axis. These reference frames are convenient in many situations. For example, the motion of wind on the Earth can be conveniently described in Earth's frame of reference, which is a rotating reference frame.

10.1 Accelerating reference frame along a straight line

Consider an inertial frame I, and a noninertial frame S that is accelerating with respect to I along the x-axis. We consider I to be fixed in space. The origin of S is denoted by $\mathbf{R}_0$. The velocity and the acceleration of the origin of the reference frame S are $\dot{\mathbf{R}}_0$ and $\ddot{\mathbf{R}}_0$ respectively. We consider the motion of a point particle, and denote the position of the particle in the reference frames I and S by $\mathbf{R} = (X, Y, Z)$ and $\mathbf{r} = (x, y, z)$ respectively. The assumptions of absolute space and time yield

$$\mathbf{R} = \mathbf{r} + \mathbf{R}_0, \tag{10.1}$$

and $t_I = t_S$, where t_I and t_S are time in reference frames I and S respectively. By taking time derivative of Eq. (10.1) we obtain

$$\dot{\mathbf{R}} = \dot{\mathbf{r}} + \dot{\mathbf{R}}_0,$$

$$\ddot{\mathbf{R}} = \ddot{\mathbf{r}} + \ddot{\mathbf{R}}_0.$$

or $\quad \mathbf{a}_I = \mathbf{a}_S + \ddot{\mathbf{R}}_0. \tag{10.2}$

Hence, the acceleration of the particle is different in I and S.

Newton's law is applicable in the reference frame I. If $\mathbf{F}$ is the force acting on the particle of mass m, then

$$m\mathbf{a}_I = \mathbf{F}. \tag{10.3}$$

Substitution of Eq. (10.2) in Eq. (10.3) yields

$$m\mathbf{a}_S = \mathbf{F} - m\ddot{\mathbf{R}}_0. \tag{10.4}$$

Thus, if we want to apply Newton's law in the noninertial frame, we would have to postulate a pseudo (unreal) force which is the mass times the negative acceleration of the reference frame S.

We illustrate the pseudo force using an example. Consider a pendulum inside a train that is accelerating with a constant acceleration A along the x-axis. After initial transients, the pendulum makes an angle θ from the vertical. Note that the bob of the pendulum is also accelerating along the x-axis with an acceleration A. The free-body diagram of the bob in an inertial frame I (e.g. platform) is shown in Fig. 10.1(a).

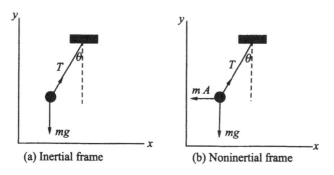

(a) Inertial frame   (b) Noninertial frame

Figure 10.1   *Free-body diagram of the bob (a) in an inertial frame; (b) in the reference frame attached to the train (noninertial frame).*

In the inertial frame I,

$$m\mathbf{a}_I = mA\hat{\mathbf{x}} = \mathbf{F}_{\text{net}}.$$

In terms of components

$$T\cos\theta = mg,$$

$$T\sin\theta = mA,$$

or $\tan\theta = \dfrac{A}{g}.$

In the train frame, which is a noninertial frame, $\mathbf{a}_S = 0$. The free-body diagram in this frame is shown in Fig. 10.1(b). Applying Eq. (10.4) we obtain

$$T\cos\theta - mg = 0,$$

$$T\sin\theta - mA = 0,$$

which again yields

$$\tan \theta = \frac{A}{g}.$$

Hence the description in both the reference frames are consistent. After this simple example, we move onto a rotating reference frame of reference. We will derive an equation of motion for an observer in the rotating frame of reference.

10.2   Reference frame rotating with a constant angular velocity

Consider a fixed inertial frame I, and a frame of reference S whose origin is the same as that of I, but which rotates with respect to I with an angular velocity Ω about the z-axis (common to both I and S). If a particle is fixed in S, then $\mathbf{a}_S = 0$, but $\mathbf{a}_I = -\Omega^2 r\hat{\mathbf{r}} = \mathbf{F}/m$ (r is the distance of the particle from the centre). Newton's law is not valid in S because it is a noninertial frame of reference. Is it possible to write an equation of motion in S? We will investigate this question in this section.

The equation of motion in rotating frames is very useful. For example, it is much more convenient to write the equation of winds, aeroplanes etc. moving on the surface of the Earth in Earth's frame of reference itself.

We denote the coordinates of the particle as (X, Y, Z) in I and as (x, y, z) in S. The unit vectors are denoted by $(\hat{\mathbf{X}}, \hat{\mathbf{Y}}, \hat{\mathbf{Z}})$ and $(\hat{\mathbf{x}}, \hat{\mathbf{y}}, \hat{\mathbf{z}})$ respectively. Therefore,

$$\mathbf{R} = X\hat{\mathbf{X}} + Y\hat{\mathbf{Y}} + Z\hat{\mathbf{Z}} = x\hat{\mathbf{x}} + y\hat{\mathbf{y}} + z\hat{\mathbf{z}}.$$

The velocity of the particle can be written as

$$\dot{\mathbf{R}} = \dot{X}\hat{\mathbf{X}} + \dot{Y}\hat{\mathbf{Y}} + \dot{Z}\hat{\mathbf{Z}}$$

$$= \dot{x}\hat{\mathbf{x}} + \dot{y}\hat{\mathbf{y}} + \dot{z}\hat{\mathbf{z}} + x\frac{d\hat{\mathbf{x}}}{dt} + y\frac{d\hat{\mathbf{y}}}{dt} + z\frac{d\hat{\mathbf{z}}}{dt}$$

Since the vectors $\hat{\mathbf{x}}$ and $\hat{\mathbf{y}}$ rotate with angular velocity $\Omega\hat{\mathbf{z}}$, we obtain

$$\frac{d\hat{\mathbf{x}}}{dt} = \hat{\boldsymbol{\Omega}} \times \hat{\mathbf{x}} = \Omega\hat{\mathbf{y}},$$

$$\frac{d\hat{\mathbf{y}}}{dt} = \hat{\boldsymbol{\Omega}} \times \hat{\mathbf{y}} = -\Omega\hat{\mathbf{x}},$$

$$\frac{d\hat{\mathbf{z}}}{dt} = \hat{\boldsymbol{\Omega}} \times \hat{\mathbf{z}} = 0.$$

Hence

$$\dot{\mathbf{R}} = \dot{\mathbf{r}} + x\Omega\hat{\mathbf{y}} - y\Omega\hat{\mathbf{x}} + 0$$

$$= \dot{\mathbf{r}} + x\boldsymbol{\Omega} \times \hat{\mathbf{x}} + y\boldsymbol{\Omega} \times \hat{\mathbf{y}} + z\boldsymbol{\Omega} \times \hat{\mathbf{z}}$$

$$= \dot{\mathbf{r}} + \boldsymbol{\Omega} \times \mathbf{r}. \tag{10.5}$$

Since the transformation rules for the components of vectors is the same as those of $\mathbf{R}$, we can generalise the above identity (Eq. (10.5)) to any vector $\mathbf{A}$:

$$\dot{\mathbf{A}} = (\dot{\mathbf{A}})_S + \boldsymbol{\Omega} \times \mathbf{A}, \tag{10.6}$$

where $(\dot{\mathbf{A}})_S$ is the rate of change of $\mathbf{A}$ in the rotating frame S. The term $\boldsymbol{\Omega} \times \mathbf{A}$ arises due to the rotation of the reference frame S. This feature is shown in Fig. 10.2. If $\mathbf{A}$ is fixed in S, then $d\mathbf{A} = d\boldsymbol{\phi} \times \mathbf{A}$ as shown in the figure. Hence $\dot{\mathbf{A}} = \boldsymbol{\Omega} \times \mathbf{A}$.

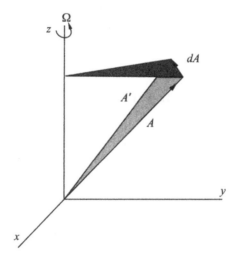

Figure 10.2   *Vector $\mathbf{A}$ is fixed in reference frame S, which is itself rotating about the z-axis with angular velocity $\boldsymbol{\Omega}$. In a small time dt, $\mathbf{A}$ is changed by $d\mathbf{A}$.*

Now we are ready to compute the acceleration.

$$\ddot{\mathbf{R}} = \ddot{x}\hat{\mathbf{x}} + \ddot{y}\hat{\mathbf{y}} + \ddot{z}\hat{\mathbf{z}} + 2(\dot{x}\dot{\hat{\mathbf{x}}} + \dot{y}\dot{\hat{\mathbf{y}}} + \dot{z}\dot{\hat{\mathbf{z}}}) + x\ddot{\hat{\mathbf{x}}} + y\ddot{\hat{\mathbf{y}}} + z\ddot{\hat{\mathbf{z}}}$$

$$= \ddot{\mathbf{r}} + 2\boldsymbol{\Omega} \times \dot{\mathbf{r}} + \boldsymbol{\Omega} \times (\boldsymbol{\Omega} \times \mathbf{r})$$

$$= \mathbf{a}_S + 2\boldsymbol{\Omega} \times \dot{\mathbf{r}} + \boldsymbol{\Omega} \times (\boldsymbol{\Omega} \times \mathbf{r}), \tag{10.7}$$

where $\mathbf{a}_S$ and $\mathbf{a}_I$ are the acceleration in the rotating and inertial frames respectively.

We could derive the above relationship by applying the rule Eq. (10.5) on itself:

$$\ddot{\mathbf{R}} = \left(\frac{d\mathbf{v}}{dt}\right)_S + \boldsymbol{\Omega} \times \mathbf{v}$$

$$= \left(\frac{d}{dt}(\dot{\mathbf{r}} + \boldsymbol{\Omega} \times \mathbf{r})\right)_S + \boldsymbol{\Omega} \times (\dot{\mathbf{r}} + \boldsymbol{\Omega} \times \mathbf{r})$$

$$= [\ddot{\mathbf{r}} + \mathbf{\Omega} \times \dot{\mathbf{r}}] + [\mathbf{\Omega} \times \dot{\mathbf{r}} + \mathbf{\Omega} \times (\mathbf{\Omega} \times \mathbf{r})]$$

$$= \ddot{\mathbf{r}} + 2\mathbf{\Omega} \times \dot{\mathbf{r}} + \mathbf{\Omega} \times (\mathbf{\Omega} \times \mathbf{r})$$

Therefore, according to Newton's laws

$$\mathbf{F} = m\ddot{\mathbf{R}} = m[\ddot{\mathbf{r}} + 2\mathbf{\Omega} \times \dot{\mathbf{r}} + \mathbf{\Omega} \times (\mathbf{\Omega} \times \mathbf{r})]. \tag{10.8}$$

That is, forces are required to provide all the three types of accelerations. The physical interpretation of all the three terms of Eq. (10.8) were discussed in Section 8.2.

The rotating frame S is a noninertial reference frame in which Newton's equation of motion $\mathbf{F} = m\mathbf{a}_S$ is not valid. The acceleration in the rotating reference frame is $\ddot{\mathbf{r}}$ which is clearly not equal to $\mathbf{F}/m$. We use Eq. (10.7) to write the equation of motion in the reference frame S:

$$\mathbf{a}_S = \mathbf{a}_I - 2\mathbf{\Omega} \times \dot{\mathbf{r}} - \mathbf{\Omega} \times (\mathbf{\Omega} \times \mathbf{r})$$

or $m\mathbf{a}_S = \mathbf{F} - 2m\mathbf{\Omega} \times \dot{\mathbf{r}} - m\mathbf{\Omega} \times (\mathbf{\Omega} \times \mathbf{r})$.

The second and the third term in the above equations are pseudo forces; they are called the *Coriolis force* and the *centrifugal force* respectively. If we want to apply Newton's law in S, we will need to include these forces in the equation of motion. We will illustrate these forces using several examples. In all these cases, we will take Ω to be positive, i.e., rotation is assumed to be counter-clockwise.

1. If a particle is at rest in S, then $\mathbf{a}_S = 0$. However, a real radially-inward force is acting on the particle. Since $\mathbf{a}_S = 0$, the observer in S says that the real force is matched by the centrifugal force that is acting radially outward. See Fig. 10.3(a) for an illustration.

2. If a particle is moving radially outward with velocity v in S, then clearly $\mathbf{a}_S = 0$. However, a real force is acting on the particle to make it possible (recall the examples discussed above). The observer in S says that the real forces F_r and F_ϕ are balanced by the Coriolis and centrifugal forces respectively as shown in Fig. 10.3(b). A similar interpretation works for a particle moving with velocity v in the tangential direction Fig. 10.3(c). Note that the direction of the Coriolis force is always towards the right if we tag along with the particle.

3. Consider a free particle shot radially outward with a constant radial velocity. In the inertial frame I, it moves straight, but in the rotating frame it turns right as shown in Fig. 10.4(a). Note that the observer is looking radially outward at points A and B. Clearly the rightward movement of the particle in S is due to the Coriolis force.

   A similar interpretation works for a free particle shot radially inward with a constant radial velocity. The observer is facing towards the centre at points A and B (Fig. 10.4(b)), hence the particle will appear to turn right for the observer. For him/her the Coriolis force pushes the particle rightward.

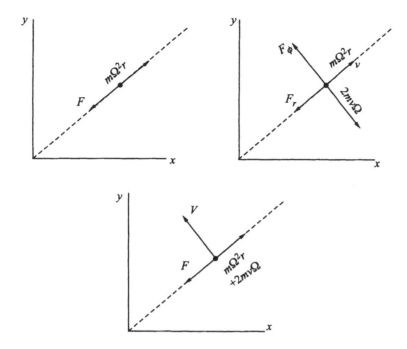

Figure 10.3   *(a) Particle is fixed in S. According to the observer in S, the real force is matched by the centrifugal force. (b) Particle is moving outward with constant velocity* **v** *in S. According to S, the real forces F_r and F_ϕ are matched by centrifugal and Coriolis force respectively. (c) For a particle moving tangential with constant velocity* **v**, *Coriolis and centrifugal forces are radially outward for an observer in S.*

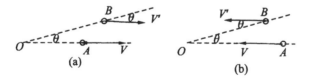

Figure 10.4   *(a) A free particle is shot radially outward. (b) A particle is shot radially inward. For an observer in S, the particle is pushed right.*

In the next section we will study the motion of a particle in Earth's reference frame for various configurations.

10.3   Earth as a reference frame

We perform most of our experiments on the Earth. Therefore, we need to understand whether the Earth is an inertial frame or not. Earth is rotating around the Sun, which in turn is rotating around the centre of the galaxy. The Earth is also spinning around its own axis. These effects arouses suspicion whether the Earth is an inertial frame or not. In the following discussion, we will study the effects of Earth's rotation on some terrestrial (on the Earth) experiments.

1. Apparent gravitational acceleration

First, we will investigate the effects of the centrifugal force due to Earth's spin. Let us consider a particle which is at rest on the surface of the Earth. We will explore all the pseudo forces acting on the particle. In Earth's frame, there is no Coriolis force on the particle because $\dot{\mathbf{r}} = 0$. However a centrifugal force of magnitude $m\Omega^2\rho\hat{\rho}$ acts on the particle as shown in Fig. 10.5. Here $\Omega = 2\pi/86400 \approx 7.27 \times 10^{-5}\text{s}^{-1}$.

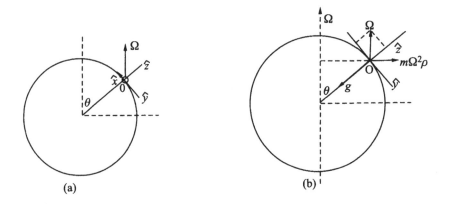

(a)   (b)

Figure 10.5   *(a) The components of the angular velocity vector $\boldsymbol{\Omega}$ in a local coordinate system. (b) Real and pseudo forces acting on a particle on the surface of the Earth.*

Hence, the effective **g** on the particle will be

$$\mathbf{g} = -g\hat{\mathbf{r}} + \Omega^2 R_e \sin\theta\,\hat{\rho},$$

where θ is the angle of the radial direction from the direction of the North Pole, and R_e is the radius of the Earth. Therefore, the horizontal (g_h) and vertical (g_v) components of **g** are

$$g_h = \Omega^2 R_e \sin\theta\cos\theta,$$

$$g_v = -g + \Omega^2 R_e \sin^2\theta.$$

The effective acceleration due to gravity is decreased due to the centrifugal force. The relative correction is of the order of

$$\frac{\Omega^2 R_e}{g} \approx \frac{(2\pi/86400)^2 \times 6400 \times 10^3 \,\text{m/s}^2}{10 \,\text{m/s}^2} \approx 0.0034,$$

which is less than one per cent. The net gravitational force acting on the particle is not radially inward, rather it makes a small angle $\tan^{-1}(g_h/g_v)$ with the vertical.

This angle is approximately $0.0034 \sin\theta \cos\theta$ rad or $0.097 \sin 2\theta$ degree.

As a result of the centrifugal forces, the acceleration due to gravity at the pole and the equator are

$$g_{\text{pole}} = -g,$$

$$g_{\text{equator}} = -g + \Omega^2 R.$$

The difference between the above two values is $\Omega^2 R \approx 0.034$ m/s^2. The experimental value of the difference these two values is however around 0.053 m/s^2. The difference between the two estimates is due to the ellipsoidal shape of the earth. The radius of the earth at the equator is larger than that at the pole. Hence g_{equator} is even smaller than the above estimate. Note that the centrifugal force is much smaller than gravity.

2. Effect of Coriolis force on terrestrial experiments

Let us estimate the magnitude of the Coriolis force in two typical situations: (a) On a cricket ball (or a base ball) when it traverses the pitch; (b) On a ballistic missile travelling from one city to another. The Coriolis force on the cricket ball is $-2\,m\mathbf{\Omega}\times\mathbf{v}$, where $\mathbf{\Omega}$ is the angular velocity of the Earth, and $\mathbf{v}$ is the velocity of a ball of mass m. We can safely take $m \approx 100$ gm, $\Omega \approx 2\pi/86400$ rad/s, and $v \approx 40$ m/s (Sohaib Akhtar's speed). Hence

$$F_{\text{Coriolis}} \approx 2 \times 0.10 \times (2\pi/86400) \times 40 \approx 6 \times 10^{-4}\,\text{N}$$

This force is quite small compared to the gravitational force.

For a missile we take, $m = 1000$ kg, and $v \approx 6000$ m/s. Using these data

$$F_{\text{Coriolis}} \approx 2 \times 1000 \times (2 \times \pi/86400) \times 6000 \approx 10^3\text{N}$$

which is comparable to the gravitational force acting on it. Designers and navigators have to take into account the Coriolis force while considering the motion of a ballistic missile.

3. Freely falling body

A particle is falling vertically downward at a latitude $\pi/2 - \theta$ (Fig. 10.5(a)). Let us examine its motion in Earth's frame (S) and in an inertial frame (I). In the reference frame I (say the Sun), the particle falls vertically downward due to gravity. However in the reference frame S, the angular velocity in the local coordinate system is $\mathbf{\Omega} = (0, -\Omega\sin\theta, \Omega\cos\theta)$ (see Fig. 10.5(b)). Note that the ratio of the centrifugal force and Coriolis force is of the order of $\Omega R/v \approx 465$ ms$^{-1}/v$. Hence centrifugal force is significant for small velocities, however it can be absorbed in the effective gravitational force as discussed above. In the coordinate system shown in Fig. 10.5(b), the Coriolis acceleration is

$$a_{\text{Coriolis}} = -2\mathbf{\Omega} \times \mathbf{v}$$

$$= -2\Omega[(-v_z \sin\theta - v_y \cos\theta)\hat{\mathbf{x}} + v_x \cos\theta\hat{\mathbf{y}} + v_x \sin\theta\hat{\mathbf{z}}].$$

Therefore, the equation of motion is

$$\dot{v}_x = -2\Omega(-v_z \sin\theta - v_y \cos\theta) \tag{10.9}$$

$$\dot{v}_y = -2\Omega v_x \cos\theta \tag{10.10}$$

$$\dot{v}_z = -g - 2\Omega v_x \sin\theta. \tag{10.11}$$

These are a reasonably complex set of equations. We solve them approximately by taking a limit $\Omega v_x/g \ll 1$. The above approximation is valid for $v_x \ll (g/\Omega) \approx 10^5$ m/s, which covers almost all terrestrial motion. Under this approximation Eq. (10.11) yields

$$v_z(t) = -gt$$

with initial condition $v_z(0) = 0$. In Eq. (10.9) we drop the second term in the right-hand side because $v_y \ll v_z$. This approximation yields

$$v_x = -\Omega \sin\theta gt^2.$$

Hence,

$$z = h - \frac{1}{2}gt^2,$$

$$x = -\Omega g \sin\theta \frac{t^3}{3}.$$

By the time the particle reaches the Earth $(T = \sqrt{2h/g})$, the deviation of the particle along the horizontal direction is

$$X = -\frac{\Omega}{3} \sin\theta \left(\frac{8h^3}{g}\right)^{1/2}$$

Note that $X/h \approx \Omega T \ll 1$. For $h = 1$ km,

$$X \approx -\sin\theta \text{ m}$$

Equation (10.10) yields

$$v_y = \frac{2}{3}\Omega^2 \sin(\theta)gt^3.$$

The ratio $v_y/v_x \sim \Omega T \ll 1$, consequently $y \ll x$.

4. Foucault pendulum

A pendulum is oscillating at a latitude of $\pi/2 - \theta$ (Fig. 10.5(a)). Let us describe the motion of the pendulum in the reference frame of the Earth. The centrifugal force can be absorbed in the effective gravitational force as discussed above. We also assume that the motion along the z direction is negligible. Under this approximation the equation of motion yields

$$\mathbf{a} = -\mathbf{g} - 2\mathbf{\Omega} \times \mathbf{v},$$

or $\dot{v}_x = 2\Omega_z v_y - \frac{g}{l} x,$

$\dot{v}_y = -2\Omega_z v_x - \frac{g}{l} y,$

where $\Omega_z = \Omega \cos\theta$ is the component of $\mathbf{\Omega}$ along the z-axis. Writing $z = x + iy$ (complex position, again this is only a mathematical construct without any physical interpretation), we obtain

$$\ddot{z} + \frac{g}{l} z + 2i\Omega_z \dot{z} = 0.$$

The solution of the above equation is

$$z(t) = \exp\left(-i\Omega_z t\right)[A\cos\left(\omega_1 t\right) + B\sin\left(\omega_1 t\right)],$$

where $\omega_1 = \sqrt{(g/l) + \Omega_z^2}$. Since $\Omega_z \ll \sqrt{g/l}$, $\omega_1 \approx \sqrt{g/l} = \omega_0$, we obtain

$$z(t) = \exp\left(-i\Omega_z t\right)[C_1 \cos\left(\omega_0 t\right) + C_2 \sin\left(\omega_0 t\right)],$$

$$x(t) + iy(t) = \exp\left(-i\Omega_z t\right) A \cos\left(\omega_0 t + \phi_0\right),$$

where A is the amplitude of the oscillation, and ϕ_0 is the initial phase. From the above expression we can easily deduce that

$$\frac{y(t)}{x(t)} = -\frac{\sin(\Omega_z t)}{\cos(\Omega_z t)} = \tan(-\Omega_z t).$$

This observation tells us that the pendulum is rotating in the clockwise direction as shown in Fig. 10.6. You can find the direction of rotation of the plane using the direction of the Coriolis force.

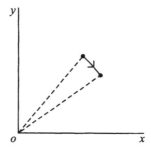

Figure 10.6   *The plane of motion of Foucault pendulum rotates clockwise with frequency Ω_z.*

At the North Pole $\Omega_z = \Omega$, hence the plane of pendulum makes a complete revolution in one day at the North Pole. At the equator however the plane of the pendulum does not rotate at all since $\Omega_z = 0$ at the equator. In Kanpur $\Omega_z = \Omega \cos 64° = 0.44\Omega$, and it would take $1/0.44 \approx 2.27$ days to complete a full revolution. It should however be kept in mind that in the Sun's reference frame, the pendulum does not rotate at all. This is where things get tricky. Let us ask a question.

QUESTION   Consider a pendulum at the North Pole. The plane of oscillation of the pendulum remains fixed in the inertial frame. The question is whether the *absolute space* holds the plane of oscillation fixed while the Earth turns below.

The answer to the above question is still uncertain. It was first addressed by Newton, and later by Mach and Einstein. We will discuss their approach to this problem very briefly in the following sections.

5. Cyclones and hurricanes

Consider a low pressure region in the atmosphere of the Earth. In the Northern Hemisphere, the wind is flowing towards the centre of a low-pressure region would turn right, and the circulation is counter-clockwise as shown in Fig. 10.7. Suppose the fluid parcel is circulating in the Northern Hemisphere at a latitude λ. In Earth's reference frame, the forces acting on the parcel are force due to pressure gradient $(dP/dr)Adr,$[1] Coriolis force $2m\Omega v \sin \lambda$ (λ is the latitude), and centrifugal force mv^2/r. Assuming that the forces are balanced in Earth's frame, we obtain

$$\frac{dP}{dr}Adr \approx 2m\Omega v \sin \lambda + m\frac{v^2}{r},$$

or $\quad \dfrac{1}{\sigma}\dfrac{dP}{dr} \approx 2\Omega v \sin \lambda + \dfrac{v^2}{r},$ $\hfill$ (10.12)

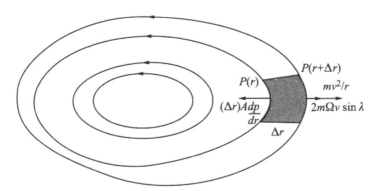

Figure 10.7   *Wind pattern near a low-pressure region*

---

[1]The pressure at two surfaces shown in the figure are $P(r)$ and $P(r + dr)$. We take a small elemental area A on these surfaces and construct a volume $|\mathbf{A} \times d\mathbf{r}|$. The force on the volume is $[P(r+dr) - P(r)]A \approx (dP/dr)Adr \approx (dP/dr)dV$.

Here Adr is the volume of the fluid parcel, and is equal to m/σ where σ is the density of the parcel. Let us estimate the velocity for a typical wind pattern. The typical parameters are $\sigma \approx 1$ kg/m^3, $dP/dr \approx 10^{-3}$ N/m^2, $\Omega \approx 10^{-4}$ s^{-1}, $r \approx 1$ km. Substitution of these values in Eq. (10.12) yield $v \approx 25$ m/s ≈ 90 km/hour.

Cyclones or hurricanes have the above patterns. In the Northern Hemisphere, the circulation is counter-clockwise, and in the Southern Hemisphere it is clockwise.

In summary, the centrifugal and Coriolis accelerations on the surface of the Earth is insignificant compared to the acceleration due to gravity in table-top experiments. The Earth can be treated as an inertial frame of reference when the accuracy required in the experiment is less than one per cent, and the length-scale of the experiment is much less than the radius of the Earth. In most elementary mechanics experiments, like air tracks, spring–mass systems, etc. we treat the Earth as an inertial frame and do not worry about centrifugal and Coriolis accelerations. For more accurate experiments, and for large-scale motion we cannot afford to ignore the above pseudo forces. For example, the Coriolis force is important for cannon balls with a range of tens of kilometres. Modern cannons are designed with built-in corrections for these forces.

What about the effect of the Earth's orbital motion around the Sun? The centrifugal acceleration due to Earth's orbital motion around the Sun is $(2\pi/(86400 \times 365))^2 \times 1.5 \times 10^{11} \approx 0.0059$ m/s^2 which is an order of magnitude smaller than the acceleration due to gravity at the surface of the Earth. Hence, both orbital motion and the spin of the Earth induce small corrections to the acceleration due to gravity.

EXAMPLE 10.1   A coin of mass m is placed on a frictional disc (coefficient of friction μ) which is rotating with constant angular speed ω. Write down the equation of motion of the coin for an observer on the disc frame. Describe the motion of the coin when the disc is frictionless ($\mu = 0$).

SOLUTION   The equation of motion of the coin in the rotating frame of reference is

$$m\frac{d^2\mathbf{r}}{dt^2} = \mathbf{F} - 2m(\mathbf{\Omega} \times \dot{\mathbf{r}}) - m\mathbf{\Omega} \times (\mathbf{\Omega} \times \mathbf{r})$$

$$= \mathbf{F} - 2m(\mathbf{\Omega} \times \dot{\mathbf{r}}) + m\Omega^2 \rho\hat{\rho}.$$

Along the vertical direction, the normal force on the coin by the disc balances the gravitational force on the coin. Along the horizontal plane, the only real force acting on the coin is the frictional force. The pseudo forces acting on the coin are the Coriolis force and the centrifugal force. The maximum value of the frictional force F_{fr} is μmg. When $F_{fr} < \mu mg$, the coin will not move relative to the disc, i.e., $\dot{\mathbf{r}} = \ddot{\mathbf{r}} = 0$. The inward frictional force will match the outward centrifugal force, i.e.,

$$F_{fr} = m\Omega^2 \rho < (F_{fr})_{\text{max}},$$

or $m\Omega^2 \rho < \mu mg$,

or $\Omega < \sqrt{\mu g/\rho} = \Omega_c.$

When $\Omega > \Omega_c$, the coin starts to slip relative to the disc. The frictional force on the coin is μmg, and its direction is opposite to the relative velocity of the coin. That is, in the rotating frame, $\mathbf{F}_{fr}$ is opposite to the direction of $\dot{\mathbf{r}} = v_r \hat{\mathbf{r}} + v_\phi \hat{\phi} = \dot{r}\hat{\mathbf{r}} + r\dot{\phi}\hat{\phi}$. The schematic diagram is shown in Fig. 10.8. We assume that the static and kinetic frictional coefficients are the same.

The components of the frictional force are

$$f_r = -\mu mg \cos \chi,$$

$$f_\phi = -\mu mg \sin \chi,$$

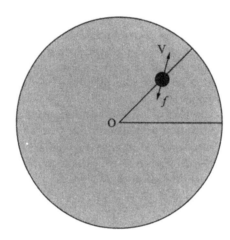

Figure 10.8   *Motion of a coin on a rotating disc.*

where χ is the angle $\dot{\mathbf{r}}$ makes with the radial direction ($\tan \chi = v_\phi / v_r$). The equation of motion of the coin in the rotating frame is

$$m(\ddot{r} - r\dot{\phi}^2) = f_r + m\Omega^2 r + 2m\Omega v_\phi,$$

$$m(\ddot{\phi} + 2\dot{r}\dot{\phi}) = f_\phi - 2m\Omega v_r.$$

The above equations can be solved numerically given initial conditions. We can then find $\mathbf{r}(t)$. Qualitatively, before the critical angular velocity $\Omega_c = \sqrt{\mu g / \rho}$, the coin appears stationary in the rotating frame. For higher angular frequencies, the coin will drift outward, and will lag behind due to friction.

Consider a case when $\mu = 0$, and the particle is at rest at $t = 0$. Here, the particle will be at rest in the inertial range forever. Therefore, in the rotating frame $\dot{r} = \ddot{r} = \ddot{\phi} = 0$, $\dot{\phi} = -\Omega$, and $v_\phi = -\Omega r$. That is, in the rotating frame, the coin is rotating in the opposite direction with angular velocity Ω. You can easily check that the above solution satisfies the equation of motion.

EXAMPLE 10.2   NEWTON'S BUCKET   Consider a bucket filled with water. We rotate the bucket about the vertical axis with constant angular frequency Ω. What is the surface profile of the water in the bucket?

SOLUTION   When we rotate the bucket, the surface is disturbed with a few ripples. The surface profile then reaches an equilibrium configuration. The configuration is shown in Fig. 10.9. A small fluid element on the surface experiences the forces shown in the figure. In the rotating frame, the fluid element is at rest. Hence

$$N\cos\theta = mg$$

$$N\sin\theta = m\Omega^2 x$$

where x is the distance of the element from the vertical axis of rotation. The above equations imply that

$$\tan\theta = \frac{dy}{dx} = \frac{\Omega^2 x}{g}.$$

Hence

$$y = \frac{\Omega^2}{2g}x^2,$$

which is the equation of a parabola. When the bucket rotates relative to the inertial frames (fixed stars), then we observe the fluid surface to be a parabola. When the bucket is not rotating, we observe a flat surface. Newton asked the following question regarding the curved surface in the bucket, "*Is it the rotation of the water through absolute space which cause the water surface to curve, or is it the relative motion between the water and the fixed stars which causes the above effect?*" In the following section we will briefly discuss the theories of Newton, Mach, and Einstein. Some of these ideas were discussed earlier in Chapter 2.

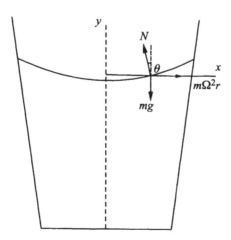

Figure 10.9   *Newton's bucket.*

10.4   Space: Newton vs. Mach

Newton's laws are invariant under Galilean transformation. Due to this symmetry we cannot determine whether a particle is at rest relative to absolute space or moving with constant velocity relative to absolute space. Newton noticed however that the acceleration with respect to absolute space could be determined by looking at the pseudo forces. If a pseudo force is present, then the reference frame at hand is noninertial.

Coming back to Newton's bucket, Newton argued that the parabolic shape of the water surface in the rotating bucket is due to the motion of the bucket relative to *absolute space*. He argued that if all the mass of the universe except the bucket rotated in the reverse direction (kinematically equivalent), the water surface will remain flat because the bucket is not rotating relative to absolute space.

In 1880 Mach countered the idea of Newton's absolute space. According to Mach, in Science, one tries to avoid unverifiable assumptions. The idea of rotating the whole universe except the bucket is unverifiable. Mach argued that the centrifugal force on the water in Newton's bucket arises due to the acceleration of water relative to the masses of the universe, not because of its motion relative to absolute space. Hence the inertial mass of a particle will change if the mass distribution of the universe changes, or if a big mass comes near the particle. These are difficult issues, and they have not yet been fully resolved.

Einstein was deeply inspired by Mach's ideas.

10.5   Einstein's notion of space

A famous experiment conducted by Michelson and Morley showed that the speed of light is the same irrespective of the relative velocity between the source and the observer. This result was in contradiction with the Galilean transformation of velocity. Recall that the Galilean transformation is based on the idea of absolute time and space. Hence, Michelson and Morley's results contradicted the notion of absolute space and time. Many brilliant scientists tried to explain Michelson and Morley's result by ingenious schemes, which would keep the framework of absolute space and absolute time intact. It was Einstein who dismissed the idea of absolute space and absolute time, and constructed the *special theory of relativity* (SR) where space and time have a different structure. Einstein's SR could explain Michelson and Morley's experiment successfully. It made many predictions which have been verified in later experiments. We will discuss these ideas in Chapters 19 and 20.

Einstein also provided a working definition of inertial frames. Consider a vertical elevator that is falling freely under gravity. A person in the elevator is holding a block of mass m. What is the net force acting on the block in the elevator's frame of reference? Using Eq. (10.4), the required force is

$$m\mathbf{a}_S = -m_G\mathbf{g} + m_i\mathbf{g}.$$

Since $m_i = m_g$, $\mathbf{a}_S = 0$. Inside the elevator, the particles at rest will remain at rest, and particles moving with constant velocities will continue to do so in the future. Hence the freely falling elevator is an inertial frame.

EXAMPLE 10.3   The acceleration due to gravity is not uniform if the size of the elevator is large. Two masses A and B separated by distance l in the vertical direction get distanced further due to non-uniformity of $\mathbf{g}$. Compute the relative acceleration and separation between the masses.

SOLUTION   The two masses A and B shown in Fig. 10.10 are falling freely under gravity.

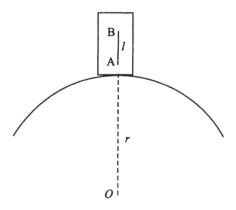

Figure 10.10   *Two masses A and B inside an elevator are falling freely under gravity.*

If the mass A is at a distance r from the centre of the Earth, then the relative acceleration between the two masses is

$$\Delta g = \frac{GM_e}{r^2} - \frac{GM_e}{(r+l)^2}$$

$$\ddot{l} = \frac{GM_e}{r^2}\left(1 - \frac{1}{(1+l/r)^2}\right)$$

$$\approx \frac{GM_e}{r^2}\frac{2l}{r} = \frac{GM2l}{r^3}.$$

This is called tidal acceleration, and it is like the acceleration of a charge particle due to an electric dipole (r^{-3}). The solution of the above equation yields

$$l(t) = c_1 \cosh\left(\frac{2GM}{r^3}t\right) + c_2 \sinh\left(\frac{2GM}{r^3}t\right).$$

If the relative velocity of the particles at $t = 0$ is zero, then

$$l(t) = l(0) \cosh\left(\frac{2GM}{r^3}t\right).$$

The above example illustrates that the elevator is an inertial frame when the size of the elevator is small such that $l(t) \approx l(0)$ and the difference $l(t) - l(0)$ is not measurable. The noninertial force inside the lift is called the *tidal force*. *The inner space of the lift is referred to as a local inertial frame if the size of the lift is small.*

The above experiment shows that a freely falling elevator is equivalent to a rocket moving in free space with constant velocity (as long as the rocket or lift is small in size). Similarly an elevator accelerating upward in free space with an acceleration **g** is equivalent to a stationary elevator placed in a gravitational field of −**g**. That is

For an observer in a reference frame S which is accelerated with constant acceleration (**a**) *relative to the inertial frame, the motion of a particle will be as if a gravitational uniform field* (−**a**) *acts in addition to the real forces.* This is called the *principle of equivalence,* and was discovered by Einstein. Naturally the principle of equivalence holds only locally (size of the lift has to be small). Note that the principle of equivalence hinges on the assumption that the inertial mass and the gravitational mass are equal.

Einstein's general theory of relativity (GR), which is the modern theory of gravitation, is based on the principle of equivalence. In GR, spacetime is curved near heavy masses, hence the spacetime structure is determined by all the matter of the universe, far and near. Test particles follow *straight paths* in the *local inertial frame*; the laws of motion are completely determined by the local spacetime. Thus, Einstein's ideas of space involves both local and global perspectives, in some sense related to the ideas of Mach.

Here we will end our discussion on noninertial frames. In the next chapter we will discuss the concept of energy and potential.

Exercises

1. Derive the equation of motion for a particle in a reference frame that is rotating with a variable angular velocity $(\dot{\Omega} \neq 0)$ with respect to an inertial frame.

2. A particle moves with a constant velocity **v** along $\hat{\phi}$ in the rotating frame of reference S; the reference frame S itself is rotating with respect to an inertial frame I with angular velocity Ω. Compute the acceleration of the particle in the inertial frame I.

3. A balloon filled with helium is connected to the floor of an open-air train by a string. The mass of the balloon is 100 grams, and the length of the string is 1 metre. What would be the configuration of the balloon when the train is moving with a constant acceleration of 1 m/s^2. Compare the results with a balloon filled with oxygen.

4. Recall Example 4.3 on the motion of a block over an accelerating wedge. In chapter 4, this problem was solved in the inertial frame. Solve the problem in the reference frame attached to the wedge.

5. Recall Example 10.1. Describe the motion of the coin in the laboratory frame with respect to which the disc is rotating with angular frequency Ω.

6. The Earth goes around the Sun. What is the centrifugal acceleration for an observer on the Earth due to this motion? Compare this acceleration with g, and with acceleration due to the spin of the Earth around its own axis.

7. The Sun rotates around the centre of the galaxy. What is the centrifugal acceleration on the solar system due to this motion? Compare this acceleration with the acceleration due to gravity on the surface of the Earth.

8. Find the apparent acceleration of gravity at the equator, the poles, and at 45° latitude. Assume Earth's shape to be spherical.

9. What should be the angular frequency of Earth's spin under which the objects at the equator will become massless? Under this situation what would be the effective gravitational accelerations at different latitudes?

10. A 400-tonne train is running due south at a speed of 100 km/hour near Kanpur station (latitude of 26 degrees). What is the horizontal force on the tracks? What is the direction of the force?

11. At the latitude of 30 degrees, three objects are shot horizontally: (a) a cricket ball with 150 km/hour; (b) a missile with 2000 km/hour; (c) an electron with $0.1c$. Estimate the maximum horizontal deviation of these objects in 1 second due to the Coriolis force. What is the approximate angle of deviation?

12. Plumb-lines are used by masons to find the local vertical direction. How much does a plumb-line deviate from the local vertical direction due to the centrifugal force on the bob as a result of Earth's rotation? Estimate the angle for Kanpur (latitude of 26 degrees).

Projects

1. Design a Foucault pendulum.

2. Design an experiment to measure the effective **g** (both magnitude and direction) on the surface of the Earth with an accuracy up to 0.1%. It is required for measuring the effect of Earth's spin. How about an experiment to measure the shift in g due to the orbital motion of the Earth around the Sun?

3. When a mass is dropped from a height, it deviates from the vertical due to Coriolis force. Design an experiment to measure this deviation.

11

Energy

11.1 Importance of conservation laws

So far we have studied the evolution of mechanical systems by solving the equation of motion, given forces and initial conditions. This procedure is called the *force method*. In this chapter we will introduce a new technique based on the *conservation of energy*. Energy is conserved for a class of systems called *conservative systems*. Using this idea, we can deduce the property of the final state of the system. We will start our discussion with a problem.

Consider the motion of a train on a smooth roller-coaster shown in Fig. 11.1.

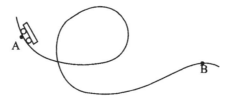

Figure 11.1 *Motion of a train on a roller-coaster.*

Suppose the train starts at point A. What is the velocity of the train at point B? The train experiences the gravitational force, and a variable normal force. The problem is very difficult to solve using the equation of motion, but is quite easy using energy conservation. We will discuss the method in this chapter.

11.2 Work and kinetic energy: Work energy theorem

The work done on a particle is related to the kinetic energy of the particle. We derive this relationship as follows. Consider a particle of mass m that experiences a force $\mathbf{F}(\mathbf{x})$ at point $\mathbf{x}$. The particle starts from point A and goes to point B along a given path. The work done on the particle by the external agency is

$$W_{AB} = \int_A^B \mathbf{F} \cdot d\mathbf{x}.$$

We can rewrite the above as

$$W_{AB} = \int_A^B m \frac{d\mathbf{v}}{dt} \cdot \frac{d\mathbf{x}}{dt} dt$$

$$= \int_A^B \frac{d}{dt} \left(\frac{1}{2} m v^2 \right) dt$$

$$= \left[\frac{1}{2} m v^2 \right]_A^B \tag{11.1}$$

$$= T(B) - T(A) \tag{11.2}$$

where $T(\mathbf{r})$ is the kinetic energy (KE) of the particle at point $\mathbf{r}$. Hence, the difference in kinetic energy of the particle at two points is equal to the work done by the external force on the particle during the transit between the two points. This is called the *work energy theorem*. The theorem holds for all kinds of forces (friction, gravity etc.). Note however that $\mathbf{F}$ is the net force acting on the particle. We illustrate the work energy theorem using several examples.

1. A particle falls under gravity from height h to the ground. Assume that the particle starts from rest, and has a velocity $\mathbf{v}$ just before hitting the ground. According to the work energy theorem

$$\frac{1}{2} m v^2 = \int_A^B \mathbf{F} \cdot d\mathbf{x} = mgh$$

yielding $v = \sqrt{2gh}$.

2. We can include air friction in the above problem. For a closed path from height $h/2$ to the ground, and back to the height $h/2$,

$$\Delta T = \int (\mathbf{F}_{\text{fric}} + m\mathbf{g}) \cdot d\mathbf{x},$$

where the integral is over the closed path. $\int m\mathbf{g} \cdot d\mathbf{x} = 0$ for the closed path, but $\int \mathbf{F}_{\text{fric}} \cdot d\mathbf{x} < 0$ because $\mathbf{F}_{\text{fric}}$ is always opposite to $d\mathbf{x}$. Hence, there is a net loss of kinetic energy over the closed path. The loss in the kinetic energy is due to friction.

3. The Earth goes around the Sun in a (almost) circular path. For a circular orbit, the work done by the gravitational force is zero because $\mathbf{F}_{\text{gravity}}$ is perpendicular to $d\mathbf{r}$. Hence, the kinetic energy remains conserved on its path.

In Chapter 3, we introduced the ideas of forces arising due to electric, magnetic, and gravitational fields. We will study the concepts of fields in a bit more detail in the following section.

11.3   Vector fields and scalar fields

So far our attention had been on the position, velocity, and acceleration of a point particle. These quantities are vectors at the point where the particle is present at that time. Now we will introduce the idea of a field. A central mass (the Sun for example) creates a gravitational field everywhere in space. At a given point, a test particle experiences a force proportional to the field at that point. So a field exists everywhere, unlike the velocity and acceleration vectors of a particle. The fields could be either vector fields or scalar fields.

A vector field is a physical quantity for which we associate a vector at every point in its domain space. The vector field at a given point transforms like a vector under the rotation of a coordinate system whose origin is at that point. Examples of vector fields are the electric field, the gravitation field, etc.

Scalar fields are quantities for which we associate a scalar at every point in its domain space. The scalar field at a given point remains unchanged under the rotation of a coordinate system whose origin is at that point. Examples of scalar fields are the temperature field, the density field etc.

In Fig. 11.2 we illustrate a vector field $\mathbf{E} = -x\hat{\mathbf{x}} - y\hat{\mathbf{y}} = -\rho\hat{\rho}$.

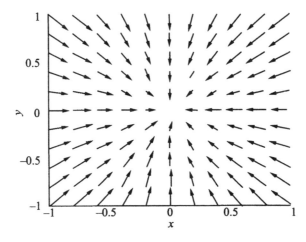

Figure 11.2   *Illustration of an electric field produced by a point charge* $\mathbf{E} = -x\hat{\mathbf{x}} - y\hat{\mathbf{y}} = -\rho\hat{\rho}$.

Here the length of the arrows denotes the magnitude of the field, while the direction of the arrows denotes the direction of the field. Sometimes the magnitude of the field is represented by colour coding, e.g., dark regions may denote intense fields, while white regions may denote weak fields.

If a particle moves in a force field, then the force field does work on the particle. The work done by the force field is $\int \mathbf{F} \cdot d\mathbf{r}$, which can be evaluated by doing the line integral. We illustrate the computational procedure using some examples.

EXAMPLE 11.1   Compute the work done on a particle by the force fields (1) $A\hat{\rho}$; (2) $(1/\rho)\hat{\phi}$; (3) $\rho\omega\hat{\phi}$ (all in cylindrical coordinates). Take the closed path given in Fig. 11.3 as the contour of integration.

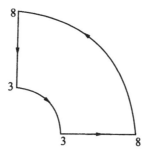

Figure 11.3   *Contour of integration for the work $\int \mathbf{F} \cdot d\mathbf{r}$.*

SOLUTION   Since the path of the particle is in the xy plane, $z = 0$. Therefore the work done by the above forces for the given closed path are

1. $\oint \mathbf{F} \cdot d\mathbf{x} = \int_3^8 A d\rho + \int_8^3 A d\rho = 0$

2. $\oint \mathbf{F} \cdot d\mathbf{x} = \int_{0,\rho=8}^{\pi/2} \frac{1}{\rho} \rho d\phi + \int_{\pi/2,\rho=3}^{0} \frac{1}{\rho} \rho d\phi = 0$

3. $\oint \mathbf{F} \cdot d\mathbf{x} = \omega \int_{0,\rho=8}^{\pi/2} \rho\rho d\phi + \omega \int_{\pi/2,\rho=3}^{0} \rho\rho d\phi = \dfrac{\omega\pi}{3 \times 2}(8^3 - 3^3) = \dfrac{485\omega\pi}{6}$.

EXAMPLE 11.2   A particle of mass 2 gm moves around in a circular orbit in the xy plane with a constant speed of 5 cm/s. The centre of the orbit is the origin, and the radius of the orbit is 5 cm. A force field $\mathbf{F} = (3x - y + z)\hat{\mathbf{x}} + (x + y - z^2)\hat{\mathbf{y}} + (3x - 2y + 4z)\hat{\mathbf{z}}$ dyne is acting on the particle. Compute the work done by the force $\mathbf{F}$? Is there any other force acting on the particle? What is the work done by the other force? What is the net work done on the particle?

SOLUTION   Work done by force $\mathbf{F}$ is

$$\oint \mathbf{F} \cdot d\mathbf{x} = \int dx(3x - y) + \int dy(x + y)$$

$$= \int dx(3x - \sqrt{25 - x^2}) + \int dy(y + \sqrt{25 - y^2})$$

$$= \tfrac{3}{2}x^2|_0^0 + 25 \int_0^{2\pi} \sin^2 \theta d\theta + \tfrac{1}{2}y^2|_0^0 + 25 \int_0^{2\pi} \cos^2 \theta d\theta$$

$$= 50\pi \text{ ergs}$$

We can solve the above integral in radial–polar coordinates as well:

$$\oint \mathbf{F} \cdot d\mathbf{x} = \oint F_\phi r d\phi = \oint (-F_x \sin \phi + F_y \cos \phi) 5 d\phi$$

$$= 5 \oint (x + y) \cos \phi d\phi - 5 \oint (3x - y) \sin \phi d\phi$$

$$= 25 \oint (\cos^2 \phi + \sin \phi \cos \phi) d\phi - 25 \oint (3 \cos \phi \sin \phi - \sin^2 \phi) d\phi$$

$$= 50\pi \text{ ergs}$$

Since the particle moves in a circle, the net force on the particle is

$$\mathbf{F}_{net} = \mathbf{F} + \mathbf{G} = -\frac{mv^2}{r} \hat{\mathbf{r}} = -10 \text{ dynes } \hat{\mathbf{r}}.$$

Therefore, the other force $\mathbf{G} = 10\hat{\mathbf{r}}$ dynes $- \mathbf{F}$. The net work done by $\mathbf{F}_{net}$ is zero. Therefore, the work done by $\mathbf{G}$ will be -50π ergs.

In the next section we will discuss a special kind of field called the conservative force field.

11.4   Conservative force fields

There are some force fields for which $\oint \mathbf{F} \cdot d\mathbf{r} = 0$ for any closed path. These special fields are called *conservative force fields*. Let us consider a path A–B–A as shown in Fig. 11.4.

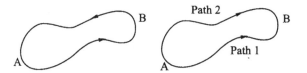

Figure 11.4   *(a) Force field* **F** *does work on the particle when the particle traverses along path A–B–A. (b) For conservative fields, the work done by the force field along path 1 is the same as the work done along path 2.*

For conservative fields $\oint \mathbf{F} \cdot d\mathbf{r} = 0$ along the path A–B–A. The integral can be rewritten as

$$\oint \mathbf{F} \cdot d\mathbf{r} = \int_{A-\text{path1}}^{B} \mathbf{F} \cdot d\mathbf{r} - \int_{A-\text{path2}}^{B} \mathbf{F} \cdot d\mathbf{r} = 0.$$

Therefore,

$$\int_{A-\text{path1}}^{B} \mathbf{F} \cdot d\mathbf{r} = \int_{A-\text{path2}}^{B} \mathbf{F} \cdot d\mathbf{r}. \tag{11.3}$$

In fact, for conservative forces $\int_{A}^{B} \mathbf{F} \cdot d\mathbf{r}$ is the same for any path; it depends only on the end-points. We take a convenient point P to be the reference point and define a *potential function* $U(\mathbf{r})$ at a point $\mathbf{r}$ using

$$U(\mathbf{r}) = - \int_{P}^{\mathbf{r}} \mathbf{F}(\mathbf{r}') \cdot d\mathbf{r}' \tag{11.4}$$

The potential function $U(\mathbf{r}) = U(x, y, z)$ is a scalar field.

We can write $\int_A^B \mathbf{F} \cdot d\mathbf{r}$ in terms of potential as follows:

$$\int_A^B \mathbf{F} \cdot d\mathbf{r} = \int_P^B \mathbf{F} \cdot d\mathbf{r} - \int_P^A \mathbf{F} \cdot d\mathbf{r},$$

$$= -U(B) + U(A). \tag{11.5}$$

Since $\int_P^B \mathbf{F} \cdot d\mathbf{r} = T(B) - T(A)$, Eq. (11.5) yields

$$T(A) + U(A) = T(B) + U(B).$$

This is the statement of *conservation of mechanical energy*. Clearly it holds only for conservative forces.

The choice of the reference point is tricky. For a potential of the form $U(r) = -\alpha/r,$ [1] we usually take the reference point to be at infinity with $U(\infty) = 0$. This reference point however is not unique as illustrated below. If we focus near the surface of the Earth, the potential function is

$$U(r_0 + h) = -\frac{\alpha}{R_e + h}$$

$$= -\frac{\alpha}{R_e}[1 + h/R_e]^{-1}$$

$$= -\frac{\alpha}{R_e} + \frac{\alpha h}{R_e^2}$$

$$= -\frac{\alpha}{R_e} + mgh, \tag{11.6}$$

since $\alpha/R_e^2 = mg$. Here R_e is the radius of the Earth.

Let us compute the potential of the same particle using $\mathbf{F} = -mg\hat{\mathbf{y}}$.

By integrating the force, we obtain the potential function as

$$U(h) = -\int_0^h F(y)dy + U(0) = mgh + U(0).$$

If we choose $U(0) = 0$ we get $U(h) = mgh,$ [2] which is the same as Eq. (11.6) apart from a constant $-\alpha/r_0$. Hence the reference point for a reference point is not always unique. However we need to keep the same reference value throughout a calculation.

---

[1] By taking radial path and integrating $\int_\infty^r \mathbf{F}(r') \cdot d\mathbf{r}'$ we can show that for $\mathbf{F}(r) = -\frac{\alpha}{r^2}\hat{r}, U(r) = -\frac{\alpha}{r}$.
[2] For large h, the formula $U(h) = mgh$ is not valid because $F \approx mg$ holds only near the surface of the Earth. We use $U(R_e + h) = -\alpha/(R_e + h)$ for computing the potential when h is of the order of R_e.

In the following simple examples we illustrate the application of the law of conservation of energy:

1. Consider a particle that falls down from height h. Suppose the particle starts with zero velocity, then at the topmost point, the particle has potential energy $U_1 = mgh$ and kinetic energy $T_1 = 0$. When the particle is about to hit the ground, $U_2 = 0$ and $T_2 = mv^2/2$. In Section 11.2 we showed that $v = \sqrt{2gh}$. When we substitute this expression of v in T_2, we obtain

$$U_1 + T_1 = U_2 + T_2,$$

thus verifying conservation of energy.

2. A charged particle $+Q$ is fixed at the origin. A test charged particle $+q$ of mass m is fired head-on towards the charged particle $+Q$ with velocity v_∞ from ∞. We can compute the distance of closest approach using the conservation of energy. Far away from the origin, $U_1 = 0$ and $T_1 = mv_\infty^2/2$, while at the point of closest approach $(r = d)$, $U_2 = Qq/(4\pi\epsilon_0 d)$ and $T_2 = 0$. Therefore, application of conservation of energy yields

$$\frac{1}{2}mv_\infty^2 = \frac{Qq}{4\pi\epsilon_0 d},$$

$$\text{or} \quad d = \frac{Qq}{2\pi\epsilon_0 mv_\infty^2}.$$

You can solve the problem by the force method too, but the energy method is much more convenient.

EXAMPLE 11.3   A planet of mass m is orbiting a star of mass M. The planet experiences a small drag force $\mathbf{F} = \eta\mathbf{v}$ due to the motion through the star's dense atmosphere. Here η is the coefficient of friction. Assuming an essentially circular orbit with radius $r = r_0$ at $t = 0$, calculate the time dependence of the radius.

SOLUTION   We had solved this problem earlier using the force method (Example .3). Here we will solve this problem using the energy method. The drag force $\eta\mathbf{v}$ dissipates a fraction of the total mechanical energy of the planet.   ote that for a circular orbit, the kinetic energy and potential energy are related by

$$\frac{1}{2}mv^2 = \frac{1}{2}\frac{\alpha}{r}. \tag{11. }$$

Therefore, the total mechanical energy of the planet would be

$$E = \frac{1}{2}mv^2 - \frac{\alpha}{r} = -\frac{1}{2}mv^2.$$

The rate of loss of mechanical energy is $\dfrac{dE}{dt} = -\mathbf{F}_{fric} \cdot \mathbf{v} = -\eta v^2$

$$\text{or} \quad mv\dot{v} = \eta v^2,$$

or $\dot{v} = \dfrac{\eta}{m}v,$

whose solution is $v(t) = v(0)\exp((\eta/m)t)$. The substitution of the above in Eq. (11.) yields

$$r(t) = \frac{\alpha}{m[v(0)]^2}\exp\left(-\frac{2\eta}{m}t\right).$$

Hence the orbit of the planet spirals toward the centre.

We find that gravitational and electrostatic forces are conservative. However, there are many forces which are not conservative. We will discuss these forces in the next section.

11.5   Nonconservative force fields

Forces for which $\oint \mathbf{F} \cdot d\mathbf{r} \neq 0$ are called *nonconservative forces*. Prominent examples of nonconservative forces are frictional force and electromotive force. Imagine that a block of mass m is sliding on a horizontal frictional surface. It starts its ourney from A with velocity v_0, rebounds at B, and then comes back to A.   n its return, the particle will have a lower speed. According to the work energy theorem, the loss in kinetic energy is equal to the work done by the frictional force in the round trip. Since the displacement is always opposite to the direction of frictional force, the work done by the frictional force is negative. This argument demonstrates the nonconservative nature of the frictional force.

   ecall that a time-varying magnetic field generates an electromotive force which accelerates the charged particles. If we confine the charged particle to a closed orbit, the speed of the particle can increase or decrease when it returns to its original position. This is due to the positive or negative work done by the electromotive force in the closed path. Hence, electromotive force is a nonconservative force.

   In the above two examples, the kinetic energy is not conserved over a closed path. For most nonconservative fields we cannot define a potential function, although in some special situations we can do so. But these definitions are not in the spirit of Eq. (11.4); we will return to this discussion in Section 11. .

   In the frictional force example described above, the loss in the mechanical energy (here kinetic energy) is equal to the gain in heat energy. Hence, the sum of kinetic and heat energy is conserved. It is found in nature that if we include all forms of energy, the total energy is always conserved. This is the *law of conservation of total energy*. The same feature is observed for the second example with the varying magnetic field; here the gain in the kinetic energy is at the expense of the electromagnetic energy, yet the sum of the kinetic energy and the electromagnetic energy is conserved. In this book we will only deal with mechanical energy.

   Equation (11.4) is used to define a potential corresponding to a given force. In the following section we will show that we can define force from the potential function. This exercise however requires some knowledge of *vector calculus* which is discussed below.

11.6 Digression to vector calculus

Mathematics of vector calculus will be covered in detail in your mathematics course. Here we will only introduce some elementary concepts of vector calculus.

11.6.1 Gradient of a potential field

In one-dimensional systems, the force field $F(x)$ is derivable from $U(x)$ using

$$F(x) = -\frac{dU}{dx}.$$

For a particle moving in two dimensions, we need to define appropriate potential and force fields. For two-dimensional systems, the potential field, a function of two independent variables x and y, is denoted by $U(x, y)$. As an example, let us consider

$$U(x, y) = \frac{1}{2}(x^2 + y^2).$$

There are several ways to visualise the potential field. One way is to plot $U(x, y)$ as a function of x, y as shown in Fig. 11.5.

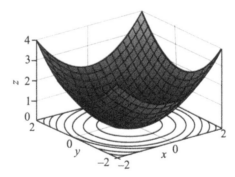

Figure 11.5   *Three-dimensional plot of scalar field $U(x, y) = (x^2 + y^2)/2$ as a function of x and y.*

Another way is to draw contours, which are lines joining all the points having the same potential. These contours are called *equipotential surfaces.* In two dimensions, the equipotential surfaces are curves, while in higher dimensions they are surfaces. See Fig. 11.6(a) for an illustration of equipotential surfaces in two-dimensions. Figure 11.6(b) shows a density plot in which the intensity of the potential is shown using a colour coding scheme. In the figure, darker regions show smaller values while the brighter regions show larger values.

Can we derive the corresponding force field at any given point (x, y) given $U(x, y)$? From the definition of the potential

$$U(X, Y) - U(x, y) = -\int_{x,y}^{X,Y} \mathbf{F}(x', y') \cdot d\mathbf{x}'.$$

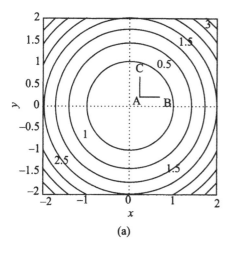
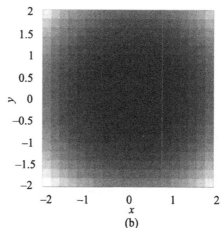

(a)   (b)

Figure 11.6   *Graphical representation of scalar field $U(x,y) = (x^2 + y^2)/2$ using (a) Contour plot of $U(x,y)$; (b) density plot of $U(x,y)$.*

For small $d\mathbf{x} = (\Delta x)\hat{\mathbf{x}} + (\Delta y)\hat{\mathbf{y}}$, the above equation becomes

$$U(X, Y) - U(x, y) \approx -F_x(x, y)\Delta x - F_y(x, y)\Delta y. \tag{11.8}$$

If we choose a small displacement only along the x-direction, i.e., $d\mathbf{x}' = (\Delta x)\hat{\mathbf{x}}$ (displacement along AB in Fig. 11.6(a)), then we obtain a slope

$$\lim_{\Delta x \to 0} \frac{U(x + \Delta x, y) - U(x, y)}{\Delta x} = \frac{\partial U}{\partial x} = -F_x$$

This is the slope of the potential function along the x-axis at point (x, y). It is also called the *partial derivative* of U along x. We have assumed that $U(x, y)$ is a continuous and differentiable function.

Similarly we can define a slope of the potential function along the y-axis, or partial derivative of U along the y-axis, by taking a displacement along the y direction (along AC in Fig. 11.6(a)), i.e.,

$$\lim_{\Delta y \to 0} \frac{U(x, y + \Delta y) - U(x, y)}{\Delta y} = \frac{\partial U}{\partial y} = -F_y$$

For arbitrary but small displacement

$$U(x + \Delta x, y + \Delta y) - U(x, y) = U(x + \Delta x, y + \Delta y) - U(x, y + \Delta y)$$

$$+ U(x, y + \Delta y) - U(x, y)$$

$$\approx \frac{\partial U}{\partial x}\Delta x + \frac{\partial U}{\partial y}\Delta y$$

Comparing the above equation with Eq. (11.8) and using the fact that Δx and Δy are arbitrary, we obtain

$$F_x = -\frac{\partial U}{\partial x}; \ \ F_y = -\frac{\partial U}{\partial y}.$$

Therefore

$$\mathbf{F}(x, y) = -\left[\frac{\partial U}{\partial x}\hat{\mathbf{x}} + \frac{\partial U}{\partial y}\hat{\mathbf{y}}\right]. \tag{11.9}$$

This is how we derive the force field from a given potential field in two dimensions. We can generalise the above definition to three dimensions as well

$$\mathbf{F}(x, y) = -\left[\frac{\partial U}{\partial x}\hat{\mathbf{x}} + \frac{\partial U}{\partial y}\hat{\mathbf{y}} + \frac{\partial U}{\partial z}\hat{\mathbf{z}}\right], \tag{11.10}$$

where

$$\frac{\partial U(x, y, z)}{\partial z} = \lim_{\Delta z \to 0} \frac{U(x, y, z + \Delta z) - U(x, y, z)}{\Delta z}$$

The bracketed-terms in the right-hand-side of Eqs. (11.9, 11.10) are called *gradient of U* in 2D and 3D. The gradient of a scalar field U is denoted by ∇U, i.e.,

$$\nabla U = \frac{\partial U(x, y, z)}{\partial x}\hat{\mathbf{x}} + \frac{\partial U(x, y, z)}{\partial y}\hat{\mathbf{y}} + \frac{\partial U(x, y, z)}{\partial y}\hat{\mathbf{z}}. \tag{11.11}$$

Hence the force is negative of the gradient of the potential function:

$$\mathbf{F} = -\nabla U$$

Suppose two points S and T are separated by a small vector $d\mathbf{r} = dx\hat{\mathbf{x}} + dy\hat{\mathbf{y}} + dz\hat{\mathbf{z}}$, then

$$\nabla U \cdot d\mathbf{r} = \frac{\partial U(x, y, z)}{\partial x}dx + \frac{\partial U(x, y, z)}{\partial y}dy + \frac{\partial U(x, y, z)}{\partial z}dz$$

$$= dU|_S^T = U(\mathrm{T}) - U(\mathrm{S}) \tag{11.12}$$

Hence, the potential difference between two adjacent points is obtained by taking a dot product of the gradient with $d\mathbf{x}$.

What is the physical interpretation of the gradient? Along equipotential surfaces $dU = \nabla U \cdot d\mathbf{r} = 0$. But $d\mathbf{r}$ is tangent to the equipotential curves, so ∇U must be perpendicular to the equipotential curves. Also $dU = \nabla U \cdot d\mathbf{r} = |\nabla U||d\mathbf{r}|\cos\theta$, where θ is the angle between ∇U and $d\mathbf{r}$. Therefore, the change in potential is maximum along the direction of ∇U ($\theta = 0$). Hence, ∇U is along the direction of the maximum change of potential (from lower potential to higher potential), and its magnitude is dU/dr. We illustrate these ideas using an example.

EXAMPLE 11.4   Consider a potential

$$U(x) = \frac{1}{2}(x^2 + y^2).$$

Compute its gradient and sketch it in the contour plot of the potential.

SOLUTION   The contour plot of the potential is shown in Fig. 11.7. The gradient of the potential is

$$\nabla U \quad = \quad \frac{\partial U}{\partial x}\hat{\mathbf{x}} + \frac{\partial U}{\partial y}\hat{\mathbf{y}}$$

$$= \quad x\hat{\mathbf{x}} + y\hat{\mathbf{y}}.$$

In Fig. 11.7 the field ∇U is depicted by arrows. The direction of the gradient is normal to the equipotential surface, and the magnitude of the gradient increases with r. Note that the force is opposite to ∇U.

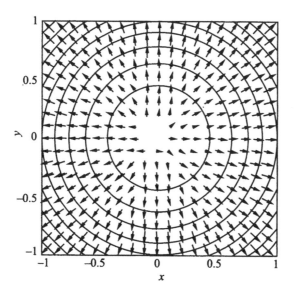

Figure 11.7   *Depiction of the gradient of the potential $U(x, y)$.*

For a three-dimensional potential

$$U(x, y, z) = \frac{k}{2}(x^2 + y^2 + z^2)$$

the equipotential surfaces are spheres, and the gradient

$$\nabla U = x\hat{\mathbf{x}} + y\hat{\mathbf{y}} + z\hat{\mathbf{z}}.$$

is perpendicular to the equipotential surface which is a sphere.

11.6.2   Derivation of the potential from a given force field

In the previous section, given the potential field, we derived the force field. Here we will describe the reverse process, i.e., derivation of a potential from a force field. We illustrate the procedure using an example. Suppose the given force field is

$$\mathbf{F}(x, y) = x\hat{\mathbf{x}} + y\hat{\mathbf{y}}.$$

Using $F_x = \partial U/\partial x$, we obtain

$$U(x, y) = -\int F_x dx + W(y) = -\frac{x^2}{2} + W(y).$$

$W(y)$ can be determined by using $F_y = -\partial U/\partial y$, which yields

$$\frac{dW}{dy} = -\frac{\partial U}{\partial y} = -y.$$

Integration of the above equation yields

$$W(y) = -y^2/2 + C,$$

where C is a constant. Hence

$$U(x, y) = -\frac{1}{2}(x^2 + y^2) + C.$$

Note that the potential is unique only up to a constant.

We can use the same procedure for obtaining the potential fields in three dimensions as well. Note however that the above procedure works only for conservative force fields. Typically, nonconservative forces do not have potentials.

EXAMPLE 11.5   The electric field due to a planar charged sheet is $\mathbf{E}(\mathbf{r}) = \hat{\mathbf{x}}\sigma/2\epsilon_0$, where σ is the surface charge density. Compute the potential function.

SOLUTION   Using the definition $F_x = -\partial U/\partial x$, we obtain

$$U(x, y, z) = -\frac{\sigma x}{2\epsilon_0} + W(y, z),$$

where x is the perpendicular distance of the point from the charged sheet. Since $F_y = F_z = 0$, $W(y, z) = $ constant that can be taken to be zero. The reference point is $x = 0$.

For many physical problems, it is easier to work in spherical–polar or cylindrical coordinates. In the following sub-section, we will define the gradient operation for these coordinate systems.

11.6.3   Gradients in spherical and cylindrical coordinates

The definition of a gradient for the Cartesian coordinate system is given in Eq. (11.11). In spherical–polar coordinates,

$$\mathbf{r} = dr\hat{\mathbf{r}} + rd\theta\hat{\theta} + r\sin\theta d\phi\hat{\phi},$$

and $$dU = \frac{\partial U}{\partial r}dr + \frac{\partial U}{\partial \theta}d\theta + \frac{\partial U}{\partial \phi}d\phi = \nabla U \cdot d\mathbf{r}.$$

Therefore, the gradient is defined as

$$\nabla U = \frac{\partial U(r,\theta,\phi)}{\partial r}\hat{\mathbf{r}} + \frac{1}{r}\frac{\partial U(r,\theta,\phi)}{\partial \theta}\hat{\theta} + \frac{1}{r\sin\theta}\frac{\partial U(r,\theta,\phi)}{\partial \phi}\hat{\phi}.$$

Using similar arguments, we derive a formula for the gradient in the cylindrical coordinate system:

$$\nabla U = \frac{\partial U(\rho,\phi,z)}{\partial \rho}\hat{\rho} + \frac{1}{\rho}\frac{\partial U(\rho,\phi,z)}{\partial \theta}\hat{\phi} + \frac{\partial U(\rho,\phi,z)}{\partial z}\hat{\mathbf{z}}.$$

Using these definitions we can work out the force fields in cylindrical and spherical–polar coordinates. Let us consider some examples:

1. For a point charge, the force field is $\mathbf{E} = \hat{\mathbf{r}}q/(4\pi\epsilon_0 r^2)$, and the potential function is

   $$U(r) = \frac{q}{4\pi\epsilon_0 r},$$

   with the reference point at ∞ with $U(\infty) = 0$.

2. For an infinitely long charged wire with radius ρ_0, the electric field is $\hat{\rho}\lambda/(2\pi\epsilon_0\rho)$, where λ is the charge density of the wire. Consequently the potential function is

   $$U(\rho) = -\frac{\lambda}{2\pi\epsilon_0}\ln(\rho/\rho_0),$$

   where ρ_0 is the radius of the charged wire. The reference point is $\rho = \rho_0$, with $U(\rho_0) = 0$.

3. The potential of an electric dipole $\mathbf{p}$ is

   $$U(\mathbf{r}) = \frac{\mathbf{p}\cdot\mathbf{r}}{r^3} = \frac{p\cos\theta}{r^2},$$

   where z-axis is along $\mathbf{p}$. The force field is

   $$\begin{aligned}
   \mathbf{F} &= -\nabla U \\
   &= -\frac{\partial U}{\partial r}\hat{\mathbf{r}} - \frac{\partial U}{r\partial\theta}\hat{\theta} - \frac{1}{r\sin\theta}\frac{\partial U}{\partial\phi}\hat{\phi} \\
   &= \frac{p}{r^3}[2\cos\theta\hat{\mathbf{r}} + \sin\theta\hat{\theta}]
   \end{aligned}$$

The equipotential surfaces are

$$C = \frac{p\cos\theta}{r^2},$$

or

$$r = \sqrt{p\left|\frac{\cos\theta}{C}\right|}.$$

The equipotential surfaces are shown in Fig. 11.8.

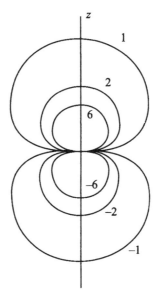

Figure 11.8   *Equipotential surfaces for an electric dipole. We take $p = 1$ and draw the equipotential lines for $C = 1, 5, 10$.*

Note that

$$\mathbf{F} = \frac{2p}{r^3}\hat{\mathbf{z}} \quad \text{for } \theta = 0,$$

and $$\mathbf{F} = \frac{p}{r^3}\hat{\theta} \quad \text{for } \theta = \pi/2.$$

The force field at a point is perpendicular to the equipotential surface at that point.

In the following section we will define another operation called the curl. The curl is a very useful operator for testing whether a force field is conservative or not.

11.6.4   Curl of a vector field

The *curl* of a vector field $\mathbf{F}$ is defined as

$$\nabla \times \mathbf{F} = \begin{bmatrix} \hat{\mathbf{x}} & \hat{\mathbf{y}} & \hat{\mathbf{z}} \\ \partial_x & \partial_y & \partial_z \\ F_x & F_y & F_z \end{bmatrix} \tag{11.13}$$

Here ∂_x is a short form of $\partial/\partial x$. Curl of a vector field can also be written in spherical–polar and cylindrical coordinates. The derivation of the curl in these coordinate systems is quite involved, so we state them without proof:

$$(\nabla \times \mathbf{F})_{\text{cylindrical}} = \begin{bmatrix} \hat{\rho} & \hat{\phi} & \hat{\mathbf{z}} \\ \partial_\rho & \partial_\phi & \partial_z \\ F_\rho & \rho F_\phi & F_z \end{bmatrix} \tag{11.14}$$

$$(\nabla \times \mathbf{F})_{\text{spherical}} = \begin{bmatrix} \frac{1}{r^2 \sin\theta}\hat{\mathbf{r}} & \frac{1}{r\sin\theta}\hat{\theta} & \frac{1}{r}\hat{\phi} \\ \partial_r & \partial_\theta & \partial_\phi \\ F_r & rF_\theta & r\sin\theta F_\phi \end{bmatrix} \tag{11.15}$$

In vector calculus there is a very nice theorem called Stokes' theorem, which state

$$\oint \mathbf{F}(\mathbf{r}) \cdot d\mathbf{r} = \int \nabla \times \mathbf{F} \cdot d\mathbf{S}, \tag{11.16}$$

where the first integral is over a closed path, and the second integral is over any *open surface* whose boundary is the same closed path (see Fig. 11.9 for an illustration). This theorem will be proved in your mathematics course. In the following section we will use this theorem to derive a test for determining whether a force field is conservative or not.

Figure 11.9   *Contour and the surface used in Stokes' theorem*

11.7 How to check if a force field is conservative or not?

How do we check whether a force field is conservative or not? Recall that a force is conservative if the work done by the force for any closed path is zero. We can use this test, but it is too cumbersome because it has to be shown to be true for infinitely many possible paths. The other method is to compute the potential. However computation of a potential function is also reasonably involved. The most straightforward test is:

A force is conservative if and only if its curl is zero everywhere.

The above statement can be proved easily using Stoke's theorem discussed above. For a conservative force, the LHS of Eq. (11.16) is zero for all possible closed paths. Hence RHS of the equation is also zero for all possible open surfaces, which requires that the integrand of the RHS must be zero. If that is not the case, then we can take a surface enclosing the nonzero curl region, which will yield a nonzero integral, a contradiction. Hence, the condition for a field to be conservative is

$$\nabla \times \mathbf{F} = 0.$$

Converse: if $\nabla \times \mathbf{F} = 0$ for a vector field everywhere, then $\oint \mathbf{F(r)} \cdot d\mathbf{r} = 0$ for all possible paths. Hence the vector field $\mathbf{F}$ is conservative.

It is easy to show that the curl of fields $x\hat{\mathbf{x}} + y\hat{\mathbf{y}}, -\hat{\mathbf{r}}/r^2$ are zero, hence these fields are conservative. An example of nonconservative field is $\mathbf{F} = \hat{\theta}/r^2$, whose curl is nonzero.

EXAMPLE 11.6   Test whether the following force fields are conservative or not by computing curls: (i) $A\hat{\mathbf{r}}$; (ii) $(1/\rho)\hat{\phi}$; (iii) $\rho\omega\hat{\phi}$ ((ii) and (iii) are in cylindrical coordinates).

SOLUTION   We compute curl of these vector fields.

1. $\mathbf{A} = A\hat{\mathbf{r}} = \frac{A}{r}(x\hat{\mathbf{x}} + y\hat{\mathbf{y}} + z\hat{\mathbf{z}})$: Simple substitution of $A_x = Ax/r, A_y = Ay/r, A_z = Az/r$ in Eq. (11.13) yields $\nabla \times \mathbf{A} = 0$. Hence $\mathbf{A}$ is a conservative field. We could also compute the curl using Eq. (11.15).

2. $\mathbf{A} = (1/\rho)\hat{\phi}$: We use Eq. (11.14) to compute the curl in the cylindrical coordinate

$$\nabla \times \mathbf{A} = \left[\frac{1}{\rho}\frac{\partial A_z}{\partial \phi} - \frac{\partial A_\phi}{\partial z} \right]\hat{\rho}$$

$$+ \left[\frac{\partial A_\rho}{\partial z} - \frac{\partial A_z}{\partial r} \right]\hat{\phi}$$

$$+ \frac{1}{\rho}\left[\frac{\partial(rA_\phi)}{\partial r} - \frac{\partial A_\rho}{\partial \phi} \right]\hat{\mathbf{z}} \tag{11.17}$$

When we substitute $A_\rho = 1/\rho$, $A_\phi = A_z = 0$ in the above formula, we obtain $\nabla \times \mathbf{A} = 0$ everywhere except the origin. Since $\mathbf{A}$ is singular at $\rho = 0$, we need to compute the curl at the origin by some other method. We apply Stokes' theorem

$$\oint \mathbf{A} \cdot d\mathbf{r} = \int (\nabla \times \mathbf{A}) \cdot d\mathbf{S}.$$

on a circular contour of radius ρ enclosing the origin. The LHS is clearly 2π. We have showed that $\nabla \times \mathbf{A} = 0$ for all the points wherever the derivatives are defined. Hence, the only place which could contribute to the integral is the origin. It turns out that the curl at the origin is a highly singular function, called the *Dirac delta function*, which is infinity at the origin, but zero everywhere else. The Delta function is a complex mathematical object, which will be discussed in more advanced mathematics and physics courses. Since $\nabla \times \mathbf{A}$ is nonzero at the origin, $\mathbf{A}$ is not a conservative force field.

3. $\mathbf{A} = \rho\omega\hat{\phi}$: We substitute $A_\phi = \rho\omega$, $A_\rho = A_z = 0$ in Eq. (11.14) and we obtain $\nabla \times \mathbf{A} = 2\omega$. Since $\nabla \times \mathbf{A} \neq 0$, $\mathbf{A}$ is a nonconservative force field.

Stokes' theorem is defined at a fixed time (snap-shot). It is not applicable for a particle trajectory when the force field is time dependent. We can test the conservativeness of time-dependent force fields by computing the work done over a closed path and checking if it is zero. We learnt earlier that the work done by an electromotive force for a closed path is nonzero. Hence EMF is a nonconservative field.

Recall that in Chapter 7, we had discussed that the symmetry of time translation is related to the law of energy conservation. In the next section we will prove this result.

11.8   Law of energy conservation from symmetry arguments

In Chapter 7 we argued the conservation of energy for one-dimensional systems from the symmetry of time translation. Here we will prove the law of energy conservation for arbitrary dimensions. The symmetry under time translation implies that the potential is not an explicit function of time. From the equation of motion we can write

$$\mathbf{v} \cdot m\frac{d\mathbf{v}}{dt} = \mathbf{v} \cdot \mathbf{F} = -\mathbf{v} \cdot \nabla U = -\frac{d\mathbf{r}}{dt} \cdot \nabla U. \tag{11.18}$$

We use a mathematical result

$$\frac{dU}{dt} = \frac{\partial U}{\partial x}\frac{dx}{dt} + \frac{\partial U}{\partial y}\frac{dy}{dt} + \frac{\partial U}{\partial z}\frac{dz}{dt} + \frac{\partial U}{\partial t}.$$

Since the potential is symmetric under time translation, $\partial U/\partial t = 0$. Consequently,

$$\frac{dU}{dt} = \frac{d\mathbf{r}}{dt} \cdot \nabla U.$$

Substitution of the above result in Eq. (11.18) yields

$$\frac{d}{dt}\frac{1}{2}mv^2 = -\frac{dU}{dt},$$

which implies that

$$\frac{1}{2}mv^2 + U = E = \text{const}$$

Thus we prove that energy is conserved for systems whose potentials are not an explicit function of time. Please keep in mind that we have implicitly assumed that the force can be written as a gradient of potential.

Energy of a particle is not an absolute number. It depends on the inertial frame in which it is measured. In the following section, we will illustrate this statement using an example.

11.9   Energy in different frames of references

In this section we will compute the energy of an object in two inertial frames. Consider a food packet of mass m being dropped from height H from an aeroplane that is moving horizontally with a constant velocity V_0. We record the kinetic energy, potential energy, and the total energy of the packet in the aeroplane and the ground reference frames at two instants of time:

- when the packet is just dropped $(t = 0)$
- when the packet hits the Earth $(t = t_f)$.

In the reference frame of the aeroplane, the values are

	KE	PE	Total
$t = 0$	0	mgH	mgH
$t = t_f$	$mv_p^2/2$	0	$mv_p^2/2$

where v_p is the speed of the packet in the plane reference frame

Conservation of energy yields $mv_p^2/2 = mgh$ or $v_p = \sqrt{2gH}$. This is consistent with the answer obtained from dynamics.

In the reference frame of the ground, the packet has horizontal velocity V_0. Hence, the values of energy are

	KE	PE	Total
$t = 0$	$mV_0^2/2$	mgH	$mV_0^2/2 + mgH$
$t = t_f$	$mv_G^2/2$	0	$mv_G^2/2$

where V_G is the speed of the packet in the ground reference frame

From the conservation of energy we obtain

$$mv_G^2/2 = mV_0^2/2 + mgH.$$

So $v_G = \sqrt{V_0^2 + 2gH} = \sqrt{V_0^2 + v_p^2}$. This result is consistent because the horizontal and vertical components of the velocity of the packet in the ground reference frame are V_0 and

v_p respectively. Note that the values of energies are different in the two reference frames, but the conservation of energy holds in both the frames.

So far we have discussed forces that could be expressed as vector fields. In the following discussion we will discuss constraint forces.

11.10   Constraints and conservation of energy

Let us consider the roller-coaster example that was introduced in the beginning of the chapter. We assume the roller-coaster to be smooth. Two forces, gravity and the normal force, are acting on the train. Gravity is a conservative force. What about the normal force? For smooth surfaces, the normal force is always perpendicular to the direction of motion of the train, so the normal force does no work on the train. We can apply conservation of energy using gravity only. The procedure is as follows.

Suppose the train starts from rest. After a complex motion it comes down by a vertical distance h. Then by conservation of energy, the final velocity of the train is

$$v = \sqrt{2gh},$$

and the direction of velocity is along the tangent of the roller-coaster. Note that it is very difficult to calculate the above velocity using Newton's equation of motion.

In all the situations when a particle is moving on a smooth surface or is guided by a smooth wire, the normal force is perpendicular to the direction of displacements. So the normal force does no work on the particle and we can apply conservation of mechanical energy to particles moving on smooth wires or on smooth surfaces.

We will end this chapter with a puzzle that illustrates the power of the conservation laws.

11.11   Conservation principle in a general framework: Process independent approach

Let us try to solve a puzzle. Consider a glass of milk (A) and a glass of water (B) both of volume V. Take out a spoon of milk (volume v) from glass A and transfer it to glass B. Mix it by stirring the liquid. Then transfer one spoon (volume v) from the mixture and transfer it back to glass A.

QUESTION: What is the ratio

$$\frac{\text{Water in glass A}}{\text{Milk in glass B}}?$$

First we compute the above ratio using the conservation principle. We use the fact that the initial and final volume of liquid in glass A and glass B are the same (V). Suppose that in the final state glass A contains α amount of water, then just from conservation of volume

	Milk in A	Water in A	Milk in B	Water in B
Initial	V	0	0	V
Final	$V - \alpha$	α	α	$V - \alpha$

Therefore the desired ratio is

$$\frac{\text{Water in glass A}}{\text{Milk in glass B}} = \frac{\alpha}{\alpha} = 1.$$

We could solve the above puzzle by another method. After the first transfer, glass B contains $V + v$ amount of fluid with V amount of water and v amount of milk. If mixing in glass B is perfect, then a spoon of mixture would contain $v^2/(V + v)$ amount of milk and $Vv/(V+v)$ amount of water. Once this spoon of mixture is transferred, glass A would contain $V - v + v^2/(V+v) = v^2/(V+v)$ amount of milk and $Vv/(V+v)$ amount of water. Glass B contains $V - Vv/(V + v) = V^2/(V + v)$ amount of water and $v - v^2/(V + v) = Vv/(V+v)$ amount of milk. Clearly the ratio is again one. Note that the above calculation assumes perfect mixing. If the mixing is not perfect, then the solution will be more complex.

When we compare the two methods we observe that the first method that is based on the conservation principle is superior because it is independent of the process. The argument used in the first method is also called a *thermodynamic argument* because it is independent of the process involved. The first method does have a limitation—we cannot compute the exact amount of fluid transfer (e.g., water in glass A) using this method.

In this chapter we have introduced the ideas of potential energy and conservation of energy. Some problems are too complex to solve using the force method, but they can be solved quite easily using the energy method. We have only discussed conservation of the sum of kinetic and potential energy. Energy, however, can also be in other forms like electrical, thermal, chemical, nuclear, etc., and the sum of all the energies for an isolated system is conserved. The law of energy conservation has been found to be strictly valid in all the experiments carried out so far.

Exercises

1. Derive potential functions for the following forces:

   (a) A constant force field

   (b) Spring force $\mathbf{F} = -k\mathbf{r}$.

   (c) Gravitational potential due to an infinite sheet of mass of constant density σ (mass per unit area).

   (d) Gravitation force $-\alpha/r^2$.

2. Compute the potential both inside and outside a spherical solid ball of constant density. Do the same calculation for a hollow spherical shell.

the centre. Forces at the three points differ by a small amount. This differential force is called the tidal force. How does this force vary with separation? What are the effects of these forces on the Earth?

4. Compute the force fields for the following potentials:

   (a) $U = k(x^2 + y^2 + z^2)$.

   (b) $U = \log \sqrt{x^2 + y^2 + z^2}$.

   (c) $U = \mathbf{r} \cdot \mathbf{g}$, where $\mathbf{g}$ is a constant vector.

   (d) $U = \mathbf{r} \cdot \mathbf{p}/r^4$, where $\mathbf{p}$ is a constant vector.

   (e) $U = \exp(-\alpha r)/r$, where α is a constant.

5. Check whether the following fields are conservative or not? Compute the potentials for the conservative forces.

   (a) $\mathbf{F} = x\hat{\mathbf{x}} + y\hat{\mathbf{y}} + z\hat{\mathbf{z}}$

   (b) $\mathbf{F} = x\hat{\mathbf{x}} - y\hat{\mathbf{y}}$

   (c) $\mathbf{F} = \phi\hat{\mathbf{r}}$

   (d) $\mathbf{F} = z\hat{\mathbf{z}}$

   (e) $\mathbf{F} = xy\hat{\mathbf{x}} + (y^2 z - x)\hat{\mathbf{y}} + z(y^2 - x^2)\hat{\mathbf{z}}$

   (f) $\mathbf{F} = \dfrac{x\hat{\mathbf{x}} + y\hat{\mathbf{y}}}{x^2 + y^2} + z\hat{\mathbf{z}}$

6. Which of the following vector fields could be generated by an isotropic source (2D or 3D):

   (a) $\mathbf{F} = x\hat{\mathbf{x}} - y\hat{\mathbf{y}} - z\hat{\mathbf{z}}$

   (b) $\mathbf{F} = x^2\hat{\mathbf{x}} + y^2\hat{\mathbf{y}} + z^2\hat{\mathbf{z}}$

   (c) $\mathbf{F} = y\hat{\mathbf{x}} - x\hat{\mathbf{y}}$

   (d) $\mathbf{F} = x^2\hat{\mathbf{x}}$

   (e) $\mathbf{F} = x\hat{\mathbf{x}}$

7. A pendulum has a bob of mass m and string of length l. The bob is given an initial velocity of u (horizontally) at the lowest point. Describe the motion of the bob for

   (a) $u \le \sqrt{2gl}$

   (b) $\sqrt{2gl} \le u \le \sqrt{4gl}$

   (c) $u > \sqrt{4gl}$.

8. Four immovable charges of magnitude $+Q$ are placed at the vertices of a square. A test charge q is moving in the field of these charges.

   (a) Derive an expression for the potential energy of the test charge q.

(b) Expand the above potential energy near the centre of the square. Keep the terms only up to second order. Note that the centre of the square is an equilibrium point.

9. A particle of mass m is resting on top of a fixed sphere. It is given a small push horizontally. At what angle from the vertical will the mass leave the surface of the sphere? Ignore initial kinetic energy of the particle.

10. Compute curl of the following vector fields:

(a) $\mathbf{F} = \rho\hat{\rho}$

(b) $\mathbf{F} = \rho\hat{\phi}$

11. Verify Stokes' law using a vector field $\mathbf{F} = \rho\hat{\phi}$, circle of radius 2 cm as a contour, and the circular region inside the contour as the open surface.

12. The equipotential surfaces for a potential field is shown in Fig. 11.10.

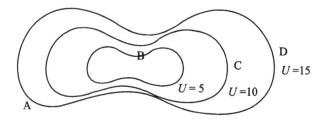

Figure 11.10   *Exercise 12*

(a) At which point among A,B,C,D is the magnitude of the force maximum?

(b) Indicate the direction of the force at all these four points.

12

Conservation of Linear Momentum and Centre of Mass

In this chapter we will introduce the second law of conservation called *conservation of linear momentum*. As discussed in the previous chapter, conservation laws are very useful in deducing the motion of mechanical systems. On many occasions, methods based on conservation laws are much simpler than the force method.

In this chapter we will also introduce the concept of centre of mass of a many-particle system. We will briefly discuss variable mass systems towards the end of the chapter.

12.1 An isolated single particle

By definition, an isolated particle is one that does not experience any external force. If $\mathbf{p}$ is the linear momentum of an isolated particle, then from Newton's second law we obtain

$$\frac{d\mathbf{p}}{dt} = \mathbf{F}_{\text{ext}} = 0,$$

which implies that

$$\mathbf{p} = \text{const},$$

or *the linear momentum of an isolated single particle is conserved.*

12.2 Isolated two-particle system

Let us consider an isolated system consisting of two interacting particles. Suppose $\mathbf{f}_{12}$ is the force acting on particle 1 due to particle 2, and $\mathbf{f}_{21}$ is the force acting on particle 2 due to particle 1. If $\mathbf{p}_1$ and $\mathbf{p}_2$ are the momenta of particles 1 and 2 respectively, then an application of Newton's laws yields

$$\frac{d\mathbf{p}_1}{dt} = \mathbf{f}_{12},$$

$$\frac{d\mathbf{p}_2}{dt} = \mathbf{f}_{21}.$$

According to Newton's third law $\mathbf{f}_{12} = -\mathbf{f}_{21}$. Hence

$$\frac{d}{dt}(\mathbf{p}_1 + \mathbf{p}_2) = 0,$$

or $\mathbf{p}_1 + \mathbf{p}_2 = const,$

which is a statement of *the conservation of total linear momentum of the two-particle system.*

12.3   System of particles and centre of mass

Let us consider a system of N particles in some reference frame I. The particles are labelled by a. The ath particle has mass m_a and is located at $\mathbf{r}_a$ as shown in Fig. 12.1(a).

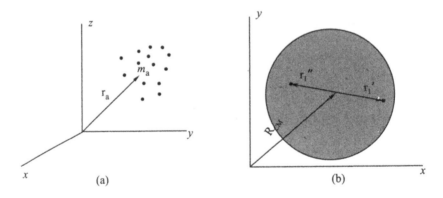

(a)   (b)

Figure 12.1(a) *A system of N particle. The ath particle located at* $\mathbf{r}_a$ *has mass* m_a. *(b) Distribution of mass around the centre of mass of the disk.*

We define $\mathbf{R}_{CM}$ as

$$\mathbf{R}_{CM} = \frac{\sum m_a \mathbf{r}_a}{\sum m_a}. \tag{12.1}$$

The point $\mathbf{R}_{CM}$ is called the *centre of mass* (CM).

In nature we have extended objects made up of a continuum of matter. For these objects the sum translates to an integral without much difficulty. Let us take a simple example. Let us find the CM of a disc of uniform density. We take a pair of points located at $\mathbf{r}_1'$ and $\mathbf{r}_1''$ on the disc (Fig. 12.1(b)) that are equidistant and opposite to each other from the centre of the disc, whose position is at $\mathbf{R}_0$. We take a symmetric patch of mass m_1 at points $\mathbf{r}_1'$ and $\mathbf{r}_1''$. We repeat this process till all the regions of the disc are covered. As a consequence we obtain

$$M\mathbf{R}_{\text{CM}} = m_1(\mathbf{R}_0 + \mathbf{r}_1') + m_1(\mathbf{R}_0 + \mathbf{r}_1'') + \dots$$

$$= M\mathbf{R}_0.$$

Since $\mathbf{r}_i' + \mathbf{r}_i'' = 0$, all the dashed coordinates get cancelled. Using similar arguments we can show that for all symmetric bodies with uniform density, the centre of mass is at the centre of the body.

We will show in the present and later chapters that the reference frame attached to the centre of mass is very useful in analysing the dynamics of mechanical systems. This reference frame is called *the centre of mass (CM) reference frame*. Let us derive some properties in the CM reference frame.

Suppose the position vector of the particle a in the reference frame I and CM reference frame are $\mathbf{r}_a$ and $\mathbf{r}_a'$ respectively. From the above arguments it is clear that

$$\sum m_a \mathbf{r}_a' = 0. \tag{12.2}$$

In a similar manner we denote the velocity of the particle a in I and CM reference frames to be $\mathbf{v}_a$ and $\mathbf{v}_a'$ respectively. By taking the time derivative of Eq. (12.2) we obtain

$$\sum m_a \mathbf{v}_a' = 0.$$

Therefore,

$$\mathbf{P} = \sum m_a \mathbf{v}_a = \sum m_a(\mathbf{V}_{\text{CM}} + \mathbf{v}_a') = M\mathbf{V}_{\text{CM}} = \mathbf{P}_{\text{CM}}.$$

Hence, the net linear momentum of a system of particles is same as the linear momentum of its CM. The internal momenta of the particles cancel each other.

Now we take a time derivative of Eq. (12.1) which yields

$$M\frac{d\mathbf{R}_{\text{CM}}}{dt} = M\mathbf{V}_{\text{CM}} = \sum m_a \mathbf{v}_a.$$

Taking another time derivative yields

$$M\frac{d\mathbf{V}_{\text{CM}}}{dt} = M\mathbf{a}_{\text{CM}} = \sum_a m_a \dot{\mathbf{v}}_a = \sum \mathbf{f}_a,$$

where $\mathbf{f}_a$ is the net force acting on m_a. We split $\mathbf{f}_a$ into internal and external forces. The internal forces $\mathbf{f}_{a,\text{int}}$ arise due to the interactions with other particles, e.g. Coulomb force between charged particles, or gravitational force between masses etc. We denote the force on the particle a due to the particle b by $\mathbf{f}_{a,b}$. The external force is caused by some external field or contact force. Hence,

$$\mathbf{f}_a = \mathbf{f}_{a,\text{ext}} + \mathbf{f}_{a,\text{int}} = \mathbf{f}_{a,\text{ext}} + \sum_b \mathbf{f}_{a,b}.$$

Now $\quad M\mathbf{a}_{\text{CM}} = \sum_a \mathbf{f}_a = \sum_a \mathbf{f}_{a,\text{ext}} + \sum_a \sum_b \mathbf{f}_{a,b}.$

From Newton's third law

$$\mathbf{f}_{a,b} + \mathbf{f}_{b,a} = 0,$$

hence all the pairs of internal forces cancel each other. Consequently

$$\frac{d\mathbf{P}}{dt} = \frac{d\mathbf{P}_{CM}}{dt} = M\mathbf{a}_{CM} = \sum_a \mathbf{f}_{a,\text{ext}} = \mathbf{F}_{\text{ext}}. \tag{12.3}$$

The rate of change of the net momentum of a system of particles is equal to the net external force. If $\mathbf{F}_{\text{ext}} = 0$, then $\mathbf{P} = $ const. This is the statement of the *conservation of linear momentum for a many-particle system.*

   The above statement of the conservation of linear momentum depends crucially on Newton's third law, which may not be valid for moving charges. It turns out that the conservation of linear momentum holds for all isolated systems if we include both particles and fields. The linear momentum of fields is reasonably complex; it will be discussed in the electromagnetic theory course.

EXAMPLES

1. EXPLODING SHELL   The net external force acting on the shell is gravity. Hence

$$\frac{d\mathbf{P}_{CM}}{dt} = \mathbf{F}_{\text{ext}} = M\boldsymbol{g}.$$

   Therefore, the trajectory of the CM of the shell is a parabola.

2. When a bullet is fired from a gun, the gun recoils back to conserve the net linear momentum of the gun and the bullet.

In earlier chapters we had confined our discussion to point particles. Using the idea of centre of mass, we can start studying extended objects. For the time being we start with two-particle systems.

12.4   Several examples of two-particle systems

In this section we will work out two examples of two-particle systems: Spring–mass system and Kepler's problem.

12.4.1   Spring–mass system

Two masses m_1 and m_2 are connected by a spring of spring constant k as shown in Fig. 12.2. The natural length of the spring is l, and the coordinates of the particles are x_1 and x_2 respectively[1].

---

[1] The displacement of the masses m_1 and m_2 are also useful coordinates. The solution of the above problem is somewhat easier in terms of these coordinates.

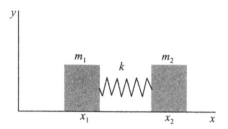

Figure 12.2 *Spring–mass system*

The equations of motion of the masses are

$$m_2\ddot{x}_2 = -k(x_2 - x_1 - l) \tag{12.4}$$

$$m_1\ddot{x}_1 = k(x_2 - x_1 - l) \tag{12.5}$$

Clearly $m_1\dot{x}_1 + m_2\dot{x}_2 = P_{CM} = \text{const}$

Physically, $P_{CM} = \text{const}$ because no external force is acting on the system. We can derive further that

$$m_1 x_1 + m_2 x_2 = M[V_{CM}t + X_{CM}(0)] = MX_{CM}(t)$$

The CM moves like a free particle with mass M and velocity V_{CM}.

By multiplying Eq. (12.4) with m_1 and Eq. (12.5) with m_2, and then by subtracting the second equation from the first, we obtain

$$\frac{m_1 m_2}{m_1 + m_2} \frac{d^2}{dt^2}(x_2 - x_1) = -k(x_2 - x_1 - l).$$

By denoting $x_2 - x_1 - l = y$ and $m_1 m_2/(m_1 + m_2) = \mu$, we obtain

$$\mu\ddot{y} = -ky, \tag{12.6}$$

which is a single particle equation. μ is called the *reduced mass*. The solution of the above equation is

$$y(t) = A\cos\left(\sqrt{k/\mu}\,t\right) + B\sin\left(\sqrt{k/\mu}\,t\right). \tag{12.7}$$

In the CM frame, $X_{CM} = V_{CM} = 0$. Therefore,

$$m_1 x_1' + m_2 x_2' = 0$$

where x_1' and x_2' are the positions of the masses m_1 and m_2 in the CM frame. Note that $x_2' - x_1' = x_2 - x_1 = y + l$.

Combining the above equation with Eq. (12.7) and using $y = x_2 - x_1 - l$, we obtain the coordinates of the masses in the CM frame:

$$x_1' = -\frac{m_2(y+l)}{m_1+m_2},$$

$$x_2' = \frac{m_1(y+l)}{m_1+m_2}.$$

Therefore, in the laboratory frame of reference

$$x_1 = -\frac{m_2(y+l)}{m_1+m_2} + X_{CM}(t),$$

$$x_2 = \frac{m_1(y+l)}{m_1+m_2} + X_{CM}(t).$$

The relative velocity is a very important quantity, which can be expressed quite nicely as

$$\dot{y} = v_2 - v_1 = v_2' - v_1'$$

$$= \frac{m_2 v_2'}{\mu} = -\frac{m_1 v_1'}{\mu}.$$

Using the idea of centre of mass, we could reduce the two-body problem to a one-body problem, which was quite easy to solve. The motion of the system can be split into two parts: (1) the motion of the CM with a constant velocity, (2) oscillatory motion with respect to the CM.

Let us take a special case when $m_1 = m_2 = m$. Here $\mu = m/2$, hence the frequency of oscillation is $\sqrt{2k/m}$. The above problem is equivalent to a single-mass problem with the same spring constant but with a reduced mass. We can argue the above result in another way—the masses oscillate keeping the CM fixed. So the CM acts as a wall. Hence, the length of the spring has been reduced to half. The quantity $F/(\Delta l/L) = K$ is a property of the material which remains the same for all sizes of the spring. However, the spring constant is defined as $k = F/\Delta l$. Hence k is inversely proportional to the length of the spring.[2] Therefore the spring constant k' of the shorter spring is $2k$, and the frequency of the new system is $\omega = \sqrt{k'/m} = \sqrt{2k/m}$.

Note that the above spring–mass system requires four initial conditions $x_{1,2}(0)$ and $\dot{x}_{1,2}(0)$, or $X_{CM}(0)$, $\dot{X}_{CM}(0)$, $y(0)$, and $\dot{y}(0)$. In the above discussion we have used the later coordinates. The equations can be solved using $x_{1,2}$ variables too, but that requires a more sophisticated approach called the matrix method, which will be covered in your mathematics course.

12.4.2   The Sun–Earth system (Kepler's problem revisited)

In our earlier discussion on Kepler's problem (Chapter 8) we had assumed that the mass of the Sun is very large so that the Sun was essentially at the CM. In the following discussion, we will lift that condition and solve the problem for an arbitrary star–planet two-body

---

[2]Physically, it is easier to stretch a longer spring than a shorter one (by the same amount). Hence the spring constant of a longer spring is smaller than that of a shorter one.

problem. The potential energy for the system is $U(r = |\mathbf{r}_1 - \mathbf{r}_2|) = -\alpha/|\mathbf{r}_1 - \mathbf{r}_2|$, where $\mathbf{r}_1$ and $\mathbf{r}_2$ are the positions of the planet and the star respectively.

In Fig. 12.3(a) we show the two masses. We work out the above problem in the CM frame. If we denote $\mathbf{r} = \mathbf{r}_1 - \mathbf{r}_2$, then the equations of motion are

$$m_1\ddot{\mathbf{r}}_1 = -\frac{\alpha}{r^2}\hat{\mathbf{r}}, \tag{12.8}$$

$$m_2\ddot{\mathbf{r}}_2 = \frac{\alpha}{r^2}\hat{\mathbf{r}}. \tag{12.9}$$

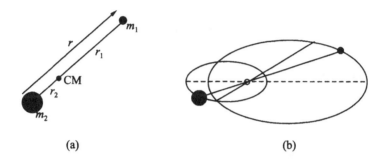

(a)   (b)

Figure 12.3   *(a) Star–planet system. (b) The orbit of the planet and the star; they are always on the opposite side of the CM.*

By multiplying Eq. (12.8) with m_2 and Eq. (12.9) with m_1, and then subtracting the second equation from the first, we obtain

$$\mu\frac{d^2}{dt^2}\mathbf{r} = -\frac{\alpha}{r^2}\hat{\mathbf{r}},$$

which is a one-body problem already solved in Chapter 8. Note that m has been replaced by the reduced mass $\mu = m_1 m_2/(m_1 + m_2)$. From the solution of $\mathbf{r}$, we can easily obtain $\mathbf{r}_1$ and $\mathbf{r}_2$:

$$\mathbf{r}_1 = \frac{m_2\mathbf{r}}{m_1 + m_2} = \frac{\mu\mathbf{r}}{m_1},$$

$$\mathbf{r}_2 = -\frac{m_1\mathbf{r}}{m_1 + m_2} = -\frac{\mu\mathbf{r}}{m_2},$$

or $m_1\mathbf{r}_1 = -m_2\mathbf{r}_2 = \mu\mathbf{r}.$

We can also deduce that $r_1/r_2 = m_2/m_1$. Since $M_S/M_E \approx 10^6$, we obtain $r_S/r_E \approx 10^{-6}$, or $r_S \approx r_E \times 10^{-6} \approx 10^8 \times 10^{-6} = 100$ km, which is quite small. Hence our assumption that the Sun is at the CM is reasonable for the Earth–Sun system.

A sketch of the motion of the star–planet system is shown in Fig. 12.3(b). The elliptical orbit of the planet is of larger size, and that of the star is of smaller size. As shown in the figure, the star and the planet are always on the opposite sides of the CM.

The above reduction of a 2-body problem to a 1-body problem illustrates the usefulness of the concept of CM. CM also finds major application in problems connected with rotation and collisions (to be discussed in later chapters).

In the earlier section we studied the procedure to compute the linear momentum for a system of particles. In the following section we will discuss kinetic and potential energies of a system of particles.

12.5   Kinetic energy and potential energy of a system of particles

12.5.1   Kinetic energy of a system of particles

The total kinetic energy of a system of particles is the sum of kinetic energies of each particle in the system. Therefore, the total kinetic energy is

$$T = \sum_a \frac{1}{2} m_a v_a^2$$

$$= \sum_a \frac{1}{2} m_a (\mathbf{V}_{\mathrm{CM}} + \mathbf{v}_a')^2$$

$$= \frac{1}{2} M V_{\mathrm{CM}}^2 + \sum_a \frac{1}{2} m_a v_a'^{\,2} + \left(\sum m_a \mathbf{v}_a' \right) \cdot \mathbf{V}_{\mathrm{CM}}$$

$$= \frac{1}{2} M V_{\mathrm{CM}}^2 + \sum_i \frac{1}{2} m_a v_a'^2 + 0 \times \mathbf{V}_{\mathrm{CM}}$$

$$= T_{\mathrm{CM}} + T_{\mathrm{int}}.$$

T_{CM} is the kinetic energy of the CM. This will be the kinetic energy (KE) if all the particles move with velocity V_{CM}. T_{int}, the internal kinetic energy, is due to the internal motion of the particles.

In the following discussion we provide some examples.

1. For the spring–mass system, $T_{\mathrm{CM}} = M V_{\mathrm{CM}}^2/2$, and the internal kinetic energy is

$$T_{\mathrm{int}} = \frac{1}{2} (m_1 v_1'^2 + m_2 v_2'^2) = \frac{1}{2\mu} m_1^2 v_1'^2 = \frac{1}{2} \mu (\dot{y})^2.$$

   In the last step we used $m_1 v_1' = -\mu \dot{y}$. Hence, the total energy is

$$T = \frac{1}{2} M V_{\mathrm{CM}}^2 + \frac{1}{2} \mu (\dot{y})^2.$$

   which is a nice expression clearly separating the CM motion and the relative motion.

2. We can derive the KE of a star–planet system using the same procedure as above. Here

$$T = \frac{1}{2}MV_{\text{CM}}^2 + \frac{1}{2}\mu v^2,$$

where $\mathbf{v} = \mathbf{v}_2 - \mathbf{v}_1 = \mathbf{v}_2' - \mathbf{v}_1'$.

3. For a spinning cricket ball, the total KE is the sum of the centre-of-mass KE and rotational KE. We will discuss rotational KE later in Chapter 17.

12.5.2 Potential energy of a system of particles

We can also define the potential energy for a system of particles. First let us think of the Earth–Sun system. The potential energy is $-GM_S M_E/R_{ES}$, where R_{ES} is the Sun–Earth distance. This is the energy required to separate the Earth from the influence of the Sun. Electrostatic potentials have a similar form. In this book we will assume that the potential energy of a pair of particles depends only on the distance between the two particles [i.e., $U_{ab} = U_{ab}(|\mathbf{r}_a - \mathbf{r}_b|)$]. For a pair of particles 1–2, the potential energy is

$$U_{12} = U_{12}(|\mathbf{r}_1 - \mathbf{r}_2|)$$

When we bring in the third particle m_3, the third particle will interact with particles 1 and 2. Hence, the potential energy will be

$$U_{123} = U_{12}(|\mathbf{r}_1 - \mathbf{r}_2|) + U_{13}(|\mathbf{r}_1 - \mathbf{r}_3|) + U_{23}(|\mathbf{r}_2 - \mathbf{r}_3|)$$

We can generalise the above argument to an N-particle system:

$$U_{\text{int}} = \frac{1}{2}\sum_a \sum_{b,a\neq b} U_{ab}(|\mathbf{r}_a - \mathbf{r}_b|).$$

The factor $1/2$ is to avoid double counting ($U_{ab} = U_{ba}$, and only one of them should be included in U_{int}). *It should be noted that the internal potential energy function is the same in all inertial frames.*

The external force, if conservative, could be expressed as a gradient of an *external potential function*. The total potential energy of a system is the sum of internal and external potential energies. The potential energy due to the external force is an explicit function of the coordinates of the particles. For example, imagine a spring–mass pair moving in the gravitational field of the Earth. Here the potential energy is the sum of the spring energy (internal) and the gravitational energy of the masses. The external potential energy of the spring–mass system is $g(m_1 y_1 + m_2 y_2)$, where $y_{1,2}$ are the vertical coordinates of the particles. To sum up

$$U = U_{\text{ext}} + U_{\text{int}} = \sum_a U_{\text{ext}}(\mathbf{r}_a) + \frac{1}{2}\sum_a \sum_{b,a\neq b} U_{ab}(|\mathbf{r}_a - \mathbf{r}_b|).$$

We can compute the external force on particle a by taking gradient $\nabla_a U_{\text{ext}}(\mathbf{r}_a)$. The net external force on the system is the sum of all the forces:

$$\mathbf{F}_{\text{ext}} = -\sum_a \nabla_a U_{\text{ext}}(\mathbf{r}_a) \tag{12.10}$$

In the next section we will consider a special case when the system is isolated.

12.6   Translation symmetry and conservation of linear momentum

When an isolated system is shifted from one place to the next, it remains unchanged, or, an isolated system is symmetric under space translation. This property is also stated as the *homogeneity of space*. Hence for an isolated system, U_{ext} cannot depend on the explicit coordinates of the particles. Consequently, $U_{\text{ext}}(\mathbf{r}_a) = \text{const}$. Thus for an isolated system $U = U_{\text{int}}$, which depends only on the difference between particle coordinates, not on their explicit values.

For an isolated system, Eq. (12.10) implies that $\mathbf{F}_{\text{ext}} = 0$. By substituting $\mathbf{F}_{\text{ext}} = 0$ in Eq. (12.10) we can immediately deduce that the total linear momentum of an isolated system is conserved. This is how we relate homogeneity of space (symmetry under space translation) to the conservation of linear momentum.

In all our discussion so far the total mass of the system is conserved. However we often encounter physics problems where it is not so. The simplest example is a rocket that loses mass through its exhaust. In the next section we will derive an equation of motion for systems with variable mass.

12.7   Newton's equation for variable mass systems

We will illustrate the derivations for Newton's equations for variable mass systems using several examples.

12.7.1   Example 1: Train + variable mass

Consider a toy train moving with a constant velocity $\mathbf{v}$. Let us place some magnets near the rails; the magnets get attached to the train while the train is moving. If we do not apply any force to the train, the train will slow down due to loading of the magnets on the train. We need to apply force to maintain the constant velocity of the train. Let us compute the required force.

We will assume that the train is moving with a constant velocity $\mathbf{v}$. If mass of the train is M, and the mass of each magnet is m, then the momentum of the train before the magnet gets attached to the train is $\mathbf{P}(t) = M\mathbf{v}$. The momentum of the train after a single magnet gets attached to the train is

$$\mathbf{P}(t + \Delta t) = (M + m)\mathbf{v}.$$

Hence, $\Delta \mathbf{P} = m\mathbf{v}$, which has to be provided by an external force. If the train picks up magnets regularly with a rate dm/dt, then the required force to move the train with a constant velocity $\mathbf{v}$ is

$$\mathbf{F} = \mathbf{v}\frac{dm}{dt}.$$

We could have derived the above equation by using

$$M\frac{d\mathbf{v}}{dt} + \mathbf{v}\frac{dm}{dt} = \mathbf{F}. \tag{12.11}$$

and setting $d\mathbf{v}/dt = 0$. The question is whether we can apply the above equation for all problems with variable mass?

Now let us think of the converse situation. A train of mass M containing sand is moving with constant velocity $\mathbf{v}$. Sand is dropped continuously from the train. Would we require any force to maintain a constant velocity for the train? If we apply Eq. (12.11) blindly, we get a nonzero force $\mathbf{F} = \mathbf{v}dm/dt$. This is incorrect! Since the dropped packet also moves with velocity $\mathbf{v}$ immediately after it is dropped, the linear momentum of the system does not change, and the train moves with the original velocity without any external force.

This asymmetry between the above two examples is because of the inertia. In the first situation, the magnets picks up momentum, while in the second case, the momentum of the sand remains unchanged. One has to be careful while applying $\mathbf{F} = d\mathbf{P}/dt$ for variable mass. We cannot simply write

$$\mathbf{F} = M\frac{d\mathbf{v}}{dt} + \mathbf{v}\frac{dm}{dt} \tag{12.12}$$

in all situations. In fact, Eq. (12.12) is not a Galilean invariant; $\mathbf{v}dm/dt$ takes different values in different inertial frames. We will show in the next example that the correct second term that is Galilean invariant is $-\mathbf{u}_{\mathrm{rel}}dm/dt$, where $\mathbf{u}_{\mathrm{rel}}$ is the relative velocity of the dm.

In the next example, we will derive the equation of motion for a rocket from first principles.

12.7.2   Example 2: Rocket

The propulsion mechanism of rockets is based on the law of conservation of linear momentum. The thrust from the exhaust pushes the rocket in the forward direction.

Suppose at a given time t, the mass of the rocket is M. In time Δt, the rocket ejects hot gas of mass Δm with relative velocity $\mathbf{u}_{\mathrm{rel}}$ to the rocket. See Fig.12.4 for an illustration.

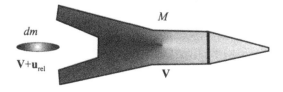

Figure 12.4   *The thrust on a rocket is achieved by pushing exhaust gas in the opposite direction.*

In the stationary frame (an inertial frame of reference), the linear momentum of the rocket + gas system is

$$\mathbf{P}(t) = M\mathbf{V},$$

$$\mathbf{P}(t + \Delta t) = (M - \Delta m)(\mathbf{V} + \Delta\mathbf{V}) + (\Delta m)(\mathbf{V} + \Delta\mathbf{V} + \mathbf{u}_{\mathrm{rel}}).$$

The change in momentum is (ignoring $\Delta m \Delta\mathbf{V}$ which is a higher-order term)

$$\Delta\mathbf{P} = M\Delta\mathbf{V} + (\Delta m)\mathbf{u}_{\mathrm{rel}}$$

Therefore,

$$\frac{d\mathbf{P}}{dt} = M\frac{d\mathbf{V}}{dt} + \mathbf{u}_{\mathrm{rel}}\frac{dm}{dt}. \tag{12.13}$$

Since the total mass is conserved, $dm/dt = -dM/dt$. Hence,

$$\frac{d\mathbf{P}}{dt} = M\frac{d\mathbf{V}}{dt} - \mathbf{u}_{\mathrm{rel}}\frac{dM}{dt}.$$

Therefore the net external force on the rocket is

$$\mathbf{F}_{\mathrm{ext}} = M\frac{d\mathbf{V}}{dt} - \mathbf{u}_{\mathrm{rel}}\frac{dM}{dt}, \tag{12.14}$$

or

$$M\frac{d\mathbf{V}}{dt} = \mathbf{F}_{\mathrm{ext}} + \mathbf{u}_{\mathrm{rel}}\frac{dM}{dt}. \tag{12.15}$$

The second term provides the boost to the rocket due to exhaust. If the rocket is moving in the positive x direction, then $\mathbf{u}_{\mathrm{rel}}$ is in the negative direction, and $dM/dt < 0$. Hence $\mathbf{u}_{\mathrm{rel}}dM/dt$ is along the positive x direction, and it provides a boost.

As an example, for a rocket coasting along the x-axis in free space, $\mathbf{F}_{\mathrm{ext}} = 0$ and $\mathbf{u}_{\mathrm{rel}} = -u\hat{\mathbf{x}} = \mathrm{const}$ (opposite direction of $\mathbf{V}$). Hence,

$$\frac{dM}{M} = -\frac{dV}{u}$$

The solution of which yields

$$V_f - V_i = -u\ln(M_f/M_i)$$

$$= u\ln(M_i/M_f)$$

If the payload is only 0.1%, then $M_i/M_f = 1000$ and

$$V_f - V_i = u\ln 1000 = 6.9u.$$

V_f depends logarithmically on M which makes it is difficult for the rocket to attain high speeds.

Let us revisit the example of the train and magnets. For the train and magnet system, $\mathbf{u}_{\rm rel} = \mathbf{v}_{\rm train}$. Hence Eq. (12.13) yields $\mathbf{F}_{\rm ext} = \mathbf{v}_{\rm train} dm/dt$. On the contrary, for the train+sand system, $\mathbf{u}_{\rm rel} = 0$. Therefore, $\mathbf{F}_{\rm ext} = 0$.

We end this chapter with a short discussion on a tricky problem involving variable mass.

12.8   Motion of chains and ropes

Lifting of a chain

The end of a chain of length L and mass density ρ, which is piled on a platform, is lifted vertically with a constant velocity v by a variable force F. Find F as a function of the height. Calculate the amount of energy loss.

SOLUTION   We apply Eq. (12.13) to the part of the chain that is above the table:

$$\mathbf{F}_{\rm ext} = M\frac{d\mathbf{v}}{dt} + \mathbf{u}_{\rm rel}\frac{dm}{dt}, \tag{12.16}$$

The motion is one dimensional. We choose the vertical coordinate axis to be the x-axis. Using $\dot{\mathbf{v}} = 0$, $\mathbf{u}_{\rm rel} = v\hat{\mathbf{x}}$, $\dot{m} = \rho v$, and $\mathbf{F}_{\rm ext} = (F - \rho g x)\hat{\mathbf{x}}$,

we obtain

$$F = \rho(gx + v^2). \tag{12.17}$$

After the whole chain has been lifted, the net work done by F is

$$W = \int_0^L F dx = \frac{1}{2}\rho g L^2 + \rho v^2 L.$$

The net gain in the potential energy is $\Delta U = MgL/2 = \rho g L^2/2$, and the net gain in the kinetic energy is $\Delta K = Mv^2/2 = \rho L v^2/2$. Therefore the net gain in the total mechanical energy is $(\rho g + \rho L v^2)/2$, which is less than the total work done by F. The balance

$$W - \Delta U - \Delta K = \rho L v^2/2.$$

goes into the inelastic interactions between the links of the chain near the table. The lost mechanical energy goes into heat and sound energy.

Lifting of a rope

Replace the open-link chain in the above problem with a flexible but inextensible rope. Compute the force required.

SOLUTION   If the bottom portion of the rope is free, then the procedure to find the force is same as above. The resultant F is the same as $\rho(gx + v^2)$ of Eq. (12.17). However there is a difference in the energy dissipation. There is no sound and heat loss in the rope. Instead there will be wiggles near the bottom, and some of the energy is converted to wave energy.

If we want to avoid the wiggles, it is necessary to impose a physical constraint at the base to guide the rope into vertical motion (Fig. 12.5). In this case, conservation of energy yields

$$F\,dx = d\left(\frac{1}{2}\rho x v^2\right) + d\left(\rho g x \frac{x}{2}\right).$$

F

Figure 12.5   *Rope, which is wound around a guide, is being pulled up.*

Hence $F = \dfrac{1}{2}\rho v^2 + \rho g x$.

After the whole rope has been lifted, the net work done by F is

$$W = \int_0^L F\,dy = \frac{1}{2}\rho v^2 L + \frac{1}{2}\rho g L^2.$$

Naturally this work is equal to the gain in total mechanical energy, unlike the chain's case. We can also compute the tension near the mouth of the guide. Since the acceleration of the rope is zero, the force balance provides us the tension near the mouth of the guide as

$$T_0 = F - \rho g x = \frac{1}{2}\rho v^2.$$

Note that in the case of a chain, the tension near the table is zero.

Exercises

1. A man moves from one end of the boat to another. The mass of the boat and the man are 100 kg and 80 kg respectively. The length of the boat is 3 metres. Assume that the speed of the boat is so slow that we can ignore the frictional force between the water and the boat. What distance would the boat move?

2. A rope of mass M and length L is lying on a frictionless table, with a short portion, l_0, hanging through a hole. Initially the rope is at rest. Find the position of the tip of the rope as a function of time for two situations:

   (a) when the rope is lying straight on the table,

   (b) when the rope is lumped near the hole.

3. A rocket is moving in interplanetary medium with a speed of 100 km/s. It gets bombarded by hydrogen ions whose density is 5 particles per cubic centimetres, and whose rms velocity is of the order of 500 km/s. What is the force exerted on the rocket due to the collisions of these particles with the rocket? Assume the mass of the rocket to be 500 kg, and the cross-section of the rocket to be 0.25 square metre.

4. Consider an arrangement shown in Fig. 12.6(a). The wedge slides on the horizontal surface without any friction, while the coefficient of friction between m_2 and the wedge is μ. For what ratio of m_1 and m_2 will the wedge remain stationary? Under what conditions will the wedge move left and right?

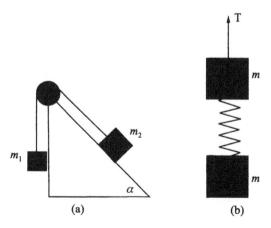

(a)   (b)

Figure 12.6   *(a) Exercise 4; (b) Exercise 8*

5. A hoop of mass M and radius R is lying horizontally on a smooth surface. A bug of mass m is moving on the hoop. Compute the trajectory of the bug and the centre of mass of the hoop.

6. A rifle recoils when a bullet is fired from it. Estimate the recoil speed of the rifle if the rifle is held very loosely.

7. Two blocks of the same mass m are connected by a spring whose unstretched length is l. The left mass rests against a wall, while the right mass is held by a stopper. The spring is compressed by length $l/2$. The stopper is removed at $t = 0$. Describe the motion of the blocks at later time.

8. A spring–mass system consisting of two masses (m each) and a spring (spring constant k) is held vertically by an experimenter (Fig. 12.6(b)). The system is let go at $t = 0$.

   (a) What is the acceleration of the CM of the spring–mass system?

   (b) Describe the motion of the particles in the CM frame.

   (c) Compute the positions of both the masses as a function of time in the laboratory frame.

9. The spring mass–system shown in Fig. 12.7(a) is being pulled with force F. Compute the position of both the blocks as a function of time.

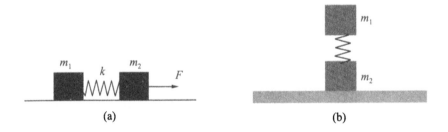

(a)   (b)

Figure 12.7   *(a) Exercises 9; (b) Exercise 10.*

10. A spring–mass system shown in Fig. 12.7(b) is resting on a horizontal surface. What force should be applied to the upper plate so that the lower one gets lifted after the pressure is removed.

11. The mass of a rocket is 40 tonnes. It contains 30 tonnes of fuel. Compute the acceleration and velocity of the rocket as a function of time. The exhaust velocity of the gases is 4 km/s and the fuel consumption is 200 kg/s.

12. A rocket is shot upward from the launch pad. The rocket expels exhaust with constant velocity v_{rel} at the rate γm, where m is the mass of the rocket, and γ is a constant.

   (a) The rocket experiences air drag $-bv$, where b is a constant and v is the speed of the rocket. Find the velocity of the rocket as a function of time.

   (b) Solve the above problem if the air drag is $-bv^2$ (turbulent drag).

13. A raindrop gains mass during its fall. The gain in mass is proportional to the instantaneous velocity and instantaneous mass of the raindrop. Compute the velocity of the raindrop assuming that it starts descending from a height h.

14. List practical examples that are based on the conservation of linear momentum.

Project

1. Read about the science involved in the mechanics of a rocket. How are modern satellites and space vehicles sent up to space? What are the intricacies of Chandrayaan (India's Moon mission)?

13

Collisions

We hear about unfortunate collisions between automobiles. These processes involve complex physics of momentum transfers, elastic properties of material, and also physical injuries. The physics of collisions is of major interest to automobile designers and engineers, and these collisions are studied using mock experiments and simulations. Realistic collisions are too complex to be studied in this book.

Physicists study collisions between elementary particles, atoms, and molecules. Rutherford bombarded α particles on to gold foils, and found that α particles get scattered by gold nuclei. This was a landmark experiment in physics. Nowadays, physicists collide elementary particles against each other, e.g., protons with protons, electrons with electrons etc. These particles move with very high speeds (close to the speed of light), so relativistic mechanics is required for their analysis. These particles are very small in size, and quantum effects dominate at these dimensions. Many particles are created during the collision process. Naturally these collision processes are quite complex too, and cannot be discussed here.

In this chapter we will study collision between particles and blocks at slow speeds (nonrelativistic). Our examples and problems essentially illustrate applications of the conservation laws. We will start with collision between two spheres in one dimension. In all our examples on collisions, bodies interact through contact forces.

13.1 One-dimensional elastic collisions

One-dimensional collisions are the simplest of all collision problems. They are observed when the bodies collide head-on.

Let us consider two spherical balls of mass m_1 and m_2. We assume m_2 to be at rest, and m_1 to be moving towards m_2 with velocity $v\hat{x}$ as illustrated in Fig. 13.1(a).[1] We assume a head-on collision between the balls. Let us denote the velocities of the two balls after the collision to be v_1' and v_2'. If the external forces acting on the system could be ignored during the collision, then the conservation of linear momentum yields

---

[1] If m_2 is not at rest, we can make a Galilean transformation to achieve this configuration. Note that the relative speed between the balls is invariant under the Galilean transformation.

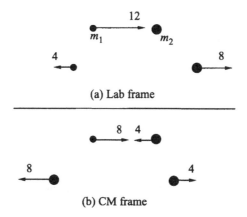

(a) Lab frame

(b) CM frame

Figure 13.1   *One-dimensional collision between two particles whose masses are 1 and 2 units in (a) laboratory reference frame; (b) centre of mass reference frame*

$$m_1 v_1' + m_2 v_2' = m_1 v. \tag{13.1}$$

We make another assumption about the collision process. We assume that the total kinetic energy of the balls is conserved. Collisions in which the total kinetic energy before and after the collisions remain the same are called *elastic collisions*. Under this approximation

$$\frac{1}{2} m_1 v_1'^2 + \frac{1}{2} m_2 v_2'^2 = \frac{1}{2} m v^2. \tag{13.2}$$

We solve the unknowns v_1' and v_2' using the above two equations. The solution after a bit of algebra is

$$v_1' = \frac{m_1 - m_2}{m_1 + m_2} v, \tag{13.3}$$

$$v_2' = \frac{2m_1}{m_1 + m_2} v. \tag{13.4}$$

The solution of the collision problems in the centre of mass (CM) reference frame is quite interesting and transparent. The CM of the particle system moves with $V_{CM} = m_1 v/(m_1 + m_2)$, and the velocities of the balls m_1 and m_2 in the CM frame are $\bar{v}_1 = v - V_{CM} = m_2 v/(m_1 + m_2)$ and $\bar{v}_2 = -m_1 v/(m_1 + m_2)$ respectively as shown in Fig. 13.1(b). The momentum of the balls m_1 and m_2 in the CM frame are $\bar{p}_1 = \bar{p} = m_1 m_2 v/(m_1 + m_2)$ and $\bar{p}_2 = -\bar{p} = -m_1 m_2 v/(m_1 + m_2)$ respectively, and the net momentum is zero.

After the collision, suppose the momentum of the balls are $\bar{p}_1' = \bar{p}'$ and $\bar{p}_2' = -\bar{p}'$. For elastic collision

$$\frac{\bar{p}'^2}{2m_1} + \frac{\bar{p}'^2}{2m_2} = \frac{\bar{p}^2}{2m_1} + \frac{\bar{p}^2}{2m_2},$$

which implies that $\bar{p}' = \pm\bar{p}$. The result $\bar{p}' = \bar{p}$ is a trivial solution that corresponds to the balls crossing each other without interactions (possible only for ghosts or imaginary particles). The other solution $\bar{p}' = -\bar{p}$ corresponds to the reversal of directions of the balls. After the collision $\bar{p}'_1 = -\bar{p}$ and $\bar{p}'_2 = \bar{p}$. Using $\mathbf{v}_i = \mathbf{p}_i/m_i$, we can compute the velocity of the balls in the CM frame, which are

$$\bar{v}'_1 = -\frac{m_2}{m_1 + m_2}v,$$

$$\bar{v}'_2 = \frac{m_1}{m_1 + m_2}v.$$

Hence the magnitude of velocity remains the same after the collision.

Now we can use transformation rules to compute the velocity of the balls in the laboratory frame. The velocity of the balls after the collision are

$$v'_1 = \bar{v}'_1 + V_{\text{CM}} = \frac{m_1 - m_2}{m_1 + m_2}v,$$

$$v'_2 = \bar{v}'_2 + V_{\text{CM}} = \frac{2m_1}{m_1 + m_2}v,$$

which are the same as our earlier calculations (Eqs. (13.3, 13.4)).

Under special cases

1. when $m_2 = m_1$, $v'_1 = 0$ and $v'_2 = v$;

2. when $m_2 \gg m_1$, $v'_1 \approx -v$ and $v'_2 \approx 0$; particle 1 simply gets reflected by the heavy particle 2.

3. when $m_2 \ll m_1$, $v'_1 \approx v$ and $v'_2 \approx v$; particle 2 is carried along by the heavy particle 1.

After this discussion on one-dimensional collisions, we will move on to two-dimensional collisions.

13.2   Two-dimensional elastic collisions

One-dimensional collisions are observed when the objects hit each other head-on. The probability of head-on collisions is very low. Typically we observe two- and three-dimensional collisions.

In the following discussion we will study collisions between two spherical balls. The scattering takes place when the balls come in contact with each other. The contact forces (normal and frictional) make the balls change directions. If the velocity vectors of the balls and the line joining the centres of the balls at the time of collision form a plane (Fig. 13.2), then the centres of the balls will remain in the same plane after the collision. This type of collision is two-dimensional, and will be analysed below. A sketch of velocity vectors during the collision in the CM frame is shown in Fig. 13.2.

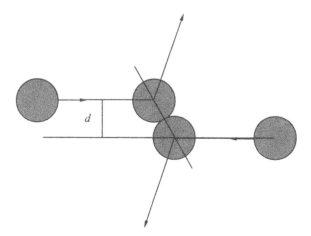

Figure 13.2   *Two-dimensional collision between two balls. The velocity vectors of the balls, and the line joining the centres of the balls at the time of collision are in a plane before and after the collision.*

If the planarity condition during the collision is not satisfied, the balls pick up momentum in the perpendicular direction, and the collision will be three-dimensional. Three-dimensional collisions are quite complex, and are beyond the scope of this book.

Suppose m_1 and m_2 are the masses of the two balls. We assume the ball m_2 to be at rest, and the ball m_1 to be moving towards m_2 with velocity $\mathbf{v}$. The perpendicular distance between the two trajectories (denoted by d in the figure) is called the *impact parameter*.

We will analyse the collision in a CM frame. The velocity of the CM is $\mathbf{V}_{\mathrm{CM}} = m_1\mathbf{v}/(m_1 + m_2)$, and its momentum is $\mathbf{P}_{\mathrm{CM}} = m_1\mathbf{v}$. Hence the velocities of m_1 and m_2 in the CM frames are $\bar{\mathbf{v}}_1 = \mathbf{v}_1 - \mathbf{V}_{\mathrm{CM}} = m_2\mathbf{v}/(m_1 + m_2)$ and $\bar{\mathbf{v}}_2 = -\mathbf{V}_{\mathrm{CM}} = -m_1\mathbf{v}/(m_1 + m_2)$. The momentum of the particles in the CM reference frame are $\bar{\mathbf{p}}$ and $-\bar{\mathbf{p}}$ respectively, where

$$\bar{\mathbf{p}} = \frac{m_1 m_2}{m_1 + m_2}\,\bar{\mathbf{v}}.$$

We assume the collision to be elastic. Hence the total kinetic energy of the systems is conserved in the collision process. If the momentum of m_1 and m_2 after the collision are $\bar{\mathbf{p}}'$ and $-\bar{\mathbf{p}}'$ respectively, then

$$\frac{\bar{\mathbf{p}}'^2}{2m_1} + \frac{\bar{\mathbf{p}}'^2}{2m_2} = \frac{\bar{\mathbf{p}}^2}{2m_1} + \frac{\bar{\mathbf{p}}^2}{2m_2}.$$

Consequently $\bar{p}' = \bar{p} = m_1 m_2 v/(m_1 + m_2)$ (Fig. 13.3(a)). This result implies that the magnitudes of the velocities of both the balls remains unchanged—only their direction changes.

Suppose the direction of scattered particle m_1 is along $\hat{\mathbf{n}}$ which makes an angle χ with the x-axis. Then the velocities of m_1 and m_2 in the CM frame after the collision will be

$$\bar{\mathbf{v}}_1' = \frac{\bar{\mathbf{p}}}{m_1} = \frac{m_2 v \hat{\mathbf{n}}}{m_1 + m_2},$$

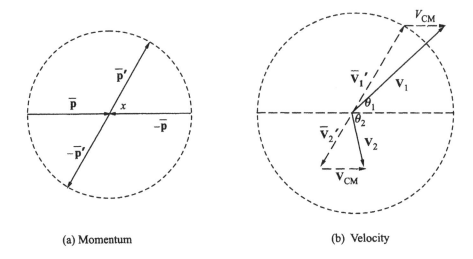

(a) Momentum   (b) Velocity

Figure 13.3   *(a) Graphical representation of momentum vectors in the CM frame. (b) The velocity vectors in CM and laboratory frame.*

$$\bar{\mathbf{v}}_2' = \frac{-\bar{\mathbf{p}}}{m_2} = -\frac{m_1 v \hat{\mathbf{n}}}{m_1 + m_2}.$$

These velocities are displayed in Fig. 13.3(b).

   We can compute the velocities of the balls in the laboratory frame by adding $\mathbf{V}_{\text{CM}}$ to $\bar{v}_{1,2}$ as shown in Fig. 13.3(b):

$$\mathbf{v}_1 = \bar{\mathbf{v}}_1' + \mathbf{V}_{\text{CM}}; \quad \mathbf{v}_2 = \bar{\mathbf{v}}_2' + \mathbf{V}_{\text{CM}}.$$

If the angles made by m_1 and m_2 from the x-axis are $\theta_{1,2}$ respectively, then

$$\tan \theta_1 = \frac{\bar{v}_1' \sin \chi}{V_{\text{CM}} + \bar{v}_1' \cos \chi} = \frac{m_2 \sin \chi}{m_1 + m_2 \cos \chi}, \tag{13.5}$$

$$\tan \theta_2 = \frac{\bar{v}_2' \sin \chi}{V_{\text{CM}} + \bar{v}_2' \cos \chi} = \frac{\sin \chi}{1 - \cos \chi} = \cot \frac{\chi}{2}. \tag{13.6}$$

Equation (13.6) implies that

$$\theta_2 = \frac{\pi}{2} - \frac{\chi}{2}.$$

The identity $\theta_2 = (\pi - \chi)/2$ can also be derived geometrically using the fact that $v_2' = V_{\text{CM}}$. For the special case when $m_2 = m_1$, we get $\theta_1 = \chi/2$. Therefore $\theta_1 + \theta_2 = \pi/2$.

   It is quite clear from the above calculation that the idea of centre of mass plays a major role in our understanding of collisions.

13.3   Inelastic collisions

In practice, kinetic energy is lost in almost all collisional processes. The lost energy goes into heat, light etc. These kinds of collisions are called *inelastic collisions*.

Let us assume that two masses m_1 and m_2 that are moving with velocities $\mathbf{v}_1$ and $\mathbf{v}_2$ respectively collide with each other. Suppose the velocities of these masses after the collision are $\mathbf{v}_1'$ and $\mathbf{v}_2'$ respectively. If external force could be ignored during the collision, then linear momentum of the system is conserved for inelastic collisions. Hence

$$m_1\mathbf{v}_1 + m_2\mathbf{v}_2 = m_1\mathbf{v}_1' + m_2\mathbf{v}_2'.$$

Suppose the energy lost during the collision process is ϵ, then

$$\frac{1}{2}m_1v_1^2 + \frac{1}{2}m_2v_2^2 = \frac{1}{2}m_1v_1'^2 + \frac{1}{2}m_2v_2'^2 + \epsilon.$$

Generally we do not test elasticity of the collision process by computing ϵ. There is a very useful constant called *coefficient of restitution (COR)* that is used to quantify the elasticity of the collisional process. COR is defined as

$$\mathrm{COR} = \frac{|\mathbf{v}_2' - \mathbf{v}_1'|}{|\mathbf{v}_2 - \mathbf{v}_1|}.$$

Note that $\mathbf{v}_2 - \mathbf{v}_1$ and $\mathbf{v}_2' - \mathbf{v}_1'$ are the relative velocities of the masses before and after the collision. For elastic collisions, COR is 1, while for completely inelastic collisions, COR is 0. We prove this result using the following arguments.

Since the relative velocity between the masses is the same in all inertial frame, we use CM reference frame for our analysis. For *two-dimensional elastic collisions* studied in the previous section, in the CM frame, the relative velocity before the collision is

$$|\bar{\mathbf{v}}_2 - \bar{\mathbf{v}}_1| = |\bar{\mathbf{p}}|\left(\frac{1}{m_1} + \frac{1}{m_2}\right),$$

where $\bar{\mathbf{p}}$ is the momentum of mass m_1 before the collision. The relative velocity after the collision is

$$|\bar{\mathbf{v}}_2' - \bar{\mathbf{v}}_1'| = |\bar{\mathbf{p}}'|\left(\frac{1}{m_1} + \frac{1}{m_2}\right),$$

where $\bar{\mathbf{p}}'$ is the momentum of mass m_1 after the collision. Therefore COR= $\bar{p}'/\bar{p}$, For elastic collisions, $\bar{p}' = \bar{p}$. Therefore, COR is 1 for elastic collisions.

In a completely inelastic collision, the relative velocity of the bodies after the collision is zero. Hence, COR is zero for a completely inelastic collision. In between these extreme cases, $|\bar{\mathbf{p}}'| < |\bar{\mathbf{p}}|$ because of the loss of kinetic energy (Exercise 1). Consequently $|\bar{\mathbf{v}}_2' - \bar{\mathbf{v}}_1'| < |\bar{\mathbf{v}}_2 - \bar{\mathbf{v}}_1|$. Therefore, COR for the intermediate situation is between 0 and 1.

In summary, inelastic collisions are characterised by a quantity called coefficient of restitution (COR). COR is 1 for elastic collisions, 0 for completely inelastic collisions, and between 0 and 1 for intermediate situations.

In the next section we will discuss collisions between extended objects and collision approximation.

13.4 Collisions between composite objects and collision approximation

Typical collisions are not between two blocks or spheres. Generally they involve composite objects with several constituent parts. The parts of the body may be loosely or tightly coupled. The coupling between the parts is a major factor in collisions. We will illustrate this statement using an example.

In an experiment, a bullet of mass m and velocity v hits a spring–mass system shown in Fig. 13.4. The mass of both the blocks of the system is M, and the spring constant of the spring is k. We investigate the state of spring–mass system after the bullet hits the left mass inelastically, and gets embedded inside it.

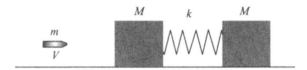

Figure 13.4   *A bullet hits the left mass of a spring–mass system*

If the spring is very stiff (i.e., spring is like a rigid rod), then the linear momentum of the bullet is transferred to both the blocks equally through the spring. An application of conservation of linear momentum yields the final velocity of the spring–mass system. If V' is the velocity of both the blocks, then

$$(2M + m)V' = mv.$$

Hence

$$V' = \frac{mv}{2M + m}.$$

Here the whole spring–mass system acts as a single mass due to extreme stiffness of the spring.

Consider the other extreme situation. Assume that during the collision process, the bullet transfers its momentum only to the left block, and spring transfers negligible momentum to the right block. Under this approximation, the velocity of the left block immediately after the collision is

$$v_1' = \frac{mv}{M + m}.$$

At this instant the right block has zero velocity. After this the CM of the spring–mass system moves with a constant velocity of $mv/(2M + m)$, and the masses oscillate in the CM frame. A similar problem was solved in Section 12.4.1.

Thus we observe that the stiffness of a spring plays a crucial role in the collision process. Typically we make either the first or the second approximation. These approximations simplify our calculation considerably. *In the second approximation, the linear momentum is not transferred across elements of the composite body. This approximation is called collision approximation.* Note however that in both the situations, the CM moves with the same velocity. The compression in the spring is rather small in the earlier situation because the spring constant is very large in that case.

Let us investigate the condition for the collision approximation. If the collision approximation hold, the velocity of the left block after the collision is $v_1' = mv/(M+m)$. Suppose that the collision time is Δt. During this time the spring gets compressed by $\Delta x \sim v_1' \Delta t$, consequently the momentum transferred by the left mass to the spring during time Δt is

$$\Delta p \sim (k \Delta x)\Delta t$$

$$\sim k v_1' (\Delta t)^2$$

$$\sim k \frac{mv}{M+m}(\Delta t)^2.$$

The collision approximation implies that the momentum transfer to the spring (Δp) is much less than the momentum of the left block, i.e.,

$$k\frac{mv}{M+m}(\Delta t)^2 \ll mv$$

or

$$\Delta t \ll \sqrt{\frac{M+m}{k}}.$$

Note that $\sqrt{(M+m)/k}$ is of the order of the time period of the oscillations in the spring–mass system. Hence collision approximations hold when the collision time is much less than the time period of the spring–mass system. When this approximation holds, we can ignore the spring and the right mass during the time the bullet collides with the left mass. The collision approximation is likely to hold for a soft spring for which the time period of the oscillation would be relatively large.

When the spring is very stiff (very large k) then the time period of oscillations is small, and the time of collision Δt could become comparable to the oscillation time scale of the spring–mass system. For these cases we need to consider the combined system during the collision. In these collisions both the masses move with the same velocity.

We should also keep in mind that Δt depends on the material property of the left mass. For example, if the block is soft, then Δt would be relatively large. The validity of the collision approximation crucially depends on the relative magnitude of Δt and the time period of oscillations of the spring–mass system. The above example shows that collision involving composite systems could be quite complex. The collision approximation may simplify the problem, in the sense that we could break up collision process piece by piece. General collisions require computer simulations for analysis.

In this chapter we discussed collisions at an elementary level. We discussed collisions between spherical balls in one and two dimensions. Our examples are only illustrative. Collisions between elementary particles studied by physicists involve high speeds where relativity and quantum mechanics are applied. In the field of engineering, collisions of composite objects (e.g. collisions between two cars) are studied; these calculations are very complex. The examples discussed in the present chapter illustrate applications of conservation laws, while the procedures described are only a stepping stone to more complex physics and engineering problems.

Exercises

1. Two masses m_1 and m_2 collide with each other.

   (a) Derive an expression for the loss in kinetic energy in the CM frame.

   (b) Show that in the CM frame the magnitude of linear momentum of each particle after the collision is always less than or equal to that before the collision.

2. A pendulum has a bob of mass M and string of length l. A bullet of mass m travelling horizontally hits the bob completely inelastically and gets embedded inside the bob. Assume the collision to be instantaneous (collision approximation). Describe the motion of the bob and bullet after the collision.

3. A particle moving parallel to the y-axis collides elastically with a parabolic mirror $y^2 = 2px$. Show that the particle will always pass through the mirror's focus irrespective of its point of impact. Locate the focus.

4. A block of mass M rests on a frictionless horizontal surface, and it is connected to a spring (spring constant k) attached to a vertical wall. A bullet of mass m hits the block inelastically with velocity v, and gets embedded inside the block. Assume the collision to be instantaneous,

   (a) Calculate the maximum compression of the spring.

   (b) Describe the final state of the block.

   (c) Obtain the numerical value when $M = 20$ kg, $m = 10$ kg, $k = 1000$ N/m, $v = 500$ m/s.

   (d) Obtain the condition for the collision to be instantaneous.

5. A spring–mass system (mass m and spring constant k) collides against a block of mass M as shown in Fig. 13.5(a). The initial velocity of mass m is **v**. Describe the motion of both the masses.

6. A spring–mass system consisting of two masses (m each) and spring constant k is falling vertically downward from height h as shown in Fig. 13.5(b). Ignore the length of the spring as compared to h. Also assume that the spring force is much stronger than the gravity forces acting on the masses and that the spring is unstretched initially.

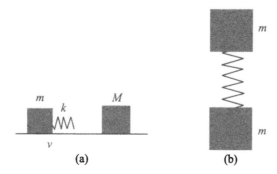

Figure 13.5   *(a) Exercise 5; (b) Exercise 6*

(a) What is the force in the spring during the fall? Describe the motion of the masses during this phase.

(b) The lower mass hits the ground inelastically, and it stays at the ground. Assume collision approximation. The upper mass compresses the spring, and it comes to rest. How much does the spring gets compressed?

(c) The spring–mass system rebounds after some time. How does the spring–mass system take off? What is the velocity of CM? What is the translational kinetic energy of the CM? What is the internal kinetic energy of the spring–mass system? How high will CM travel upward before coming to rest.

(d) Under what condition is the collision approximation valid?

(e) What would be the motion of the system if the lower mass collided *elastically*, and the momentum is transferred to the upper mass instantly (very stiff spring)?

(f) What would be the motion of the system if the lower mass collided *elastically*, but the momentum of the lower block is transferred to the upper block only after the spring is fully compressed.

7. A test charge $+q$ is shot towards a fixed charge $+Q$ from far with an initial velocity of v_∞. The velocity of the test particle is in the same direction as the line joining the centres of the charge particle. Describe the motion of the test particle. Assume that the mass of charge particles q and Q are m and M respectively.

8. [LONG PROBLEM]   Solve problem 7 for an impact parameter d. You would need to compute the trajectory of the test particle in similar lines as Kepler's problem. What would be the trajectory of the test particle if it is a negative charge?

9. A ball of mass M containing chemicals is travelling with a constant velocity of v_0. This ball explodes into three pieces whose masses are $M/4$, $M/4$, and $M/2$. During the explosion, the amount of energy released is E. The masses $M/4$ move with speed v at an angle of $45°$ and $-45°$ from the original direction, while $M/2$ moves in the original direction with velocity v'. Compute v and v'.

10. A steel ball of mass 10 gm falls from a height of 1 metre on to a horizontal surface. Assume that the velocity of the ball decreases by a factor of 1.25 after every impact. Find the total momentum transfer after n collisions. What is the height achieved by the ball after the nth collision. Compare your result with that when it is an elastic collision.

11. List practical situations where you encounter collisions between bodies.

14

Simple Harmonic Motion

In this chapter we will discuss a very important and generic feature of stable systems. If a stable system is disturbed slightly, it oscillates around the stable equilibrium position. We start our analysis with a study of a typical one-dimensional stable system. If the equilibrium position of the system is q_0, the form of potential energy near q_0 is

$$U(q) = U(q_0) + \frac{(q - q_0)^2}{2} U''(q_0) + ...,$$

where ... represents the higher order terms. Note that $U'(q_0) = -F(q_0) = 0$ at the equilibrium position. We can choose $U(q_0) = 0$. The condition of stability demands that $U''(q_0) > 0$ because the restoring force has to always point towards the equilibrium position. We denote $U''(q_0)$ by k. For small displacements around the equilibrium position $(q - q_0)$, we can ignore the higher order terms. We denote $q - q_0$ by x. For small x, the potential energy is

$$U(x) = \frac{1}{2} k x^2 \tag{14.1}$$

with $k > 0$.

If the mass of the particle is m, then the equation of motion will be

$$m\ddot{x} = -kx, \tag{14.2}$$

whose solution is rather straightforward

$$x(t) = c_1 \cos \omega_0 t + c_2 \sin \omega_0 t$$

or

$$x(t) = A \cos (\omega_0 t + \alpha)$$

where $\omega_0 = \sqrt{k/m}$ is called the *natural frequency* of the oscillator. The constants $c_{1,2}$ or A, α are determined using initial conditions. The functions sin and cos are called simple harmonic functions. Because of this nomenclature, the above motion is called *simple harmonic motion (SHM)*. This is a generic feature for all stable systems near an equilibrium point.

As discussed in Chapter 5, we obtain periodic motion for all convex potentials. However, harmonic motion occurs only when the potential is quadratic (x^2), which is typically valid for small displacements near a minima of a potential.

The kinetic and potential energy of the oscillator are

$$T = \frac{1}{2}m\dot{x}^2 = \frac{1}{2}m\omega_0^2 A^2 \sin^2(\omega_0 t + \alpha),$$

$$U = \frac{1}{2}kx^2 = \frac{1}{2}m\omega_0^2 A^2 \cos^2(\omega_0 t + \alpha).$$

Note that the total energy

$$E = \frac{1}{2}m\dot{x}^2 + \frac{1}{2}kx^2 = \text{const}, \tag{14.3}$$

which is not surprising because the force $-kx$ is conservative.

Examples of SHM

1. **Spring–mass system**
   If the spring constant is k, and the mass of the block is m, then $U = kx^2/2$ which is of the form of Eq. (14.1).

2. **Pendulum**
   A pendulum with a bob of mass m and rod length l has potential energy $mgl(1 - \cos(\theta))$, where θ is the angle of the pendulum from the stable position (bottom-most position). For small θ, $U(\theta) \sim mgl\theta^2/2$, which is of the same form as Eq. (14.1). Note that $\theta = \pi$ is the unstable equilibrium position of the pendulum.

3. **A piston above a compressed gas**
   At equilibrium, the weight of the piston is balanced by the pressure force. This is a stable position. The motion of the piston will be SHM for small displacements (Exercise 7).

4. **Swing of a tree branch**
   If θ is the angular displacement of the tree branch, then the elastic energy is proportional to $k\theta^2/2$ which is of the same form as Eq. (14.1). The motion of the tree branch is SHM.

5. **LC circuit**
   See Fig. 14.1. The capacitance of the capacitor is C, and the inductance of the inductor is L. If the charge contained in the capacitor is Q, and the current flowing in the circuit is $I = \dot{Q}$, then the total energy of the LC system is

   $$E = \frac{1}{2}L\dot{Q}^2 + \frac{Q^2}{2C},$$

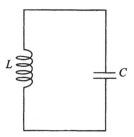

Figure 14.1   *LC circuit*

which is equivalent to Eq. (14.3), with Q analogous to x, and $\dot{Q}$ analogous to v. The dynamical equation is

$$L\ddot{Q} + \frac{Q}{C} = 0,$$

which is equivalent to Eq. (14.2). The inductor has inertia (like mass), and the capacitor has stiffness (like spring).

We shall make a remark before going to the next section. If $k < 0$, then the potential function $-kx^2/2$ is concave at $x = 0$, and the equilibrium point is unstable. As discussed in Section 5.1.4. the solution of the equation of motion for this system is

$$x(t) = c_1 \exp(Bt) + c_2 \exp(-Bt) \tag{14.4}$$

with $B = \sqrt{|k|/m}$. Clearly $x \to \infty$ as $t \to \infty$. In an exceptional case when $c_1 = 0$, $x(\infty) = 0$. An example of an unstable equilibrium is the uppermost point of the pendulum $(\theta = \pi)$.

EXAMPLE 14.1   Consider the motion of a particle in a potential $U(x) = -\frac{1}{2}x^2 + \frac{1}{4}x^4$. Find the equilibrium points of the potential and the frequency of oscillation of the particle wherever SHM is possible.

SOLUTION   The potential function is plotted in Fig. 14.2. The force is zero when $U'(x) = 0$, i.e., at $x = 0, \pm 1$. These are the equilibrium points. We expand the potential near the equilibrium points. Near $x = 0$, $U(x) \approx -\frac{x^2}{2}$ hence $x = 0$ is an unstable point, and no SHM is possible near $x = 0$. Near $x = \pm 1$,

$$U(x) \approx U(x_0) + \frac{1}{2}U''(x_0)(x - x_0)^2 = -\frac{1}{4} + \frac{1}{2}2(x - x_0)^2.$$

Hence the potential is convex near $x = \pm 1$ with $k = 2$. Consequently the particle performs an SHM near $x = \pm 1$ with a frequency of $\sqrt{2/m}$.

In the next section we will discuss SHM in higher dimensions.

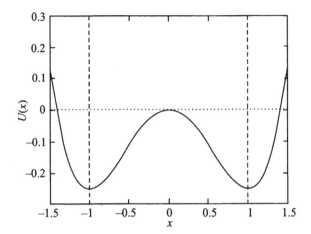

Figure 14.2   *Plot of $U(x) = -\frac{1}{2}x^2 + \frac{1}{4}x^4$ vs. x*

14.1   Oscillations in higher dimensions

Let us explore the possibility of oscillations in two-dimensional systems. If (x_0, y_0) is an equilibrium point of a two-dimensional system, we can expand the potential energy $U(x, y)$ near the equilibrium point as

$$
\begin{aligned}
U(x, y) &= U(x_0, y_0) + \frac{1}{2}\frac{\partial^2 U}{\partial x^2}\bigg|_{(x_0, y_0)}(x - x_0)^2 + \frac{1}{2}\frac{\partial^2 U}{\partial y^2}\bigg|_{(x_0, y_0)}(y - y_0)^2 \\
&+ \frac{\partial^2 U}{\partial x \partial y}\bigg|_{(x_0, y_0)}(x - x_0)(y - y_0) + ..
\end{aligned}
$$

At the equilibrium point, the partial derivatives $\partial U/\partial x = \partial U/\partial y = 0$. We make a change of variable $x' = x - x_0$ and $y' = y - y_0$, and set $U(x_0, y_0) = 0$. We also ignore the higher order term for small x', y'. The potential energy as a function of new variables is

$$
U(x', y') = \frac{1}{2}k_1 x'^2 + \frac{1}{2}k_2 y'^2 + k_3 x' y'.
$$

Using a coordinate transformation, we can reduce the above potential to

$$
U(X, Y) = \frac{1}{2}k_1 X^2 + \frac{1}{2}k_2 Y^2.
$$

The proof of the above result is a part of matrix analysis, and is beyond the scope of this book. Anyway, when $k_1, k_2 > 0$, then the potential function is a paraboloid with a minimum at $(0, 0)$. The equation of motion is

$$
m\ddot{X} = -k_1 X,
$$

$$m\ddot{Y} = -k_2 Y.$$

The motion of the particle is an SHM along both X and Y directions with frequencies $\omega_1 = \sqrt{k_1/m}$ and $\omega_2 = \sqrt{k_2/m}$ respectively. If ω_1/ω_2 is a rational number, then the orbit will be a closed one (Exercise 6).

When $k_{1,2} < 0$, then the motion is unstable. The particle will move away from the equilibrium point. An interesting feature arises when one of the coefficients is positive, and the other is negative. For example, if $k_1 > 0$ and $k_2 < 0$, then the motion is SHM only along the x-axis and unstable along the y-axis. A test particle in the electrostatic field has such potential near equilibrium points (Exercises 8, 9).

We can generalise the above analysis to three and higher dimensions, however we will focus on one dimension. In the next section we will discuss the motion of an oscillator under the influence of external and frictional forces.

14.2   Forced oscillations

In physics we encounter many situations when an oscillator is forced by an external agency. One such example is the motion of an electron of an atom that is forced by an electromagnetic field. The equation of motion for a forced oscillator is

$$\ddot{x} + \omega_0^2 x = F(t)/m,$$

where $F(t)$ is the external forcing function. The above equation is not easily solvable for arbitrary $F(t)$. The forced oscillation by simple harmonic functions (sin and cos) is most common, for example the motion of an electron in an atom forced by electromagnetic waves.

Let us take

$$F(t) = F_0 \cos{(\omega_f t)}.$$

For $w_0 \neq \omega_f$, the solution is straightforward:

$$x(t) = A\cos{(\omega_0 t)} + B\sin{(\omega_0 t)} + \frac{F_0}{m(\omega_0^2 - \omega_f^2)}\cos{(\omega_f t)}. \tag{14.5}$$

The first two terms are the homogeneous terms, and the last term is the particular term discussed in Chapter 5.

For the initial condition $(x(0), v(0))$, we obtain

$$x(t) = x(0)\cos{(\omega_0 t)} + \frac{v(0)}{\omega_0}\sin{(\omega_0 t)} + \frac{F_0}{m(\omega_0^2 - \omega_f^2)}[\cos{(\omega_f t)} - \cos{(\omega_0 t)}]. \tag{14.6}$$

The solution $x(t)$ indicates oscillations in time with two frequencies, the natural frequency of the oscillator w_0, and the forcing frequency ω_f. Figure 14.3(a) illustrates $x(t)$ for a generic initial condition.

The behaviour however changes when the forcing frequency matches the natural frequency. We can obtain the solution of this case by taking the limit $\omega_f \to \omega_0$ in $x(t)$.

$$x(t) = x(0) \cos{(\omega_0 t)} + \frac{v(0)}{\omega_0} \sin{(\omega_0 t)} + \lim_{\omega_f \to \omega_0} \frac{F_0(\cos{(\omega_f t)} - \cos{(\omega_0 t)})}{m(\omega_0^2 - \omega_f^2)},$$

$$= x(0) \cos{(\omega_0 t)} + \frac{v(0)}{\omega_0} \sin{(\omega_0 t)} + \frac{F_0 t}{2m\omega_0} \sin{(\omega_0 t)} \qquad (14.7)$$

Application of L'Hospital's rule provides the final answer. For large t, $F_0 t/(2m\omega_0)$ is the most dominant term. Hence $x(t)$ oscillates with increasing amplitude in time. This phenomena is called *resonance*. Figure 14.3(b) shows $x(t)$ vs. t plot under the resonance condition.

Let us understand the phase relationship between the forcing function $F(t)$ and response $x(t)$. Note that when $\omega_0 \neq \omega_f$, $x_{\text{particular}}(t)$ and $F(t)$ are in phase, and $\dot{x}_{\text{particular}}$ and $F(t)$ are out of phase. The average power input by the forcing function is

$$\langle P \rangle = \frac{1}{T} \int_0^T F(t)\dot{x}(t)dt,$$

where T is the time period of the oscillator. The function is crucially dependent on the phase relationship between $F(t)$ and $\dot{x}(t)$. When $\omega_0 \neq \omega_f$, the above integral is zero for both homogeneous and particular parts. Consequently $\langle P \rangle = 0$.

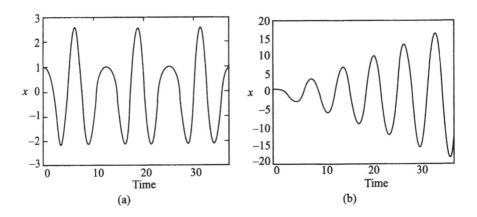

(a)   (b)

Figure 14.3   *(a) Plot of $x(t)$ vs. t for a forced oscillator. We take $\omega_0 = 1, \omega_f = 1.5, F_0 = m = 1$, $x(0) = 1$, and $v(0) = 0$. (b) Plot of $x(t)$ vs. t for a forced oscillator under resonance condition. We take $\omega_0 = \omega_f = 1.0, F_0 = m = 1, x(0) = 1$, and $v(0) = 0$.*

At resonance, the velocity is

$$\dot{x}(t) = -A\omega_0 \sin{(\omega_0 t + \alpha)} + \frac{F_0}{2m\omega_0} \sin{(\omega_0 t)} + \frac{F_0 t}{2m} \cos{(\omega_0 t)}.$$

Note that the last term of the velocity is in phase with the forcing, and the force pumps in energy to the oscillator. The power supplied by the forcing is

$$\langle P \rangle = \frac{1}{T} \int_0^T F(t) \dot{x}(t) dt$$

$$= \frac{F_0^2}{T 2m} \int_0^T t \cos^2(\omega_0 t) dt$$

$$= \frac{F_0^2 T}{4m}.$$

Since $\langle P \rangle \neq 0$, the amplitude of the oscillator grows with time. In fact, the input power is proportional to T. The homogeneous solution will be in phase with $F(t)$ only when $\alpha = \pi/2$. However such fine tuning in phases is unlikely in generic systems.

At resonance, when the amplitude becomes large, the small x approximation breaks down, and we need to keep higher order terms of potential or force. These terms typically saturate the growth. Another major approximation is the neglect of friction that damps the growth. We will study the effect of friction in the next section.

14.3   Damped oscillation

We find that the oscillations in a pendulum, tree branch, LRC circuit etc. get damped after sometime. This is due to the frictional force acting on the system. In the present section we will consider frictional oscillations with and without external forcing.

First we will analyse frictional oscillations without forcing. We will assume that the frictional force is proportional to the velocity of the particle. This assumption is valid when the mass moves slowly in a fluid. The equation of motion for the system is

$$m\ddot{x} = -kx - b\dot{x}.$$

We denote $b/m = 2\gamma$. This equation has been solved in Chapter 5. For a trial function $\exp(at)$, the values of a are

$$a_{1,2} = -\gamma \pm \sqrt{\gamma^2 - \omega_0^2}.$$

When $\gamma \neq \omega_0$, the general solution is

$$x(t) = \exp(-\gamma t) \left[c_1 \exp\left(-\sqrt{\gamma^2 - \omega_0^2} t\right) + c_2 \exp\left(\sqrt{\gamma^2 - \omega_0^2} t\right) \right] \tag{14.8}$$

For $\gamma < \omega_0$, the bracketed term is oscillatory, and the overall amplitude decreases with time. The general solution can be written as

$$x(t) = A \exp(-\gamma t) \cos\left(\sqrt{\omega_0^2 - \gamma^2} t + \alpha\right)$$

This type of motion is called *under-damped oscillation*. Figure 5.1 illustrates a typical under-damped oscillation.

When $\gamma > \omega_0$, $-\gamma \pm \sqrt{\gamma^2 - \omega_0^2} < 0$, hence the motion has no oscillation. This type of motion is called *over-damped oscillation*. A typical over-damped oscillation is also illustrated in Fig. 5.1

$\gamma = \omega_0$ is a special case in which $a_1 = a_2 = -\gamma$, and the solution is

$$x(t) = c_1 \exp(-\gamma t) + c_2 t \exp(-\gamma t).$$

For initial conditions $x(0) = A$ and $\dot{x}(0) = 0$, the solution is

$$x(t) = A \exp(-\gamma t)[1 + \gamma t],$$

which is illustrated in Fig. 5.1. This type of motion is called *critically-damped oscillation* and is intermediate between under-damped and over-damped oscillations.

14.4 Forced-damped oscillation

Now let us introduce forcing on a damped oscillator. The equation of motion of the oscillator is

$$\ddot{x} + 2\gamma \dot{x} + \omega_0^2 x = (F_0/m) \cos (\omega_f t). \tag{14.9}$$

In the following discussion we will derive the steady-state solution of this oscillator.

14.4.1 Steady-state solution

The homogeneous solution is provided in Eq. (14.8). The particular solution unfortunately involves both $\cos \omega_f t$ and $\sin \omega_f t$, which makes the calculation quite cumbersome. We follow a simpler approach. We rewrite Eq. (14.9) as

$$\ddot{x} + 2\gamma \dot{x} + \omega_0^2 x = \Re[(F_0/m) \exp (i\omega_f t)].$$

Compare this equation with

$$\ddot{z} + 2\gamma \dot{z} + \omega_0^2 z = (F_0/m) \exp (i\omega_f t),$$

whose particular solution is

$$z_p(t) = \frac{(F_0/m) \exp (i\omega_f t)}{-\omega_f^2 + 2i\gamma\omega_f + \omega_0^2}$$

Clearly, $x(t) = \Re[z(t)]$, where $\Re$ stands for the real part of the argument, is the desired solution. Hence, the particular solution of the oscillator is

$$x_p(t) = \Re \frac{(F_0/m)\exp{(i\omega_f t)}}{-\omega_f^2 + 2i\gamma\omega_f + \omega_0^2} = \frac{(F_0/m)\cos(\omega_f t - \theta)}{\sqrt{(\omega_0^2 - \omega_f^2)^2 + 4\gamma^2\omega_f^2}},$$

where $\tan\theta = \dfrac{2\gamma\omega_f}{\omega_0^2 - \omega_f^2}.$ (14.10)

The homogeneous solution, called transients, gets damped in several time periods of the oscillator. Therefore, for $t \gg 2\pi/\omega_0$

$$x(t) \approx x_p = \frac{(F_0/m)\cos{(\omega_f t - \theta)}}{\sqrt{(\omega_0^2 - \omega_f^2)^2 + 4\gamma^2\omega_f^2}}$$

$$= \frac{F_0\cos{(\omega_f t - \theta)}}{m\omega_0^2}\frac{1}{\sqrt{(\omega_r^2 - 1)^2 + 4\omega_r^2\gamma_r^2}},$$

where $\omega_r = \omega_f/\omega_0$ and $\gamma_r = \gamma/\omega_0$. This is the steady-state solution of the oscillator. The amplitude of the oscillator is $A = F_0/[(m\omega_0^2)\sqrt{(\omega_r^2 - 1)^2 + 4\omega_r^2\gamma_r^2}]$ which is finite, in contrast to that for a undamped-forced oscillator where the amplitude increases linearly with time.

We will analyse x_p for weak damping $\gamma \ll \omega_0$. Figure 14.4 shows a plot of the amplitude $A/(F_0/m\omega_0^2)$ and phase θ vs. ω_f. For low driving frequency ($\omega_f \ll \omega_0$), the amplitude $A \approx F_0/(m\omega_0^2)$ and $\theta \approx 0$, i.e., the response is in phase with the forcing. For high driving frequency ($\omega_f \gg \omega_0$), the amplitude $A \approx F_0/(m\omega_f^2)$ and $\theta \approx \pi$; hence, the response and the forcing have a phase difference of π.[1] At resonance $\omega_f = \omega_0$, the phase is $\pi/2$, and the amplitude $F_0/(2m\gamma\omega_0)$.

The maxima of the amplitude A occurs when

$$f(\omega_r) = (\omega_r^2 - 1)^2 + 4\omega_r^2\gamma_r^2$$

is a minima, which is at

$$\omega_{r0} = \sqrt{1 - 2\gamma_r^2}$$

with $f_{\min} = 4\gamma_r^2(1 - \gamma_r^2) \approx 4\gamma_r^2$. Consequently the maximum value of amplitude A is

$$max(A) = \frac{F_0}{2m\omega_0^2\sqrt{f_{\min}}} \approx \frac{F_0}{2m\omega_0^2\gamma_r}.$$

---

[1] The above phenomena can be nicely illustrated by an oscillator made up of a rubber band and a key chain (attached at the bottom of the rubber band). This system behaves as a forced oscillator if we shake the base of the oscillator vertically up and down (Exercise 12). When the forcing frequency is small, the forcing function and displacement of the key chain are approximately in the same phase. However for large forcing frequency, these two oscillations are out of phase. The reader is encouraged to do this experiment himself/herself.

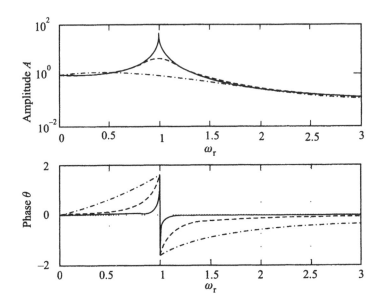

Figure 14.4 *Plot of amplitude* $A/(F_0/m\omega_0^2)$*(upper plot) and phase (lower plot) of a damped-forced oscillator as a function of* ω_r*. The solid, dashes, and chained lines represent amplitudes and phases for* $\gamma_r = 0.01$ *(solid), 0.1 (dashed), and 0.5 (chained) respectively.*

We measure the width Δ_1 of the half maximum of A using the definition $A(\omega_f - \Delta_1/2) = max(A)/2$. Taylor expansion of f yields

$$f(\omega_{r0} + \frac{\Delta_{1,r}}{2}) \approx f(\omega_{r0}) + \frac{\Delta_{1,r}^2}{8} f''(\omega_{r0}) + ...,$$

or $4f(\omega_{r0}) \approx f(\omega_{r0}) + \Delta_{1,r}^2.$ (14.11)

The solution of Eq. (14.11) yields $\Delta_{1,r} = \gamma_r 2\sqrt{3}$ or $\Delta_1 = \Delta_{1,r}\omega_0 \approx \gamma 2\sqrt{3}$. Hence the width Δ_1 is proportional to the frictional coefficient. For small frictional forces ($\gamma \ll \omega_0$), the relative width Δ/ω_0 is rather small. The half-width of an oscillator is a measure of sharpness of the resonance. Hence, sharper resonance is observed for smaller γ_r.

We can also analyse the power absorption by the forced-damped oscillator.

14.4.2   Power absorption

Power input to the oscillator by the external force is

$$P = F\dot{x} = -F_0 \cos(\omega_f t) A\omega_f \sin(\omega_f t - \theta)$$

$$= -\frac{AF_0\omega_f}{2} [\sin(2\omega_f t - \theta) - \sin\theta]$$

We take the time average of P over a time period of the forcing function. Since $\langle \sin x \rangle = 0$, the average value of P will be

$$\langle P \rangle = \frac{AF_0\omega_f}{2}\sin\theta$$

$$= \frac{F_0^2}{4\gamma m}\frac{4\gamma^2\omega_f^2}{(\omega_0^2 - \omega_f^2)^2 + 4\gamma^2\omega_f^2}$$

$$\langle P \rangle = \frac{F_0^2}{4\gamma m}\frac{4\gamma_r^2\omega_r^2}{(1 - \omega_r^2)^2 + 4\gamma_r^2\omega_r^2}$$

In Fig. 14.5 we plot $\langle P \rangle/(F_0^2/4\gamma m)$ as a function ω_r. $\langle P \rangle$ peaks at $\omega_r = 1$ or $\omega_f = \omega_0$, which is the resonance condition. Note that at resonance, the position x_p lags behind the force by a phase difference of $\pi/2$, but the velocity is in phase with the force. Consequently, the external force pumps energy into the oscillator most efficiently. The maximum value of P is $P_{\max} = F_0^2/(4\gamma m)$; for other values of ω_f, there is only a partial absorption of energy. For $\omega_f \ll \omega_0$ or $\omega_f \gg \omega_0$, the energy absorption is negligible. Note that for a frictionless oscillator, $\langle P \rangle$ is zero except at the resonance.

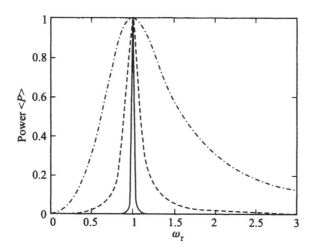

Figure 14.5 *Plot of normalised power absorbed by an oscillator $\langle P \rangle/(F_0^2/4\gamma m)$ as a function of ω_r. The solid, dashes, and chained lines represent power for $\gamma_r = 0.01$, 0.1, and 0.5 respectively.*

Let us look at the energy balance. At the steady state, the energy input from the external force must be balanced by the frictional losses, which is

$$D = 2m\gamma\dot{x}^2$$

$$= 2m\gamma A^2\omega_f^2\cos^2(\omega_f t - \theta)$$

Since $\langle \cos^2(\omega t - \theta)\rangle = 1/2$, we obtain

$$\langle D \rangle = m\gamma\omega_f^2 \frac{(F_0/m)^2}{(\omega_0^2 - \omega_f^2)^2 + 4\gamma^2\omega_f^2}$$

which is the same as $\langle P \rangle$. Hence, the energy fed by the external force gets dissipated by the frictional force, and there is no gain or loss of energy by the oscillator. This is consistent with the fact that the oscillator is in a steady state.

14.4.3   Resonance width and Q factor

In Section 14.4.1 we computed the half-width of an oscillator resonance using the oscillator amplitude. We can also quantify the resonance half-width in terms of power absorbed by the oscillator. Let us compute the resonance width Δ_2 using $P(\omega_f + \Delta_2/2) = P_{\max}/2$. The required condition is

$$(\omega_0^2 - \omega_f^2)^2 + 4\gamma^2\omega_f^2 = 8\gamma^2\omega_f^2$$

or $\omega_f^2 - \omega_0^2 = \pm 2\gamma\omega_f.$

For small γ, the above condition translates to

$$\omega_f - \omega_0 \approx \pm\gamma.$$

Therefore, at $\omega_f = \omega_0 \pm \gamma$, the absorption reduces to half of the maximum value. The width at the half maximum of the absorption curve of Fig. 14.5 is $2\gamma/\omega_0$. The damping constant thus determines the width of the absorption line.

We define an important quantity called Q *factor* or *quality factor* that characterises the resonance property of an oscillator. The quality factor or Q factor is defined as the ratio of the resonance frequency to the resonance width:

$$Q = \frac{\omega_0}{\Delta_2} = \frac{\omega_0}{2\gamma}.$$

A large Q (lower γ) indicates a narrow resonance. Large Q oscillators are more accurate and better tuned. Typical mechanical systems have $Q = 10^3$ and antennas have $Q = 10^7$.

There is another way to define the Q factor. For a damped oscillator, it is defined as π times the ratio of decay time and the period of the oscillator, which is $\omega_0/(2\gamma)$. Clearly both the definitions are equivalent.

In the following discussion, we will discuss two examples of forced-damped oscillations.

14.4.4   LRC circuit

Let us consider an LRC circuit driven by an AC drive as shown in Fig. 14.6. The equation for the circuit is

$$L\frac{dI}{dt} + RI + \frac{Q}{C} = V_0 \cos(\omega_f t). \tag{14.12}$$

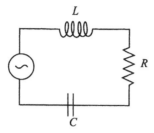

Figure 14.6   *LRC circuit*

The general solution of the equation is a sum of homogeneous and particular solutions. The homogeneous solution however dies down in the time scale of L/R, and only the particular solution survives. To compute the particular solution we use the same trick as described in Section 14.4.1. The external voltage is the real part of $V_0 \exp{(i\omega_f t)}$. Let us try a particular solution of the form

$$I(t) = \Re(I_0 \exp{(i\omega_f t)}).$$

So

$$Q(t) = \int I dt = \Re \frac{I_0}{i\omega_f} \exp{(i\omega_f t)}.$$

The substitution of the above form of $I(t)$ and $Q(t)$ in Eq. (14.12) yields

$$I(t) = \Re \frac{V_0 \exp{(i\omega_f t)}}{R + (i\omega_f L + \frac{1}{i\omega_f C})}$$

$$= \frac{V_0 \cos{(\omega_f t - \theta)}}{\sqrt{R^2 + (\omega_f L - \frac{1}{\omega_f C})^2}},$$

with

$$\tan \theta = \frac{\omega_f L - \frac{1}{\omega_f C}}{R}.$$

We call $R + (i\omega_f L + \frac{1}{i\omega_f C})$ as the complex impedance Z. The impedance of the resister, inductor, and capacitor are R, $i\omega_f L$, and $1/(i\omega_f C)$ respectively. The current lags behind the voltage by an angle θ.

Differentiation of the above equation also yields

$$L \frac{d^2 I}{dt^2} + R \frac{dI}{dt} + \frac{I}{C} = -V_0 \omega_f \sin{(\omega_f t)}.$$

This equation is analogous to Eq. (14.9) when we map $R/L = 2\gamma$, $1/(LC) = \omega_0^2$, and $V_0\omega_f/L = F_0/m$. The inductor provides inertia, the resistance provides dissipation, and the capacitor provides the restoring force.

LRC circuits will be discussed thoroughly in your electrical engineering course.

14.4.5   Electrons oscillating in an electromagnetic field

When we apply an electromagnetic field to a medium, typically the motion of an electron in the medium can be described by the following equation:

$$m\frac{d^2y}{dt^2} + m\gamma\frac{dy}{dt} + m\omega_0^2 y = qE_0 \cos\left(\omega_f t\right)$$

where y is the position of the electron, e is the charge of the electron, and E_0 and ω_f are the amplitude and frequency of the external electric field respectively. This equation is analogous to Eq. (14.9) described before. We can write down the solution using the procedure described above. The details of the solution will be discussed in your electromagnetic theory course.

EXAMPLE 14.2   Two pendulums of equal length l and bob mass m are connected together by a spring (spring constant k). Find the frequencies of oscillations of the system. Solve for the coordinates of the pendulum if the initial conditions are (1) $\theta_1(0) = 0.1$, $\theta_2(0) = -0.1$, and zero velocities; (2) $\theta_1(0) = \theta_2(0) = 0.1$ and zero velocities, and (3) $\theta_1(0) = 0.1$, $\theta_2 = 0$, and zero velocities. Interpret the above results. What happens when $g = 0$ in case (3).

SOLUTION   Let the (anti-clockwise) angular displacements of the two pendulums be θ_1 and θ_2. We assume these angles to be small. Under this condition, the equations of motion for the horizontal direction are

$$ml\ddot{\theta}_1 = -mg\theta_1 - kl(\theta_1 - \theta_2), \tag{14.13}$$

$$ml\ddot{\theta}_2 = -mg\theta_2 - kl(\theta_2 - \theta_1). \tag{14.14}$$

Addition and subtraction of the above equations yields

$$\ddot{\zeta} = -\frac{g}{l}\zeta,$$

$$\ddot{\phi} = -\frac{g}{l}\phi - \frac{2k}{m}\phi,$$

where $\zeta = \theta_1 + \theta_2$ and $\phi = \theta_1 - \theta_2$. The solutions of the above equations are

$$\zeta(t) = A\cos\left(\omega_0 t\right) + B\sin\left(\omega_0 t\right),$$

$$\phi(t) = C\cos\left(\omega_1 t\right) + D\sin\left(\omega_1 t\right),$$

with $\omega_0 = \sqrt{g/l}$ and $\omega_1 = \sqrt{(2k/m) + (g/l)}$. The solution of the pendulum for various initial conditions are given below.

1. The initial condition $\theta_1(0) = 0.1, \theta_2(0) = -0.1$ and zero initial velocity yields $A = B = D = 0$, and $C = 0.2$. Therefore,

$$\phi(t) = 0.2\cos(\omega_1 t), \quad \zeta(t) = 0.$$

In terms of θ_1 and θ_2 the solution is

$$\theta_1 = 0.1\cos(\omega_1 t),$$

$$\theta_2 = -\theta_1.$$

The bobs have a phase difference of π. The CM remains fixed, and masses move either towards each other or away from each other as shown in Fig. 14.7(a).

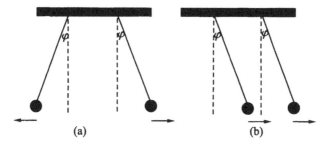

(a)   (b)

Figure 14.7   *Sketch of different types of motion exhibited by the spring–pendulum of Example 14.2: (a) the pendulums have a phase difference of π; (b) pendulums are in phase with each other.*

2. When $\theta_1(0) = \theta_2(0) = 0.1$ and $v_1(0) = v_2(0) = 0$, the constants are $B = C = D = 0$ and $A = 0.2$. Hence $\zeta(t) = 0.2\cos(\omega_0 t)$ and $\phi(t) = 0$. Consequently $\theta_1 = \theta_2$. The solutions in terms of θ_1 and θ_2 are

$$\theta_1(t) = \theta_2(t) = 0.1\cos(\omega_0 t).$$

Pictorially the motion is illustrated in Fig. 14.7(b).

3. For the initial condition $\theta_1(0) = 0.1, \theta_2(0) = 0$ and $\dot\theta_1(0) = \dot\theta_2(0) = 0$, $B = D = 0$ and $A = C = 0.1$. Hence,

$$\zeta(t) = 0.1\cos(\omega_0 t),$$

$$\phi(t) = 0.1\cos(\omega_1 t).$$

Therefore the solution in terms of θ_1 and θ_2 are

$$\theta_1(t) = 0.05(\cos(\omega_0 t) + \cos(\omega_1 t)),$$

$$\theta_2(t) = 0.05(\cos(\omega_0 t) - \cos(\omega_1 t)).$$

If $g = 0$, then $\omega_0 = 0$ and $\omega_1 = \sqrt{2k/m}$. In this case the motion of the pendulums is the same as that of a double mass connected to a spring. For the initial condition (3), the solution is

$$\theta_1(t) = 0.1 \cos^2(\omega_1 t/2),$$

$$\theta_2(t) = 0.1 \sin^2(\omega_1 t/2).$$

The solution $\theta_{1,2}$ have been plotted in Fig. 14.8. When θ_1 is fully stretched, $\theta_2 = 0$, and vice versa.

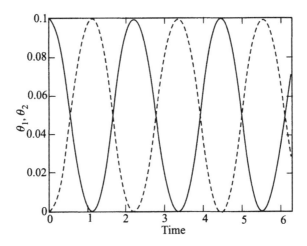

Figure 14.8   *Plot of $\theta_1(t)$ (solid line) and $\theta_2(t)$ (dashed line) as a function of t for initial condition (3) of Example 14.2.*

EXAMPLE 14.3   A bead of mass m is constrained to move on a hoop of radius b. The hoop rotates with constant angular velocity Ω around a vertical axis which coincides with the diameter of the hoop. Neglect friction between the hoop and the bead.

1. Write down the equation of motion.

2. Find the equilibrium positions. Determine their nature.

SOLUTION   We solve the problem in the rotating frame. The free-body diagram of the bead is shown in Fig. 14.9(a). Here C is the centrifugal force. The Coriolis force is perpendicular to the plane of the hoop, and it is balanced by the normal force exerted by the hoop. The equation of motion for the bead along the θ direction is

$$m b \ddot{\theta} = C \cos\theta - mg \sin\theta$$

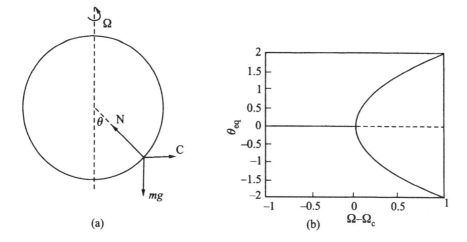

Figure 14.9   *(a) A bead of mass m moving on the hoop. (b) Plot of θ_{eq} as a function of Ω; the solid limes represent the stable solution and the dashed line represents the unstable solution.*

or $a\ddot{\theta} = -g\sin\theta + a\Omega^2 \sin\theta\cos\theta$.   (14.1.5)

Along the radial direction

$$m\dot{\theta}^2 R = N - mg\cos\theta - C\sin\theta$$

When the particle is in equilibrium, the net forcing acting on it is zero. Hence at equilibrium position the angular acceleration is zero. Using Eq. (14.15) we obtain equilibrium points as

$$\theta_{eq} = 0, \text{ or } \cos\theta_{eq} = \frac{g}{a\Omega^2}.$$ (14.16)

The first solution $\theta_{eq} = 0$ is the bottom-most point. Since $\cos\theta \leq 1$, the second solution appears only when $\Omega \geq \Omega_c = \sqrt{g/a}$. Now we will investigate the nature of these equilibrium points.

Expanding Eq. (14.15) near $\theta = 0$ we obtain

$$a\ddot{\theta} = -\theta[g - a\Omega^2],$$

which indicates that $\theta = 0$ is a stable equilibrium point for $\Omega < \Omega_c$, and the particle executes SHM with frequency $\sqrt{(g - a\Omega^2)/a}$. For $\Omega > \Omega_c$, the equilibrium point $\theta_{eq} = 0$ becomes unstable. To study the nature of stability of $\theta_{eq} = \cos^{-1}[g/(a\Omega^2)]$, we substitute $\theta = \theta_{eq} + \phi$ in Eq. (14.15) which yields

$$a\ddot{\phi} = -a\Omega^2[\sin^2\theta_{eq}]\phi.$$

Hence the bead performs SHM near $\theta_{eq} = \cos^{-1}[g/(a\Omega^2)]$ with frequency $a\Omega^2 \sin^2\theta_{eq}$.

It is interesting to observe the variation of θ_{eq} with Ω. Denoting $\Omega = \Omega_c + \Delta\Omega$ and expanding Eq. (14.16) near $\theta_{eq} = 0$ (for $\Omega = \Omega_c$) we obtain

$$\theta_{eq}^2 = \frac{4\Delta\Omega}{\Omega_c}.$$

Hence θ_{eq} increases gradually from 0. In Fig. 14.9(b) we illustrate the stable equilibrium point by a solid line and the unstable equilibrium point by a dashed line.

In summary, for frequencies $\Omega < \Omega_c$, the bead performs SHM near $\theta = 0$. At $\Omega = \Omega_c$, a new stable equilibrium point appears. For $\Omega \geq \Omega_c$, the bead performs SHM near $\theta = \theta_c$.

We end our discussion on SHM here.

Exercises

1. Review Exercises 4 and 5 from Chapter 5.

2. Compute the average power input to the forced-oscillator by an external forcing for both damped and undamped cases? When is the power input by the external forcing to the oscillator maximum?

3. A dissipation-less linear spring is forced using the following forces. Derive an expression for the amplitude of the oscillator whose initial condition is $(x(0), v(0))$ for each of these conditions.

   (a) $F = F_0$ for $t > 0$ and zero for $t < 0$.

   (b) $F = at$ for $t > 0$ and zero for $t < 0$.

4. Describe the motion of a spring–mass system in an incline (see Fig. 14.10(a)). The left end of the spring is fixed at the end of the wedge, which is fixed on the ground. What is the frequency of oscillation?

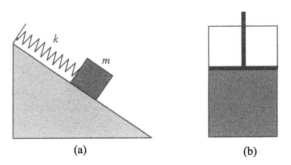

(a)   (b)

Figure 14.10   *(a) Exercise 4; (b) Exercise 7.*

5. Imagine a tunnel from one end of the Earth to the other end going through the centre of the Earth. Compute the potential of a particle that is moving in this tunnel. If this particle is released in the tunnel, what would be its motion?

6. Sketch the trajectory of a two-dimensional oscillator for which $\omega_x/\omega_y = 1, 2, \sqrt{2}$. Make the simplest assumptions on the initial condition. Is the motion periodic? Derive a general condition for the motion of a two-dimensional oscillator to be periodic.

7. A cylinder contains gas at a constant temperature T, and the gas is covered by a piston of mass m as shown in Fig. 14.10(b). The cross-section area of the cylinder is A. Describe the motion of the piston when the piston is displaced slightly from its equilibrium position. Neglect friction between the piston and the wall.

8. Two immovable charges of magnitude $+Q$ are placed at a distance R apart. A test charge $+q$ of mass m is moved around.

    (a) Find the position(s) of equilibrium point(s).

    (b) Compute the potential near the equilibrium point. Keep terms only up to the second order.

    (c) What is the nature of the equilibrium point? Describe the motion of the test particle.

    (d) Do the same analysis for negative q.

9. Four immovable charges of magnitude $+Q$ are placed at the vertices of a square. Do the same analysis as in Exercise 8.

10. A particle is moving under the influence of a central potential $U(r) = -A/r^n$, where $A > 0$.

    (a) Find circular orbits for this potential. (Recall the discussion in Chapter 8)

    (b) Classify the stability of these circular orbit in terms of n.

11. Describe the motion of a charged particle with charge q and mass m whose initial condition is $\mathbf{r}(0) = 0$ and $\mathbf{v}(0) = v_0\hat{\mathbf{x}}$, and is subjected to an external electric field $\mathbf{E} = E_0 \cos(\omega t)\hat{\mathbf{x}}$.

12. A mass of 50 gm (e.g., a small key) is suspended from a support with a rubber band of spring constant 0.1 Nm^{-1}.

    (a) Describe the nature of oscillation of the above system.

    (b) The support of the rubber band starts vibrating up and down harmonically with a frequency of 1 Hz and an amplitude of 1.0 cm. Write down the equation of motion for the mass. Describe the motion of the mass.

13. List practical examples of simple harmonic oscillations.

14. Compute the spring constant of a rubber band and spring using an appropriate experiment. What is the advantage of having the spring curled instead of being straight?

Projects

1. Learn about the suspensions in a car.

2. Design a tuner for a medium-wave radio.

3. How do mechanical clocks work? Refer to Wikipedia.

15

Nonlinear Oscillations and Chaos

In the last chapter we discussed the motion of unforced and forced oscillators. In all of these cases, the response ($x(t)$) was proportional to the forcing (stimulus). Therefore, these oscillators are said to be *linear*. The highest power of $x, \dot{x}, \ddot{x}$ in these equations of the oscillator were one, and the corresponding potentials were quadratic. However when the response becomes significantly large, the higher order terms start to become significant, making the equation of motion *nonlinear;* here the highest power of one or more of the variables $x, \dot{x}, \ddot{x}$ is more than one. You can easily verify that the superposition principle does not hold for nonlinear equations (Exercise 1).

The study of nonlinear systems forms a vast subject called nonlinear physics and chaos. Nonlinear equations are quite complex to solve. Most of the time they do not have analytical solutions, and we need to resort to numerical solutions. It is impossible to introduce the field of nonlinear oscillations in just a few pages. Here we will only illustrate the richness of the field by several simple examples.

15.1 Driven damped pendulum

A pendulum has a massless rod of length l and a bob of mass m. The frictional force acting on the bob is bv, where v is the velocity of the bob. The pendulum is driven by a force $F\cos(\omega t)$. We denote the counter-clockwise displacement of the pendulum by ϕ. The equation of the pendulum is

$$ml\ddot{\phi} + bl\dot{\phi} + mg\sin\phi = F\cos(\omega t). \tag{15.1}$$

The normalisation of the above equation yields

$$\ddot{\phi} + 2\gamma\dot{\phi} + \omega_0^2 \sin\phi = A\omega_0^2 \cos(\omega t) \tag{15.2}$$

where $b/m = 2\gamma$, $l/g = \omega_0^2$, and $F/ml = (F/mg)(g/l) = A\omega_0^2$.

Equation (15.2) is nonlinear, and does not have an analytic solution. We will study this problem by solving Eq. (15.2) numerically using a computer. The method is described in Chapter 5. We solve Eq. (15.2) for various values of forcing amplitudes A. For our study we will take $\omega = 2\pi$ (time period equal to 1), $\omega_0 = 1.5\omega$, $\gamma = \omega_0/4$, and the initial condition as $\phi = \dot{\phi} = 0$ at $t = 0$.

For small A, the amplitude of the pendulum is small, and we can approximate $\sin\phi \approx \phi$. Under this condition, the equation of the pendulum is linear and the solution was studied in Section 14.4.1. The motion is simple harmonic till A is around 0.2. For large A (till around 1), the motion remains periodic, but is not simple harmonic. The amplitude is a superposition of $\sin(n\omega t)$ $(n \geq 1)$.

Interesting patterns emerge when we increase A even further. For $A = 1.5$, the amplitude of the pendulum as a function of time is shown in Fig. 15.1(a).

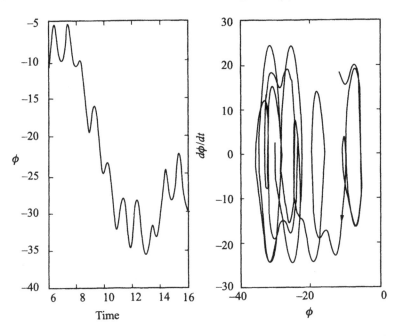

Figure 15.1   *The left figure is a plot of $\phi(t)$ vs. t, while the right figure is a plot of ϕ vs. $\dot{\phi}$ for $A = 1.5$. The other parameters are $\omega = 2\pi$, $\omega_0 = 1.5\omega$, $\gamma = \omega_0/4$.*

In Fig. 15.1(b), we plot the phase space $[(\phi, \dot{\phi})$ space] trajectories. These trajectories show that the motion of the pendulum is random. Technically this motion is called *chaotic*. Before the pendulum becomes chaotic, it exhibits some interesting motion, but the features of this motion are quite technical. We refer the reader to standard books on chaos (Strogatz 1994; Taylor 2005).

Chaotic systems have many interesting features. We will discuss one of them here. Let us observe the motion of two pendulums which start with slightly different initial conditions. For example, let us take the initial condition to be (a) $[\phi = 0.5, \dot{\phi} = 0.0]$ and (b) $[\phi = 0.5, \dot{\phi} = 0.00001]$. The evolution of ϕ for these systems are plotted in Fig. 15.2 as solid and dashed lines respectively.

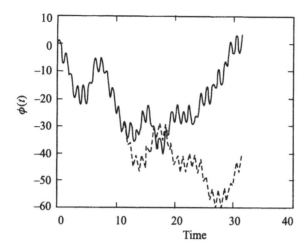

Figure 15.2   *Plot of $\phi(t)$ vs. t for $A = 1.5$ and two different initial conditions: (a) $[\phi = 0.5$, and $\dot{\phi} = 0.0$] (solid line) and (b) $[\phi = 0.5; \dot{\phi} = 0.00001$] (dashed line). The other parameters are $\omega = 2\pi$, $\omega_0 = 1.5\omega$, $\gamma = \omega_0/4$. We observe that the future for case (a) and (b) are very different.*

We find that the two solutions are close for a short while, after which they start to deviate strongly. After $t \approx 10$, the evolution for the two cases are very different. This property is a generic feature of chaotic systems, and is called *sensitivity to initial conditions*.

Even if we bring the two initial conditions closer, say (a) $[\phi = 0.5, \dot{\phi} = 0.0]$ and (b) $[\phi = 0.5; \dot{\phi} = 0.0000001]$, the two solutions deviate from each other. However the deviations starts a bit later. Hence, if two chaotic systems differ even by a small amount in the beginning, then the configuration of the systems will be very different after a finite time. This feature is seen for all initial conditions however close they might be.

In the next section we will describe another system called a *double pendulum* which exhibits chaos.

15.2   Double pendulum

In this section we will describe an experimental system called a *double pendulum* which shows chaos. The system is exhibited in Fig. 15.3. There are two double pendulums side by side. The top pendulum consists of two aluminum bars hinged at a horizontal rod. The two bars are joined together at the bottom by a hinge; a single aluminum bar hangs freely from the bottom hinge. Both the hinges are free. We study the double pendulum for various initial conditions.

For small displacements of the pendulum (the angles of both the pendulums are less than 60 degrees or so), the motion is periodic and regular. However when the displacement is large (e.g. greater than $\pi/2$), then the motion of the bar become chaotic. There is no periodicity in the motion of the pendulums. If we test the sensitivity to initial conditions of the double pendulum by starting the two double pendulums from the vertical position, we find that the state of both the pendulums is very different after a short time. This result shows that the motion of a double pendulum is sensitive to initial conditions.[1]

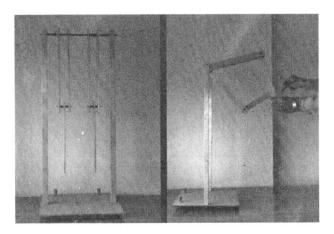

Figure 15.3   *A picture of double pendulum.*

15.3   Three-body system

In Chapter 8 we studied the motion of a planet under the influence of a central star. The orbit of the planet was found to be relatively simple and orderly (ellipse). However the situation is very different for a planet under the influence of a binary star. This system is called a three-body system, and has been studied by many physicists and mathematicians. The trajectory of the planet for most initial conditions is known to be chaotic similar to the trajectories of the driven pendulum or double pendulum. The system is also quite sensitive to initial conditions; a slight change in the initial condition of the planet changes the future of the planet dramatically. For some initial conditions, the planet may escape to infinity. These results show that a three-body system is much more complex than the two-body system studied in Chapter 8. The interested student can refer to Wikipedia for further details and references.

The above systems show chaos for a range of parameters. Scientists have found many other systems that show chaos. Some prominent examples are nonlinear oscillators, population dynamics of a species, heart, neurons, stock markets etc.

Chaotic systems exhibit unexpected and nontrivial behaviour. Sensitivity to initial conditions is of major relevance to the predictions of physical systems. Recall that according to Newton's laws of motion, one can predict the future of a mechanical system given its dynamical equations and its initial condition. Around 300 years back scientists believed that the whole universe is deterministic, hence we can predict the future of every system including our universe. Laplace conjectured that our lives are governed by physical laws,

---

[1]Refer to the website http://brain.cc.kogakuin.ac.jp/~kanamaru/Chaos/e/ for simulation and video display of a double pendulum.

hence we cannot have *free will*. In the nineteen fifties, John von Neumann argued that computers would be able to perform long-term weather predictions, and we could accordingly plan our vacation. Chaos theory in some sense demolishes these conjectures because of the sensitivity to initial conditions in chaotic systems.

Many physical systems including the weather system is known to be chaotic. For the future prediction of these systems, we record their initial conditions to a certain accuracy. These initial conditions are fed into a computer that solves the equation of motion of these systems. The recorded value always differs from the real value by a small value, however accurate our measurement may be. This small error in the initial conditions will get amplified after a short time, hence the computer solution will differ from the real system after a short time. This is the reason why long-term predictions of chaotic systems is impossible. As an example, even the best available computer of today cannot predict the weather for more than five days.

Here we will end our introductory discussion on chaos. Chaos is a frontier area of physics research with many interesting and tantalising results. The interested reader can refer to Strogatz (1994) and Taylor (2005).

Exercises

1. Consider a nonlinear equation

$$\ddot{x} + x + x^3 = 0.$$

If $x_1(t)$ and $x_2(t)$ are the two independent solutions of the above equation, then show that $c_1 x_1(t) + c_2 x_2(t)$ does not satisfy the above equation for an arbitrary set of constants c_1 and c_2. This exercise illustrates that the linear superposition principle does not hold for general nonlinear equations.

2. A quadratic map is a famous example of a chaotic system. In this example, the system is described by a number $x_n \in [0, 1]$, and the evolution of the system is given by

$$x_{n+1} = f(x_n) = a x_n (1 + x_n).$$

   (a) Plot $f(x_n)$ and $x_{n+1} = x_n$. The intersection points x^* are called the *fixed points*. Show that
   $$x^* = f(x^*).$$

   (b) Show using numerical simulation or analytic calculation that for $0 \le a < 1$, $x_n \to x^* = 0$ as $n \to \infty$ irrespective of initial conditions. You will find the convergence to be quite fast.

   (c) Show that for $1 \le a < 3$, $x_n \to x^* = (1/a) - 1$ as $n \to \infty$.

   (d) Show that for $3 \le a \le 3.449$, x_n jumps between two values for large n. This phenomena is called period doubling.

(e) Show that for $3.449 \leq a \leq 3.54$, x_n jumps between four values for large n. This is called period 4. If you increase a a bit more, you will see period 8 and period 16. The window of a however decreases for higher periods.

(f) For a greater than 3.57, x_n is random. This state is called a chaotic state.

(g) In the chaotic region, check sensitivity to initial conditions by exploring the future of the system from two nearby initial conditions.

3. Explore the dynamics of a driven-damped pendulum described in this chapter near $A = 1.1$ $\omega = 2\pi$, $\omega_0 = 1.5\omega$, $\gamma = \omega_0/4$. You may observe period doubling.

Projects

1. Study systems that exhibit chaos. Some well-known examples are Duffing oscillator, double pendulum, nonlinear LRC circuit, and three-body problem.

16

Waves: Oscillations in Continuous Media

We have studied oscillations in lumped systems, where kinetic and potential energies are stored in different elements. For example, in a spring–mass system, the mass contains the kinetic energy and the spring contains the potential energy. In this chapter we will study oscillations in a continuous medium where each element of the medium contains both kinetic and potential energy. For small disturbances in the medium, oscillations can be expressed in terms of simple harmonic motion. We will illustrate this phenomena by two examples: (1) longitudinal waves in a solid medium, and (2) transverse waves in a string.

16.1   Longitudinal waves in a solid medium

We will formulate the wave motion in a solid medium (say a rod) by making an analogy with the dynamics of an infinite chain of a spring–mass system (Fig. 16.1). We assume that all the blocks have the same mass m, and all the springs have spring constant k. The displacement of the ith mass is denoted by ζ_i. The force on the ith mass due to the left and right springs are $k(\zeta_{i+1} - \zeta_i)$ and $k(\zeta_i - \zeta_{i-1})$ respectively. Therefore, the equation of motion of the ith mass is

$$m\ddot{\zeta}_i = k(\zeta_{i+1} - \zeta_i) - k(\zeta_i - \zeta_{i-1}). \tag{16.1}$$

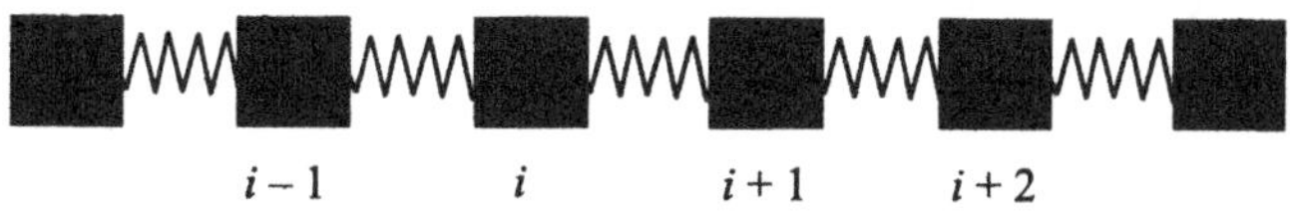

$$i-1 \qquad i \qquad i+1 \qquad i+2$$

Figure 16.1   *An infinite chain of a spring–mass system.*

The equations for the other masses have a similar form. By a somewhat complex algebra it can be shown that if the amplitude of disturbances of ζ_is are small compared to the length of the springs, the oscillations of the mass are harmonic. When ζ_is become large, higher terms like $(\zeta_{i+1} - \zeta_i)^2$ appear in the right-hand side of Eq. (16.1), and the harmonic functions are no more solutions of the equation. We make an analogy of the waves in the spring–mass system with that in a solid rod.

A solid rod is not a chain of mass-spring systems. Each element of the rod has both elastic and inertial properties, so it can simultaneously take the role of spring and mass. We will divide the rod into small segments of size Δx each as shown in Fig. 16.2.

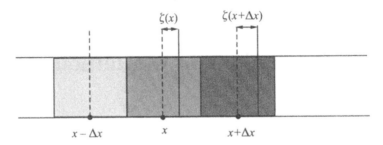

Figure 16.2   *Wave in a solid rod.*

Let us focus our attention on the centres of the segments whose positions are $(x-\Delta x)$, x, $(x+\Delta x)$, ... in the equilibrium position. When we apply a small disturbance to the rod, the material at point x moves to $x + \zeta(x)$, hence the displacement at the point x is $\zeta(x)$. Similarly the displacements at points $(x - \Delta x)$ and $(x + \Delta x)$ are $\zeta(x - \Delta x)$ and $\zeta(x + \Delta x)$ respectively. These displacements cause compression and stretch on the elements due to the elastic properties of the material. These forces act on the shaded region that has mass $\rho A \Delta x$, where ρ is the density of the material, and A the cross-section area of the rod. We can write the equation of the shaded region as

$$\rho A \Delta x \frac{\partial^2}{\partial t^2} \zeta(x) = k(\zeta(x + \Delta x) - \zeta(x)) - k(\zeta(x) - \zeta(x - \Delta x)),$$

where k is corresponding spring constant of the rod. Using Taylor's expansion, the above equation simplifies to

$$\rho A \Delta x \ddot{\zeta}(x) = k \left\{ \left[\frac{\partial \zeta}{\partial x} \right]_{x+\Delta x} - \left[\frac{\partial \zeta}{\partial x} \right]_x \right\} \Delta x$$

or

$$\frac{\partial^2 \zeta}{\partial t^2} = \frac{Y}{\rho} \frac{\partial^2 \zeta}{\partial x^2} \qquad (16.2)$$

where $Y = k\Delta x/A$ is the Young's modulus of the rod. The above equation is called *the wave equation*. Since the oscillation is along the length of the rod, these types of waves are called *longitudinal waves*. Note that sound propagates in the medium through this wave.

We will try out the following solution for Eq. (16.2)

$$\zeta(x, t) = A \sin(kx - \omega t). \qquad (16.3)$$

Substitution of this solution in Eq. (16.2) yields

$$\omega^2 = \frac{Y}{\rho}k^2.$$

Hence $\omega/k = \sqrt{Y/\rho} = C_s$ is the speed of the wave in the rod. A disturbance in the medium travels with this speed. We prove this result using the following argument.

At time t, $\zeta(x,t)$ (Eq. (16.3)) is shown by the solid line in Fig. 16.3. We focus on position x where the wave has maximum amplitude (point A in the figure). After a small time interval Δt, the peak of the wave moves to the right, to position $(x + \Delta x)$ (point B in the figure). Since points A and B have the same phase, we have

$$kx - \omega t = k(x + \Delta x) - \omega(t + \Delta t).$$

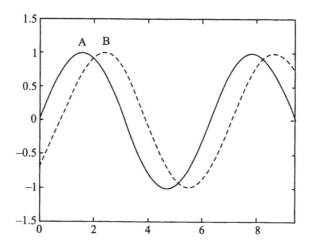

Figure 16.3   *Profile of the wave ζ at two different times t and $t + \Delta t$.*

Therefore

$$\frac{\Delta x}{\Delta t} = \frac{\omega}{k} = C_s.$$

Hence we have shown that the wave $\zeta(x,t)$ is moving to the right with speed C_s.

If we focus on any fixed point x' and observe the amplitude of the disturbance as a function of time, we find the motion to be

$$\zeta(x',t) = A\sin(kx' - \omega t),$$

a simple harmonic motion (SHM) in time with period $T = 2\pi/\omega$. Note that ω is called frequency of the wave. The motion is harmonic because of the small amplitude of the disturbances. For large amplitudes, the above approximation breaks down, and the solution typically becomes much more complex.

We will take a snapshot of the wave profile at a given time (e.g. the solid line of Fig. 16.3). How far apart are the two closest points that have the same phase? We call this distance the *wavelength*. According to Eq. (16.3)

$$k(x + \lambda) - \omega t - (kx - \omega t) = 2\pi.$$

Therefore the wavelength of the wave is $\lambda = 2\pi/k$.

After having studied waves on a solid rod, we will now study waves on strings.

16.2   Transverse waves on a string

In the previous section we discussed longitudinal waves in a solid rod. In the present section we will derive the wave equation for a stretched string whose mass density (mass per unit length) is μ, and tension is T. We will focus on a small segment of the string, and write down Newton's law for that segment.

For a small displacement of the string, the tension at all the points of the string is the same as the first approximation. Now we will focus on a small element (shown in Fig. 16.4) of the string.

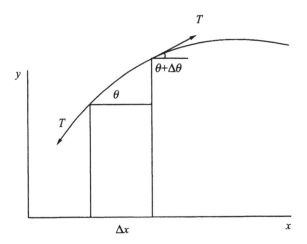

Figure 16.4   *Small element of a vibrating string.*

The size of the element is Δx, and its displacement from the equilibrium position is $\zeta(x, t)$ in the transverse direction (perpendicular to the length of the string). The net transverse force on the string element is $T \sin(\theta + \Delta\theta) - T \sin\theta$. Consequently the equation of motion is

$$(\mu\Delta x)\frac{\partial^2}{\partial t^2}\zeta(x) = T \sin(\theta + \Delta\theta) - T \sin\theta$$

For small θ, $\sin\theta \approx \theta \approx \tan\theta = d\zeta/dx$. Therefore,

$$(\mu\Delta x)\frac{\partial^2}{\partial t^2}\zeta(x) = T\left[\frac{\partial\zeta}{\partial x}\right]_{x+\Delta x} - T\left[\frac{\partial\zeta}{\partial x}\right]_x$$

or

$$\frac{\partial^2\zeta}{\partial t^2} = c^2\frac{\partial^2\zeta}{\partial x^2} \tag{16.4}$$

where $c = \sqrt{T/\mu}$ is the speed of the wave propagation on the string.
We can verify that

$$\zeta(x,t) = A\sin(kx - \omega t)$$

is a solution of the above equation with similar interpretation as in the previous example. The wave is moving to the right with speed $c = \sqrt{T/\mu}$. This solution is valid for an infinite string where a wavefront is always moving to the right.

In practical situations we encounter strings that are clamped either at both the ends or at one end. In the following discussion we will focus our attention on a string that is clamped at both the ends as shown in Fig. 16.5. At the ends $\zeta(0,t) = 0$ at all times. This boundary condition and Eq. (16.4) is satisfied by a function of the form

$$\zeta(x,t) = \sin(kx)\cos(\omega t + \phi) \tag{16.5}$$

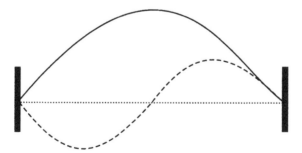

Figure 16.5   *Waves on a clamped string.*

with $\omega/k = \sqrt{T/\mu}$ and

$$\sin(kL) = 0 \quad \text{or} \quad k = \frac{n\pi}{L}.$$

We denote the above k and ω by k_n and ω_n respectively. Clearly infinitely many (k_n, ω_n) are possible given the above conditions. Therefore a general solution for the clamped string is

$$\zeta(x,t) = \sum_n A_n\sin(k_nx)\cos(\omega_nt + \phi_n)$$

where $k_n = \dfrac{n\pi}{L}$, and $\omega_n = ck_n$ (Exercise 3).

Now we come to an important point. When we pluck a string and let it go, we will never observe a pure sine wave of the form given in Eq. (16.5). The solution will always be a combination of many sine waves. We will compute the wave solution using the following procedure.

Suppose the initial configuration is a triangular form with zero velocity. In the following discussion we will solve for the configuration of the string $\zeta(t)$ at any arbitrary time. The transverse velocity of any portion of the string is given by

$$\frac{\partial \zeta(x,t)}{\partial t} = - \sum_n A_n \omega_n \sin(k_n x) \sin(\omega_n t + \phi_n).$$

At $t = 0$, $\frac{\partial \zeta}{\partial t} = 0$ for all x. Therefore $\phi_n = 0$ and the general solution with the above initial condition is

$$\zeta(x,t) = \sum_n A_n \sin(k_n x) \cos(\omega_n t). \tag{16.6}$$

Our next objective is to determine A_n, which is obtained by the initial condition:

$$\zeta(x,0) = \sum_n A_n \sin(k_n x) \tag{16.7}$$

or

$$A_n = \frac{2}{L} \int_0^L \zeta(x,0) \sin(k_n x).$$

We substitute A_n in Eq. (16.6), and that gives us the final solution. Thus we solve the time evolution of a plucked string. The final solution is a combination of many sine waves.[1] Note that waves in a finite solid rod can also be solved using the above procedure.

It is interesting to note that the $\zeta = 0$ is the stable equilibrium configuration for both the systems. The wave configuration oscillates around the stable configuration just like the pendulum bob oscillates around its equilibrium position. The major difference is that the waves are displacements in the continuous media, while the displacement of the pendulum bob is the displacement of a lumped system.

In the above two examples, the wave was transmitted in a continuous medium (rod and string respectively). In the next section we will describe electromagnetic waves that do not require a medium. Yet the propagation of electromagnetic waves have certain similarities with the waves in the rod and the string in terms of the governing equation. We present very basic features of electromagnetic waves here to contrast different scenarios in wave phenomena.

---

[1] The above solution uses a very important mathematical theorem—any continuous function $\zeta(x,0)$ can be expanded using $\sin(k_n x)$ as in Eq. (17.7)). This theorem is due to a mathematician named Fourier.

16.3   Electromagnetic waves

We can generate electromagnetic waves by shaking charged sheets. The details cannot be discussed here because to comprehend it, we would need a lot of background information which is not relevant to the rest of the text here. For certain configurations we will generate plane waves travelling perpendicular to the charged planes. If the charged sheets are along the yz-plane, and if the sheets are shaken along the y-direction in a harmonic fashion, then we have

$$\frac{\partial^2 E_y}{\partial t^2} = c^2 \frac{\partial^2 E_y}{\partial x^2},$$

$$\frac{\partial^2 B_z}{\partial t^2} = c^2 \frac{\partial^2 B_z}{\partial x^2},$$

where E_y and B_z are the electric and magnetic fields respectively, and c is the speed of light. The solution of the above equations are

$$E_y = E_{y0} \sin(kx - \omega t),$$

$$B_z = B_{z0} \sin(kx - \omega t),$$

with $\omega/k = c$. Clearly both electric and magnetic fields are perpendicular to the wave vector which is along the x-direction; hence the wave is transverse.

The above wave is one of the simplest example of an electromagnetic (EM) wave. Since the wave is moving in a plane, it is called a plane wave. Electromagnetic waves appear in many different forms. In the following, we will list some of the important features of electromagnetic waves:

1. EM waves are generated by the motion of charged particles. Stationary charge particles do not generate waves.

2. The propagation of EM waves does not require any medium. They can travel in vacuum. The light (an EM wave) from the Sun and stars travel through empty space (devoid of material particles).

3. For a plane wave, the three vectors, electric field, magnetic field, and wave number ($\mathbf{E}$, $\mathbf{B}$, and $\mathbf{k}$ respectively) are perpendicular to each other. In fact they are connected by the following relations:

$$\hat{\mathbf{k}} \times \hat{\mathbf{E}} = \hat{\mathbf{B}},$$

$$\hat{\mathbf{k}} \times \hat{\mathbf{B}} = -\hat{\mathbf{E}},$$

$$\hat{\mathbf{E}} \times \hat{\mathbf{B}} = \hat{\mathbf{k}},$$

where $\hat{\mathbf{E}}, \hat{\mathbf{B}}$, and $\hat{\mathbf{k}}$ are unit vectors along the electric field, magnetic field, and the wavevector respectively. The above relations can be derived from Maxwell's equations.

4. In the above example the electric field vector is always in the xy-plane, and the magnetic field vector is in the xz-plane. Hence these waves are called *plane polarised waves*. In many situations, the electric and magnetic field vectors circulates in the yz-plane. These waves are called *circularly polarised waves*. Matter manipulates light as it propagates through it, and brings in rich features like converting a plane polarised light to a circularly polarised light.

5. In the above example, plane EM waves were generated by a charged sheet. A vibrating point charge generates spherical EM waves.

6. In EM waves, the electric and magnetic fields help each other in the propagation of waves. Recall that tension provides the restoring force for the waves in the string. The analogous restoring forces is not that apparent in EM waves. Naturally, EM waves are more complex and richer compared to the waves on the string.

EM waves are one of the most important waves in physics and one of the most ubiquitous waves. We receive signals in our television and mobile phones through these waves. EM waves will be discussed in more detail in advanced courses in electromagnetism.

In this chapter we have discussed wave motion at a very introductory level. We encounter waves in a vast range of physical phenomena, e.g., water waves, pressure waves, tsunami, etc. These waves are discussed in advanced courses.

Exercises

1. Discuss various examples of wave motion around you.

2. Verify that

$$\zeta(x, t) = \sin(kx) \cos(\omega t - \phi)$$

is a solution of

$$\frac{\partial^2 \zeta}{\partial t^2} = c^2 \frac{\partial^2 \zeta}{\partial x^2}.$$

3. [SUPERPOSITION PRINCIPLE]   Verify that the general solution of a clamped string is

$$\zeta(x, t) = \sum_n A_n \sin(k_n x) \cos(\omega_n t + \phi_n)$$

where $k_n = n\pi/L$ and $\omega_n = ck_n$.

4. Study the following wave solutions. Here x is in metres and t is in seconds. (i) Sketch the wave at time $t = \pi/2$. (ii) What is the wavelength and frequency of the wave?

(iii) What is the speed of the wave? (iv) What is the direction of propagation of the wave? (v) Describe the motion at $x = 3$.

(a) $\sin(3x + 9t)$

(b) $\sin(3x - 6t)$

(c) $\sin(5x)\cos(10t)$

5. Write down the general solution of a one-metre string that is clamped at both ends. The initial velocity of the string $\partial\psi/\partial t(t = 0)$ is zero, and the initial configuration $\psi(t = 0)$ is a symmetric triangle with height in the middle of the string as 5 mm. Compute A_ns of Eq. (16.7) for the above string.

6. A radio station emits electromagnetic waves at 1 MHz frequency. What is the speed and wavelength of the wave? What is the nature of the wavefront?

7. Estimate the speed of waves generated in the strings of a veena.

8. Estimate the speed of longitudinal waves in a slab of air, water, iron.

17

Angular Momentum

So far we have studied the motion of a single point particle, or that of a system of particles. The equation of motion and properties of the system were characterised by energy and linear momentum. In the present chapter and the next we will study rotation of a mechanical system. We will introduce the concept of angular velocity, angular momentum, rigid body, torque etc. The equations reveal a certain analogy between angular velocity and linear velocity, angular momentum and linear momentum, and torque and force. However there are some crucial differences between these quantities.

In the present chapter we will derive relationships between the angular velocity and the angular momentum of a rigid body. This relationship is a part of rotational kinematics. We can study mechanical systems using rotational kinematics. As an example we will study the motion of a free top. But at first we will define basic quantities like angular momentum and angular velocity.

17.1 Angular momentum of a single particle

We will first define angular momentum of a particle of mass m about a point O. If the position of the particle is $\mathbf{r}$, then the angular momentum of the particle about O is given by $\mathbf{L} = \mathbf{r} \times \mathbf{p}$, where $\mathbf{p}$ is the linear momentum of the particle. Angular momentum is always defined about a point in any frame of reference, inertial or noninertial.

Let us take the time derivative of $\mathbf{L}$:

$$\frac{d\mathbf{L}}{dt} = \mathbf{r} \times \frac{d\mathbf{p}}{dt}.$$

In an inertial frame $d\mathbf{p}/dt = \mathbf{F}$, hence,

$$\frac{d\mathbf{L}}{dt} = \mathbf{r} \times \mathbf{F}.$$

We call $\mathbf{r} \times \mathbf{F}$ *torque* and denote it by $\mathbf{N}$. Hence, in an inertial frame $\frac{d\mathbf{L}}{dt} = \mathbf{N}$. *If the torque, $\mathbf{N}$, on the particle is zero, then the angular momentum $\mathbf{L}$ is a constant.* This law is called the *conservation of angular momentum*.

The force is radial in Kepler's problem. If we choose the origin of the central force as the reference point, then the torque on the planet about the origin is zero. Consequently angular momentum measured with respect to the central point is conserved for the central potentials.

17.2 Angular momentum of a system of particles

Angular momentum is an additive quantity like energy and linear momentum. Therefore, for a system of particles, the angular momentum is

$$\mathbf{L} = \sum_a \mathbf{r}_a \times \mathbf{p}_a,$$

where $\mathbf{r}_a$ and $\mathbf{p}_a$ denote the position vector and the linear momentum of the ath particle.

In earlier chapters we discussed the usefulness of the centre of mass (CM). Let us first derive the angular momentum of an object about its CM. Let us denote $\mathbf{r}_a$ and $\mathbf{p}_a$ as the position and the linear momentum of the ath particle of the body in the inertial frame I, and $\mathbf{r}'_a$ and $\mathbf{p}'_a$ as the corresponding quantities in the CM frame. In the inertial frame I, the total angular momentum of the body about its origin is

$$\mathbf{L} = \sum_a \mathbf{r}_a \times \mathbf{p}_a$$

$$= \sum_a (\mathbf{R}_{\text{CM}} + \mathbf{r}'_a) \times (\mathbf{P}_{\text{CM}} + \mathbf{p}'_a)$$

$$= \mathbf{R}_{\text{CM}} \times \mathbf{P}_{\text{CM}} + \sum \mathbf{r}'_a \times \mathbf{p}'_a$$

$$= \mathbf{R}_{\text{CM}} \times \mathbf{P}_{\text{CM}} + \mathbf{L}_{\text{CM}}, \tag{17.1}$$

where $\mathbf{L}_{\text{CM}}$ is the angular momentum of the body about the CM. Hence the total angular momentum of a system of particles is the sum of the angular momentum of the CM and the angular momentum wrt the CM. We will illustrate this theorem using several examples.

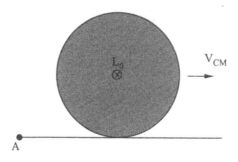

Figure 17.1   *Rolling wheel: The angular momentum L_0 of CM about point A points into the page*

Examples

1. **A rolling wheel**
   When a wheel rolls without slipping on a straight line, its angular momentum can be split into two parts: the angular momentum of the CM of the wheel, and the angular momentum of the wheel about the CM. If we take point A shown in Fig. 17.1 as the reference point, then the angular momentum of the CM about A is $\mathbf{L}_0 = -MRV_{\text{CM}}\hat{\mathbf{z}}$, where M and R are the mass and the radius of the wheel respectively. The angular momentum of the wheel wrt its CM will be computed later in this chapter.

2. **Sun–Earth system**
   Let us write down the angular momentum of the Sun–Earth system with respect to the centre of galaxy in the galaxy reference frame (inertial frame). If the coordinate and linear momentum of the CM of the Sun–Earth system is $\mathbf{R}_{\text{CM}}$ and $\mathbf{P}_{\text{CM}}$ respectively, and the plane of motion of the Earth–Sun is the xy-plane, then the angular momentum of the Sun–Earth system wrt the centre of the galaxy in the galaxy frame is $\mathbf{L} = \mathbf{R}_{\text{CM}} \times \mathbf{P}_{\text{CM}} + \mu r^2 \dot{\phi}\hat{\mathbf{z}}$, where μ is the reduced mass of the system, r is the distance between the Sun and the Earth, and $\dot{\phi}$ is the angular velocity of the Earth around the Sun. The second term has been derived in Chapter 8. Here spin of the Earth has been ignored

The time derivative of the angular momentum of a system of particles is

$$\frac{d\mathbf{L}}{dt} = \sum \mathbf{r}_a \times \mathbf{F}_a$$

$$= \sum_a \mathbf{r}_a \times \mathbf{F}_{a,\text{ext}} + \sum_a \sum_b \mathbf{r}_a \times \mathbf{f}_{a,b}$$

$$= \sum_a \mathbf{r}_a \times \mathbf{F}_{a,\text{ext}} + \sum_a \sum_b (\mathbf{r}_a \times \mathbf{f}_{a,b} + \mathbf{r}_b \times \mathbf{f}_{b,a}), \tag{17.2}$$

where $\mathbf{f}_{a,b}$ is the internal force on the ath particle due to the bth particle as shown in Fig. 17.2. According to Newton's third law, $\mathbf{f}_{a,b} = -\mathbf{f}_{b,a}$. Therefore,

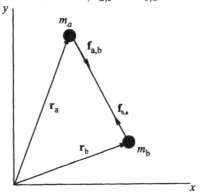

Figure 17.2   *Internal forces between particles a and b.*

$$\frac{d\mathbf{L}}{dt} = \sum_a \mathbf{r}_a \times \mathbf{F}_{a,\text{ext}} + \sum_a \sum_b (\mathbf{r}_a - \mathbf{r}_b) \times \mathbf{f}_{a,b}.$$

We assume that $\mathbf{f}_{a,b}$ is along $\mathbf{r}_{ab}$. This condition is called the *strong form of Newton's third law*; the action and reaction forces are not only equal and opposite, but they are also along the line joining the two particles. Under this condition, the total internal torque (the second term) vanishes.

$\sum_a \mathbf{r}_a \times \mathbf{F}_{a,\text{ext}}$ is the total external torque on the system. Hence,

$$\frac{d\mathbf{L}}{dt} = \mathbf{N}_{\text{ext}}. \qquad (17.3)$$

The internal torques do not contribute to the net torque. If $\mathbf{N}_{\text{ext}} = 0$, then $\mathbf{L}$ is a constant. That is, *if the net torque on a system of particles is zero, then the total angular momentum of the system is conserved. This is the statement of the conservation of angular momentum for a system of particles.*

17.3 Rotation symmetry and conservation of angular momentum

An isolated system is unchanged when it is rotated by a certain angle. The net external torque on an isolated body is also zero. If the net external torque was nonzero, then the direction of external torque would be special, which is not the case. No direction in space is special. This property is also stated as *isotropy of space*.

Since the torque on an isolated system is always zero, the angular momentum $\mathbf{L}$ of the isolated system is a constant. Thus the total angular momentum of an isolated system is conserved. This is how we relate isotropy of space (symmetry under space rotation) to the conservation of angular momentum.

The discussion up till now has been on a system of particles. An important sub-class of a system of particles is the rigid body that will be discussed below.

17.4 Rigid body

A rigid body is one in which the relative distance between any pair of points remains constant. The shape and size of this body is unchanged irrespective of the force applied. A large force can change the dimension of any real body due to elasticity. However for rigid bodies these changes are assumed to be negligible. An example of a rigid body is a steel ball subjected to a moderate external force.

Any motion of a rigid body can be split into two parts: (a) translation of the CM and (b) rotation about the CM. During the translation, all the points of the body move by a constant distance. As an example consider a ball thrown by a bowler. The motion consists of translation of the CM of the ball, and the rotation of the ball about the CM.

The specification of CM requires three coordinates say (X, Y, Z), and the rotation about CM requires three angles. Visualisation of rotation of a rigid body is a bit complex. One way to specify rotation is the following: First locate the axis of rotation, which requires two angles θ and ϕ relative to the original coordinate system (see Fig. 9.4). θ is the angle between the rotation axis and the vertical z-axis, and ϕ is the angle between the projection of the rotation axis and the x-axis. After fixing the rotation axis at the above configuration, we rotate the rigid body about the rotation axis by an angle ζ. These three angles are called *Euler angles*. Thus a general motion of a rigid body can be specified by six variables $(X, Y, Z, \theta, \phi, \zeta)$. Note that the rotation about a fixed axis can be specified by one angle. However, when the rotation axis itself revolves, then the angles θ and ϕ provide the orientation of the rotation axis, and ζ provides the angle of rotation about the axis.

We need six coordinates to specify the configuration of a rigid body: three for space translation, and three angles for the orientation of the rigid body. The angular velocity of a rigid body is a combination of time derivatives of the above angles. In general the rotation of a rigid body could be quite complex. In the following discussion we will provide a definition of angular velocity.

17.5 Angular velocity

We begin by considering the rotation of a rigid body about a fixed axis. Imagine a planar body rotating about an axis passing through point O as shown in Fig. 17.3. We mark a line on the body, say OA. The configuration of the body is specified by an angle between line OA and a reference axis, say the x-axis. In Fig. 17.3 this angle is denoted by ϕ. The angular velocity of the body is quantified by $\dot{\phi}$, and its direction is in the direction of the axis of rotation.[1] By convention the counter-clockwise rotation is considered to be positive.

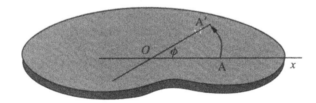

Figure 17.3 *A planar body is rotating about an axis passing through O. The rotation of the body is measured by an angle a line on the body makes with respect to a reference axis (here x-axis).*

The displacement of a point on the rigid body is a function of the angular velocity ω. Suppose a rigid body is rotating with angular velocity ω wrt the z-axis as shown in Fig. 17.4. For the point P in Fig. 17.4, the angle traversed $d\phi$ in a small time dt is ωdt. The displacement of the point P in a small time interval dt is $d\phi \times \mathbf{r} = (\omega \times \mathbf{r})dt$. Therefore the instantaneous linear velocity of point P is $\omega \times \mathbf{r}$.

---

[1]The reason for considering rotation to be a vector is a bit tricky. For the above definition of rotation, large rotations do not obey the commutative property of addition of two vectors. However small rotations do obey the commutative property and hence only small rotations are vectors. Refer to Halliday and Resnick (2004), and Kleppner and Kolenkow (1973) for details.

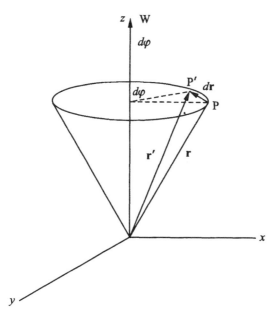

Figure 17.4   *Displacement of a point P on a rigid body under rotation.*

The angular velocity of a rigid body is the same for all points on the rigid body. This theorem can be proved by the following arguments. Suppose a rigid body is rotating about a vertical axis passing through point O of Fig. 17.5. Let us consider two points on the rigid body O' and A as shown in the figure. The velocity of point A is

$$\mathbf{V}_A = \omega_O \times \mathbf{r} = \omega_O \times (\mathbf{a} + \mathbf{r'}) = \mathbf{V}_{O'} + \omega_O \times \mathbf{r'}. \tag{17.4}$$

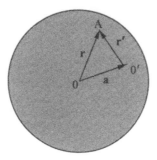

Figure 17.5   *Angular velocity ω of a rigid body about any point is the same. Here we show the equality of ω for two reference points O and O'.*

But the velocity of point A is the sum of velocities of O' and the velocity of point A wrt O'. That is,

$$\mathbf{V}_A = \mathbf{V}_{O'} + \omega_{O'} \times \mathbf{r}', \tag{17.5}$$

where $\omega_{O'}$ is the angular velocity of the body wrt point O'. Comparing Eqs. (17.4, 17.5) we can deduce that $\omega_O = \omega_{O'}$. That is, the angular velocity about both the points O and O' are the same. Physically, in time dt the line OA rotates by an angle $\omega_O dt$ wrt a fixed reference axis. The line O'A rotates by the same angle in time dt wrt the fixed reference axis (convince yourself). Thus we prove that angular velocity about any point in a rigid body is the same.

In many situations the rotation axis is not fixed, for example, for a moving cricket ball, rolling wheel etc. For a wheel rolling on a straight line, the axis moves in a straight line. In these systems too, the angular velocity of the rigid body about any point in the body is the same. As an example, the angular velocities of a wheel about the CM and about the bottom-most point are the same.

In many situations we find that the velocity of all the points on a straight line is zero. For the rotation about a fixed axis, this line happens to be the axis itself. It turns out that we can find this line for any moving rigid body. For example, for a rolling wheel, the axis passing through the bottom-most point has zero velocity. This axis is called the *instantaneous axis of rotation*.

For the rotation about one axis, the angular velocity is along the axis of rotation. A general rotation however could be more complex. A frisbee spins as well as wobbles. If we choose the axis perpendicular to the frisbee plane as the z-axis, and the wobble axis as the x-axis as shown in Fig. 17.6 then

$$\mathbf{\Omega} = \Omega_x \hat{\mathbf{x}} + \Omega_z \hat{\mathbf{z}}.$$

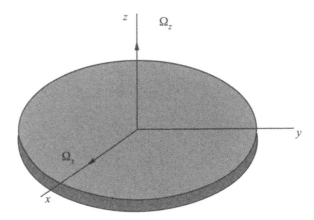

Figure 17.6   *A wobbling frisbee has angular velocity Ω_z along the z-axis, and Ω_x along the x-axis.*

Sometimes the angular velocity of a body can be quite confusing. We will illustrate several important examples to clarify salient concepts regarding angular velocity.

17.6   Examples illustrating angular velocity

1. **A rolling wheel**
   Assume the radius of the wheel to be R. The motion of the wheel can be seen as a combination of translation of the wheel by a distance a, and rotation about the CM by an angle a/R.

2. **A coin (C_2) rolling over another coin (C_1) of the same radius**
   When coin C_2 moves from configuration α to configuration β (Fig. 17.7), line O_2A has moved an angle π. We can divide this rotation in two parts: (a) the line joining centres of C_1 and C_2 rotate by an angle $\pi/2$. This part is called an orbital rotation. If coin C_2 slided rather than rolled over, we would have had only orbital rotation, and lines O_1A and O_2A would have moved by only an angle $\pi/2$. (b) Due to the rolling of the coin C_2, line O_2A makes an additional rotation of angle $\pi/2$. This is called spin. Hence the net rotation of coin C_2 is the sum of the orbital rotation and the spin.

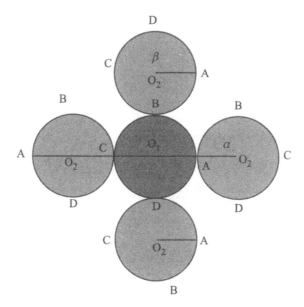

Figure 17.7   *Coin C_2 rolling over coin C_1 without slipping.*

Since the radius of both the coins are the same, we obtain $\theta_{\text{orbit}} = \theta_{\text{spin}}$ (prove it). The total angle traversed by line O_2A is

$$\theta_{\text{net}} = \theta_{\text{orbit}} + \theta_{\text{spin}} = 2\theta_{\text{orbit}}.$$

When coin C_2 returns to the original spot after $\theta_{\text{orbit}} = 2\pi$, line O_2A has covered $\theta_{\text{net}} = 4\pi$, which corresponds to two complete revolutions of coin C_2. See Fig. 17.7 for an illustration of the motion of four points ABCD and line O_2A.

3. Earth–Sun system

In one year (365.24 solar days), the centre of the Earth returns to its original position. Hence in one year the Earth completes an orbital motion of 2π radian, and spin of $2\pi \times 365.24$ radian. Consider a point P on the surface of the Earth (Fig. 17.8). It is the closest point on the Earth to the Sun in the first configuration. After one solar day it will again be the closest to the Sun. The duration is 24 hours or a solar day. This is how a solar day is defined. Note however that in one solar day, the line OP has rotated by

$$\theta = 2\pi + \frac{2\pi}{365.24}.$$

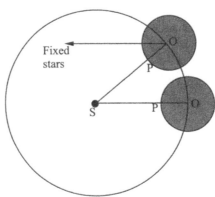

Figure 17.8   *Motion of the Earth around the Sun.*

Hence the angular velocity of the Earth is

$$\omega = \frac{\theta}{24 \text{ hours}} = \frac{2\pi \times 366.24}{365.25 \times 86400}$$

$$= 7.29 \times 10^{-5}/\text{s}.$$

A sidereal day is the time required for the line OP to rotate by 2π. Using the above discussion we conclude that

$$T_{\text{sidereal}} = \frac{365.24}{366.24} \times 24 \text{ hours} \approx 23\,\text{h}\,56\,\text{m}.$$

After discussing angular velocity, we move on to the computation of kinetic energy of a rigid body.

17.7   Kinetic energy of a rigid body, and moment of inertia

As discussed in Chapter 12, the total kinetic energy of a mechanical system is the sum of the kinetic energy of the CM and the kinetic energy wrt the CM. If a rigid body is rotating with angular velocity $\mathbf{\Omega}$, then the kinetic energy of the rigid body is

$$T = \frac{1}{2}MV_{CM}^2 + T_{rot},$$

where $T_{rot} = \frac{1}{2}\sum m_a(\mathbf{\Omega} \times \mathbf{r}_a)^2$.

For rotation about a fixed axis, say z-axis

$$T_{rot} = \frac{1}{2}\sum m_a(\mathbf{\Omega} \times \mathbf{r}_a)^2$$

$$= \frac{1}{2}\sum m_a(y_a\hat{\mathbf{y}} - x_a\hat{\mathbf{x}})^2\Omega_z^2$$

$$= \frac{1}{2}(\sum m_a(y_a^2 + x_a^2))\Omega_z^2$$

$$= \frac{1}{2}I_{zz}\Omega_z^2,$$

where $\sum m_a(y_a^2 + x_a^2) = I_{zz}$ is called the moment of inertia about the z-axis. The above expression is valid only for rotation about the z-axis. For an arbitrary rotation with angular velocity $\mathbf{\Omega} = \Omega_x\hat{\mathbf{x}} + \Omega_y\hat{\mathbf{y}} + \Omega_z\hat{\mathbf{z}}$ the rotational kinetic energy is

$$T_{rot} = \frac{1}{2}\sum m_a(\mathbf{\Omega} \times \mathbf{r}_a)^2$$

$$= \frac{1}{2}\sum m_a[(\Omega_y z_a - \Omega_z y_a)^2 + (\Omega_z x_a - \Omega_x z_a)^2 + (\Omega_x y_a - \Omega_y x_a)^2]$$

$$= \frac{1}{2}\sum_i\sum_j I_{ij}\Omega_i\Omega_j$$

where i, j can take values x, y, z, and

$$I_{xx} = \sum m_a(y_a^2 + z_a^2)$$

$$I_{yy} = \sum m_a(x_a^2 + z_a^2)$$

$$I_{zz} = \sum m_a(x_a^2 + y_a^2)$$

$$I_{xy} = I_{yx} = -\sum m_a x_a y_a$$

$$I_{yz} = I_{zy} = -\sum m_a y_a z_a$$

$$I_{xz} = I_{zx} = -\sum m_a x_a z_a.$$

The quantity I_{ij} is called the *moment of inertia* of the rigid body. It is a nine component object with $I_{ij} = I_{ji}$. This object is an example of a second-rank tensor, and has been discussed in more detail in Appendix E.

If we choose our axes to be the symmetry axes of the rigid body, then $I_{xy} = I_{yz} = I_{zx} = 0$, and we obtain

$$T_{\text{rot}} = \frac{1}{2}I_{xx}\Omega_x^2 + \frac{1}{2}I_{yy}\Omega_y^2 + \frac{1}{2}I_{zz}\Omega_z^2.$$

Because of the above simplification, the symmetry axes of a rigid body are very important and convenient for calculations. These axes are also called the *principal axes*. We state a theorem without proof:

We can always find three symmetry axes for any rigid body.

As an example we will illustrate the principal axes of a uniform cylinder in Fig. 17.9.

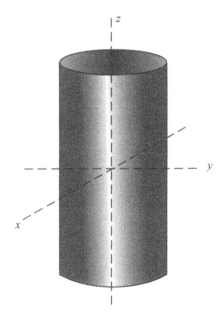

Figure 17.9   *Axes x, y, z are the symmetry axes of a cylinder.*

One of the axes is the axis of the cylinder, and the other two axes are horizontal axes passing through the CM.

Let us compute the moment of inertia of a cylinder about its principal axes. Suppose that the mass density of the cylinder is σ, the radius and the height of the cylinder are R and H respectively. About the vertical axis, the moment of inertia I_{zz} is

$$I_{zz} = \int \sigma(x^2 + y^2)d\mathbf{r}$$

$$= \sigma \int (\rho^2)\rho d\rho d\phi dz$$

$$= \sigma 2\pi H \frac{R^4}{4}$$

$$= \frac{1}{2} M R^2,$$

where M is the mass of the cylinder.

Similarly

$$I_{xx} = \int \sigma(y^2 + z^2) d\mathbf{r}$$

$$= \sigma \int (\rho^2 + z^2 - \rho^2 \cos^2 \phi) \rho d\rho d\phi dz$$

$$= \sigma \int d\rho \rho^3 d\phi \sin^2 \phi dz + \sigma \int z^2 \rho d\rho d\phi dz$$

$$= \sigma \pi H \frac{R^4}{4} + \sigma \frac{1}{12} \pi R^2 H^3$$

$$= \frac{1}{4} M R^2 + \frac{1}{12} M H^2.$$

Due to x–y symmetry

$$I_{yy} = I_{xx}$$

and

$$I_{xy} = I_{yz} = I_{zx} = 0.$$

A disc is a special case of a cylinder where $H \ll R$. Hence for a disc

$$I_{zz} = \frac{1}{2} M R^2$$

$$I_{xx} = I_{yy} = \frac{1}{4} M R^2.$$

The moment of inertia of a thin cylinder ($R \ll H$) is

$$I_{zz} = 0$$

$$I_{xx} = I_{yy} = \frac{1}{12} M H^2.$$

Using a similar calculation, we can show that the moment of inertia of a sphere about its symmetric axis is $(2/5)MR^2$, where M and R are the mass and the radius of the sphere. The moment of inertia of a cube about any of its symmetry axes is $(1/6)Ma^2$, where a is the side of the cube.

If a disc is spinning about the vertical axis passing through its CM with angular frequency Ω, then its kinetic energy will be

$$T = \frac{1}{2} I_{zz} \Omega^2 = \frac{1}{4} M R^2 \Omega^2.$$

For a frisbee, the angular velocity in general is

$$\boldsymbol{\Omega} = \Omega_x \hat{\mathbf{x}} + \Omega_y \hat{\mathbf{y}} + \Omega_z \hat{\mathbf{z}},$$

hence its kinetic energy is

$$T = \frac{1}{2} \left[\frac{1}{4} M R^2 (\Omega_x^2 + \Omega_y^2) + \frac{1}{2} M R^2 \Omega_z^2 \right].$$

EXAMPLE 17.1   A plank of length l and mass m is standing vertically on a frictionless surface. The plank starts to fall at $t = 0$. Assuming conservation of energy (to be proved in the next chapter), find the angular velocity of the plank as a function of time.

SOLUTION   We solve the above problem by applying conservation of energy. At $t = 0$, the total energy of the plank is $mgl/2$. Since there is no horizontal force on the plank, its CM will fall down vertically. Consider a configuration when the CM of the plank is at y (Fig. 17.10).

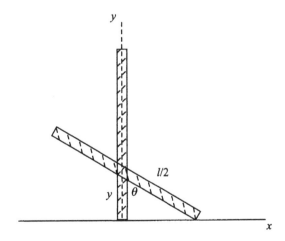

Figure 17.10   *A plank falling under gravity.*

Since $y = (l/2) \cos \theta$, we obtain

$$\dot{y} = -\frac{l}{2} \sin \theta \dot{\theta}. \tag{17.6}$$

The total kinetic energy of the plank is the sum of kinetic energy of CM and rotational kinetic energy that is

$$KE = \frac{1}{2}m\dot{y}^2 + \frac{1}{2}I\dot{\theta}^2$$

$$= \frac{1}{2}m\dot{y}^2 + \frac{1}{2}\frac{ml^2}{12}\dot{\theta}^2$$

Now we apply conservation of energy, which yields

$$mg\frac{l}{2} = mgy + \frac{1}{2}\frac{ml^2}{12}\dot{\theta}^2 + \frac{1}{2}m\dot{y}^2.$$

Using eq. (17.6) we obtain

$$\dot{y}^2 = mg(\frac{l}{2} - y)\frac{6\sin^2\theta}{(1 + 3\sin^2\theta)}.$$

The above expression provides the vertical velocity of the CM of the plank.

17.8   Parallel axis theorem

In the last section we studied the moment of inertia about the symmetry axes of a rigid body. However on many occasions we need to compute the moment of inertia about some other axis. The parallel axis theorem provides us the moment of inertia about any axis parallel to the principal axis. According to the *parallel axis theorem*,

The moment of inertia about an axis (z') that is a distance a away and parallel to the symmetry axis (say z-axis) is

$$I'_{zz} = I_{zz,\text{CM}} + Ma^2, \tag{17.7}$$

where $I_{zz,\text{CM}}$ is the moment of inertia about the axis passing through the CM, and I'_{zz} is the moment of inertia about the new axis.

PROOF   Without any loss of generality we can choose the new axis to pass through a point O' on the x-axis as shown in Fig. 17.11. The moment of inertia about the new axis is

$$I'_{zz} = \sum m_a(x_a'^2 + y_a'^2)$$

$$= \sum m_a[(x_a - a)^2 + y_a^2]$$

$$= \sum m_a(x_a^2 + y_a^2) + a^2 \sum m_a - 2a \sum m_a x_a$$

$$= I_{zz,\text{CM}} + Ma^2.$$

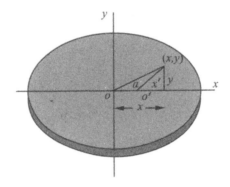

Figure 17.11   *Parallel axis theorem*

Hence proved!

As an example, the moment of inertia of a wheel about the bottom-most point is

$$I_3' = \frac{1}{2}MR^2 + MR^2 = \frac{3}{2}MR^2.$$

In the last two sections we discussed moment of inertia and kinetic energy of a rigid body. In the next section we will describe the angular momentum of a rigid body.

17.9 Angular momentum of a rigid body

The angular momentum of a rigid body about a reference point is

$$\mathbf{L} = \sum \mathbf{r}_a \times \mathbf{p}_a$$

$$\sum \mathbf{r}_a \times m_a(\mathbf{\Omega} \times \mathbf{r}_a)$$

$$= \sum m_a[\mathbf{\Omega} r_a^2 - \mathbf{r}_a(\mathbf{\Omega} \cdot \mathbf{r}_a)]$$

where $\mathbf{\Omega}$ is the angular velocity of the rigid body. Note that the reference point need not be the CM. In terms of components

$$L_x = I_{xx}\Omega_x + I_{xy}\Omega_y + I_{xz}\Omega_z, \tag{17.8}$$

$$L_y = I_{yx}\Omega_x + I_{yy}\Omega_y + I_{yz}\Omega_z, \tag{17.9}$$

$$L_z = I_{zx}\Omega_x + I_{zy}\Omega_y + I_{zz}\Omega_z, \tag{17.10}$$

where I_{ij} is the moment of inertia of the rigid body defined in the previous section. We can also write Eqs. (17.8–17.10) in tensor notation as

$$L_i = \sum_j I_{ij}\Omega_j.$$

The above formulas are valid for any point of reference. The centre of mass is often a convenient reference point for many angular momentum calculations. As discussed in Section 17.7, I_{xy}, I_{xz}, I_{yz} are zero for symmetric axes. Hence, with the choice of symmetric axes as our reference axes, the angular momentum of the rigid body is

$$\mathbf{L} = I_{xx}\Omega_x\hat{\mathbf{x}} + I_{yy}\Omega_y\hat{\mathbf{y}} + I_{zz}\Omega_z\hat{\mathbf{z}}, \tag{17.11}$$

where $\hat{\mathbf{x}}$, $\hat{\mathbf{y}}$, and $\hat{\mathbf{z}}$ are the unit vectors along the principal axes. Equation (17.11) is the equation of rotational kinematics. We can derive many important properties of a mechanical system using this law.

The angular momentum $\mathbf{L}$ is in general not parallel to $\mathbf{\Omega}$. They are parallel only when

1. $I_{xx} = I_{yy} = I_{zz} = I$ (valid for a sphere or a cube) where $\mathbf{L} = I\mathbf{\Omega}$

2. $\mathbf{\Omega}$ is along one of the principal axis. For example, if $\Omega_x = \Omega_y = 0$ and $\Omega_z \neq 0$, then $\mathbf{L} = I_{zz}\mathbf{\Omega}$.

The form of Eqs. (17.8–17.10) is similar to $p_i = mv_i$. However, there are several crucial differences between these two equations. (1) The proportionality constant between the angular momentum and the angular velocity is a tensor, while the proportionality constant between the linear momentum and the linear velocity is a scalar (mass). (2) The angular momentum depends on the choice of origin, which is not the case for linear momentum.

The above differences are very important. We have to be careful while solving rotation problems. In the following discussion we will illustrate angular momentum calculations using several examples.

17.10 Examples of angular momentum calculations

1. A rolling wheel
Imagine a wheel of mass M and radius R rolling without slipping with angular velocity Ω. The angular momentum of the wheel about its CM is

$$L_{\text{CM}} = I\Omega = \frac{1}{2}MR^2\Omega.$$

where I is the moment of inertia of the wheel about the axis passing through the CM (centre of the wheel). Therefore the angular momentum of the wheel about point A of Fig. 17.1 is

$$\mathbf{L}_A = \mathbf{R}_{\text{CM}} \times \mathbf{P}_{\text{CM}} + \mathbf{L}_{\text{CM}}$$

$$\text{or } \mathbf{L}_A = \left(MR^2\Omega + \frac{1}{2}MR^2\Omega\right)\hat{\mathbf{z}} = \frac{3}{2}MR^2\Omega\hat{\mathbf{z}}.$$

The angular momentum of the wheel about the bottom-most point is the same as above. We can also compute L_B about the bottom-most point using the parallel axis theorem:

$$L_B = I_B \Omega = (MR^2 + \frac{1}{2}MR^2)\Omega = \frac{3}{2}MR^2\Omega.$$

We have used the theorem that the angular velocity of all the points on a rigid body is the same.

2. **A coin rolling over another coin of the same radius**
See Fig. 17.7 for an illustration. The angular momentum of coin C_2 about the centre of the fixed coin (C_1) is

$$\mathbf{L} = \mathbf{R}_{CM} \times \mathbf{P}_{CM} + \mathbf{L}_{CM}$$

$$= (2R \times M \times 2R\Omega_{\text{orbit}} + \frac{1}{2}MR^2\Omega_{\text{spin}})\hat{\mathbf{z}},$$

where R is the radius of each coin. Since $\Omega_{\text{orbit}} = \Omega_{\text{spin}}$, the net angular momentum is

$$\mathbf{L} = \frac{9}{2}MR^2\Omega_{\text{spin}}\hat{\mathbf{z}}.$$

We can also derive the above expression using $\mathbf{L} = \sum \mathbf{r} \times \mathbf{p}$ with $\mathbf{v} = \mathbf{V}_{CM} + \mathbf{v}'$ (Exercise 1).

3. **Earth–Sun system**
The angular momentum of the Earth wrt the centre of the Sun is

$$\mathbf{L} = mr^2\omega\hat{\mathbf{z}} + \frac{2}{5}mR^2\Omega_S\hat{\mathbf{s}},$$

where m is the mass of the Earth, r is the Earth–Sun distance, ω is the orbital angular velocity, R is the radius of the Earth, Ω_S is the spin angular velocity of the Earth. Note that the spin angular momentum is inclined by 23.5 degrees from the perpendicular to the orbital plane. From the observed data

$$\omega = \frac{2\pi}{365.25 \times 86400}\text{rad/s}$$

$$\Omega_S = \frac{2\pi}{86400}\text{rad/s}.$$

4. **Rod making an angle θ with the rotating axis**
The configuration of the rod is shown in Fig. 17.12. The angular velocity of the rod is Ω along the AA' axis. To compute the angular momentum of the rod we resolve Ω along the principal axes as

$$\mathbf{\Omega} = \Omega_x\hat{\mathbf{x}} + \Omega_z\hat{\mathbf{z}}.$$

Therefore the angular momentum of the rod is

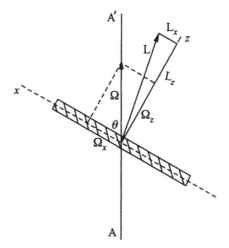

Figure 17.12   *A rod, which is inclined from the vertical axis by an angle θ, is rotating about the vertical axis with angular velocity $\mathbf{\Omega}$.*

$$\mathbf{L} = I_{xx}\Omega_x\hat{\mathbf{x}} + I_{zz}\Omega_z\hat{\mathbf{z}}.$$

Clearly $\mathbf{L}$ and $\mathbf{\Omega}$ are *not* in the same direction.

5. Torque-free precession of a wobbling top and a frisbee

A sphere or cube has $I_{xx} = I_{yy} = I_{zz}$. However this symmetry is not obeyed by many rigid bodies. The rigid bodies for which $I_{xx} = I_{yy} \neq I_{33}$ are called *tops*. Imagine that a top is spinning with angular velocity $\mathbf{\Omega}_S$ in free space, and has angular momentum $\mathbf{L}$. If the angular momentum and angular velocity are in the same direction, the top will spin happily in the direction of $\mathbf{L}$. However if the direction of $\mathbf{L}$ and $\mathbf{\Omega}_S$ are not in the same direction, then the top starts to precess due to the reason discussed below.

As shown in Fig. 17.13(a), the rigid body has angular momentum $\mathbf{L}$ along the z-axis, and $\mathbf{\Omega}$ is inclined with respect to the z-axis. The precession angular velocity is $\mathbf{\Omega}_P$, hence

$$\mathbf{\Omega} = \mathbf{\Omega}_S + \mathbf{\Omega}_P$$

$$= (\Omega_S + \Omega_P\cos\theta)\hat{\mathbf{z}} + \Omega_P\sin\theta\hat{\mathbf{x}}$$

$$= \Omega_z\hat{\mathbf{z}} + \Omega_x\hat{\mathbf{x}}.$$

The net angular momentum can be resolved along the principal axes as

$$L_x = L\sin\theta = I_{xx}\Omega_x$$

$$L_z = L\cos\theta = I_{zz}\Omega_z$$

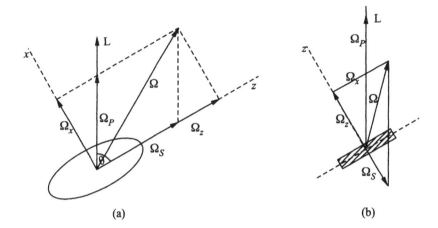

Figure 17.13   *(a) Torque-free precession of a wobbling top. (b) Torque-free precession of a wobbling frisbee.*

The above equations yield

$$\Omega_x = \frac{L_x}{I_{xx}} = \frac{L \sin \theta}{I_{xx}},$$

$$\Omega_z = \frac{L_z}{I_{zz}} = \frac{L \cos \theta}{I_{zz}}$$

Also,

$$\Omega_P = \frac{\Omega_x}{\sin \theta} = \frac{L}{I_{xx}}, \tag{17.12}$$

$$\Omega_S = \Omega_z - \Omega_P \cos \theta = \Omega_z \left[1 - \frac{I_{zz}}{I_{xx}} \right], \tag{17.13}$$

or $$\Omega_S = \Omega_P \cos \theta \left[\frac{I_{xx}}{I_{zz}} - 1 \right]. \tag{17.14}$$

Note that $\Omega_S \neq \Omega_z$.

There are two types of tops: (a) thin top: $I_{xx} > I_{zz}$ (e.g., the tops children play with), (b) flat top: $I_{xx} < I_{zz}$ (e.g., frisbee). For tops of type (a), according to Eq. (17.14), Ω_P and Ω_S have the same sign, while for the tops of type (b), Ω_P and Ω_S have different signs (Fig. 17.13(b))

When we apply the above equations to a spinning frisbee ($I_{xx} = I_{zz}/2$), we find that

$$\Omega_S = -\frac{\Omega_P \cos \theta}{2}.$$

For small θ, $\Omega_S = -\Omega_P/2$; that is, the frisbee precesses twice as fast as it spins. The sense of precession and spin are also opposite. For $\theta = 45°$,

$$\Omega_S = -\frac{L}{2I_{xx}}\cos\theta = -\frac{L}{2\sqrt{2}I_{xx}}$$

$$\Omega_z = -\Omega_S.$$

With this we close our discussion on angular momentum. Our discussion also included various aspects of rotational kinematics. One of the main points to remember is that the angular momentum depends on the point of reference, and need not be in the same direction as the angular velocity.

In the next chapter we will discuss the dynamics of rigid bodies.

Exercises

1. In Example 2 of Section 17.6, a coin (C_2) of radius R is rolling over another coin (C_1) of the same radius. Compute the angular momentum of coin C_2 wrt the centre of coin C_1 from first principles, i.e., using the definition $\mathbf{L} = \sum \mathbf{r} \times \mathbf{p}$ with $\mathbf{v} = \mathbf{V}_{\mathrm{CM}} + \mathbf{v}'$.

2. Compute the moment of inertia of a (a) sphere about an axis passing through the centre; (b) cube about one of its symmetry axes; (c) cube about one of its edges.

3. Two discs of moment of inertia I_1 and I_2 are rotating about the same concentric axis with angular frequencies ω_1 and ω_2 respectively. The top disc falls onto the bottom disc, and they start moving with the same velocity after a while.

   (a) Compute the common angular velocity of the discs.

   (b) Compute the kinetic energy lost in the process. Where does the energy go?

4. A fixed gear A whose moment of inertia is I_1 and radius is R_1 is rotating with angular frequency ω_1. A second gear B touches gear A and rotates without slipping about an axis parallel to the axis of gear A. The radius and moment of inertia of the second gear are R_2 and I_2 respectively. Compute the angular momentum of the whole system about the axis of gear A.

5. A cycle wheel whose moment of inertia is I and radius is R is connected to a vertical axis through a horizontal shaft of length a as shown in Fig. 17.14(a). The wheel rotates freely about the vertical axis with angular frequency Ω.

   (a) Compute the net angular velocity of the wheel.

   (b) Compute the angular momentum of the wheel wrt the hinge.

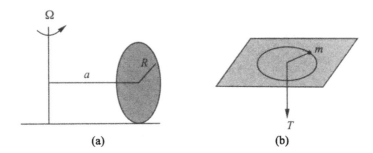

Figure 17.14   *(a) Exercise 5; (b) Exercise 6*

6. A particle of mass m, connected to one end of a string as shown in Fig. 17.14(b), is rotating around in a circle of radius r_0 with speed u on a frictionless table. At $t = 0$, the string is pulled through the hole in the middle with a force such that the radius of the circle decreases at a constant rate.

   (a) What is the force on the string?

   (b) What are the linear and angular velocities of the mass?

7. A cubical block of size d rests on top of a fixed cylindrical drum of radius R. What is the largest value of d beyond which the block will become unstable.

8. A thin rod of length l is resting on top of a fixed sphere of radius R. Is the configuration of the rod stable?

18

Rigid Body Dynamics

In the last chapter we studied rotational kinematics. In the present chapter we will study the dynamical equation for a rigid body. The motion of a rigid body involves movement of the CM of the rigid body, and the rotation of the rigid body about the CM. We will set up dynamical equations for the rigid body in terms of the applied forces and torques on the body. Study of these equations form the subject of rotational dynamics.

When a rigid body rotates about a fixed or a single axis, the motion is relatively simple. However a general rotation when the rotation axis itself rotates is quite complex. In this chapter we will discuss both these cases, but our discussion for the latter class of rotational motion will be at an elementary level. Study of a full-fledged rotation with all the three axis moving around is very complex, and is beyond the scope of this book.

18.1 Equation of motion of a rigid body

As discussed in the previous chapter, configuration of a rigid body can be specified by its centre-of-mass (CM) and the three Euler angles. To describe the motion of a rigid body, we need to write down dynamical equations for these six quantities. The equation of motion for the centre of mass of the rigid body in an inertial frame is

$$\frac{d\mathbf{P}_{\mathrm{CM}}}{dt} = \mathbf{F}_{\mathrm{ext}} \tag{18.1}$$

where $\mathbf{F}_{\mathrm{ext}}$ is the external force acting on the rigid body.

For the angles we use the equation for angular momentum. Recall that the total angular momentum is given by

$$\mathbf{L} = \mathbf{R}_{\mathrm{CM}} \times \mathbf{P}_{\mathrm{CM}} + \mathbf{L}_{\mathrm{CM}}.$$

Therefore

$$\frac{d\mathbf{L}}{dt} = \mathbf{R}_{\mathrm{CM}} \times \mathbf{P}_{\mathrm{CM}} + \frac{d\mathbf{L}_{\mathrm{CM}}}{dt} = \mathbf{R}_{\mathrm{CM}} \times \mathbf{F}_{\mathrm{ext}} + \mathbf{N}_{\mathrm{CM,ext}},$$

where $\mathbf{N}_{\text{CM,ext}}$ is the torque with respect to the CM. The rate of change of angular momentum about the CM is given by

$$\frac{d\mathbf{L}_{\text{CM}}}{dt} = \mathbf{N}_{\text{CM,ext}}. \tag{18.2}$$

The torque $\mathbf{R}_{\text{CM}} \times \mathbf{F}_{\text{ext}}$ accounts for the orbital motion of the CM.

Let us take the well-studied Sun–Earth system as an example. The motion of the Earth involves both orbital motion and spin. If we take the Sun as the origin, then for the Earth the torque $\mathbf{R}_{\text{CM}} \times \mathbf{F}_{\text{ext}}$ is zero, hence the orbital angular momentum is a constant, and the trajectory of the Earth is an ellipse. To a good approximation, the torque on the Earth wrt its CM, $\mathbf{N}_{\text{CM,ext}}$, is also zero, and hence the Earth continues to spin with constant angular velocity.[1]

Equation (18.2) is applicable in any inertial frame. It is also valid for noninertial frames as long as we take the reference point to be the CM. This is because the pseudo forces pass through the CM, and hence do not contribute to the torque.

18.2   Rotation about a single axis

It is easier to describe the motion of a rigid body when it is rotating about a single axis. Note that the axis of rotation need not be fixed. The angular momentum is also always along the angular velocity for these problems. For such cases we can completely solve for the CM and angular configuration of the rigid body using Eqs. (18.1) and (18.2). We will illustrate the procedure using several examples.

EXAMPLE 18.1   A cylinder rolls down an inclined plane without slipping. Describe the motion of the cylinder.

SOLUTION   The cylinder is rolling down the inclined plane without slipping as shown in Fig. 18.1. Gravity and frictional forces are acting on the cylinder. The frictional force opposes the tendency to slip. Hence, the frictional force is along $-\hat{\mathbf{x}}$. Let a be the acceleration of the cylinder, and α its angular velocity. We take the CM of the cylinder to be the reference point. We can do so even though the CM is accelerating because the pseudo force passes through the CM, and it causes no additional torque. Therefore, the equation of motion of the cylinder is

$$ma = mg \sin\theta - f, \tag{18.3}$$

$$I\alpha = fR. \tag{18.4}$$

The constraint that the cylinder rolls down without slipping provides the third equation:

$$a = \alpha R$$

---

[1] The moon and the Sun exert a torque on the Earth due to its oblate shape. Because of this torque the axis of the Earth precesses. See French (1971) for details.

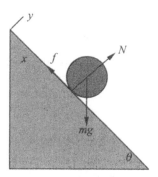

Figure 18.1   *A cylinder rolling down on an incline*

Now we have three equations and three unknowns a, α and f. Solving these equations yields

$$a = \frac{mg \sin \theta}{m + I/R^2},$$ (18.5)

and $f = \dfrac{Ia.}{R^2}.$ (18.6)

For the cylinder $I = MR^2/2$, hence $a = (2g/3) \sin \theta$. Thus the cylinder rolls down with a constant acceleration. We can easily determine the position of the CM and the angular position of the cylinder using a and α. We will observe some important points about the motion of the cylinder.

1. For a block sliding down an incline, we have only one unknown a and one equation $ma = F_{\text{net}}$. For the rolling cylinder, we have three equations (force, torque, and constraint equations) that help us determine three unknowns: a, α, and f. This would be a typical scenario for a one-dimensional motion with rotation about a single axis.

2. Compare the acceleration of the cylinder with a sliding block without friction ($a = g \sin \theta$). The cylinder falls slower than the block.

3. For a rigid body $I = kMR^2$, where k is a constant that depends on the geometrical shape of the body (e.g., for all cylinders $k = 1/2$). Therefore, the acceleration of a rolling rigid body in terms of k is

   $$a = \frac{g \sin \theta}{1 + k}.$$

   Hence the acceleration of the cylinder a does not depend on the mass and radius of the rigid body. So cylinders of different masses, radii, and densities will roll down with the same acceleration. Newton used this result to disprove Aristotle according to whom a heavy mass falls faster that a lighter one.

4. Frictional force given by Eq. (18.6) is required for the cylinder to roll down. If the surface cannot provide this frictional force, then the cylinder will slip.

5. The frictional force does not do any work while the cylinder is rolling. $\int \mathbf{f}.d\mathbf{x}$ is zero because the point of contact does not slip ($d\mathbf{x} = 0$).

6. Since the frictional force does not dissipate any energy, we can also solve the above problem using conservation of total energy. When the cylinder comes down by a distance x along the incline, the conservation of energy yields the following equation

$$mgx \sin \theta = \frac{1}{2}MV_{\text{CM}}^2 + \frac{1}{2}I\omega^2 .$$

The LHS is the loss in potential energy, and the RHS is the gain in KE. The time derivative of the above equation with $V_{\text{CM}} = \omega R$ gives the same result as above.

7. The point of contact is at instantaneous rest, so it can also be chosen to be a reference point for the torque equation that yields

$$I'\alpha = mg \sin \theta R$$

with $I' = I + MR^2$. This equation yields the same result as expected.

EXERCISE 18.2   A cylinder is pushed up an incline. Consequently the cylinder rolls up the incline without slipping. Describe the motion of the cylinder.

SOLUTION   When the cylinder is rolling up, the frictional force is still upward to oppose the tendency of the cylinder to slip down the incline. Hence the equation of motion is exactly the same as Eqs. (18.3) and (18.4) and hence the cylinder slows down by a deceleration given by Eq. (18.5). The motion for this case is the time-reversed motion of Exercise 18.1.

It is interesting to note that the time-reversal symmetry is respected in this problem even in the presence of frictional force. This is because the rolling frictional force does not dissipate energy, hence it does not break the time-reversal symmetry.

EXAMPLE 18.3   A homogeneous sphere of mass M and radius R is thrown horizontally on the floor with a speed v_0 such that it slides initially. Calculate the speed at which rolling will occur? Describe the motion.

SOLUTION   At $t = 0$, the ball has linear velocity, but no angular velocity. So it slides as well as rolls as described below. Let F be the magnitude of the *constant* frictional force acting on the ball, then the equations of motion of the ball are

$$Ma = -F,$$

$$\frac{2}{5}MR^2\alpha = FR$$

In this example we treat clock-wise rotation as positive. Note that $a < 0$ and $\alpha > 0$. The linear velocity of the CM at time t is

$$v = v_0 + at,$$

and the angular velocity of the ball is

$$\omega = \alpha t.$$

Hence the ball skids as well as rotates. Note that $a \neq \alpha R$ because the ball is skidding. The condition $a = \alpha R$ is valid only in cases where there is rolling without slipping. The velocities at different points are shown in Fig. 18.2.

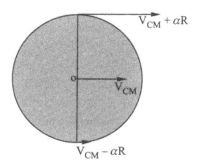

Figure 18.2   *Velocities of various points of the skidding sphere.*

The ball will start rolling without slipping when the velocity of the bottom-most point vanishes, i.e., $v_{\text{roll}} = \omega_{\text{roll}} R$. The condition is

$$\frac{5Ft}{2M} = v_0 - \frac{Ft}{M}.$$

Therefore,

$$Ft = \frac{2}{7} v_0 M.$$

Using $v_{\text{roll}} = \omega_{\text{roll}} R = \alpha t R$, we obtain

$$v_{\text{roll}} = \frac{5}{7} v_0.$$

The above problem can also be solved using the conservation of angular momentum. If we take the angular momentum about the bottom-most point, then the torque due to the frictional force is zero. Hence the angular momentum of the ball about the bottom-most point is conserved. The initial angular momentum is $M v_0 R$, and the final angular momentum is $M v_{\text{roll}} R + I\omega$. The final configuration is when the ball rolls without slipping. Conservation of angular momentum implies that

$$M v_0 R = M R^2 \omega + \frac{2}{5} M R^2 \omega.$$

Hence

$$v_{\text{roll}} = \omega R = \frac{5}{7} v_0,$$

the same result as above.

In summary, the frictional force produces a torque which increases the angular velocity. Meanwhile the linear velocity decreases due to the frictional force. When $v_{\text{CM}} = \omega R$, the sphere starts to roll without slipping. After this, the frictional force vanishes, and the sphere continues to roll with a constant angular velocity (Section 18.4). We observe this phenomena during the touchdown of an aeroplane; at first the wheels of the aeroplane skid and roll, and then they start to roll without skidding.

18.3   Rotation with constraints

Rotation typically involves constraints. One such constraint for a rigid body rotation is that the distance between any two points does not change. As a result of this constraint, *the net work done by internal forces is zero*. The proof of this statement is as follows.

PROOF   We will consider the internal forces in pairs and focus on two point masses a and b on whom the action and reaction forces are $\mathbf{f}_{a,b}$ and $\mathbf{f}_{b,a}$ respectively. If the displacements of the particles a and b are $\delta\mathbf{r}_a$ and $\delta\mathbf{r}_b$ respectively, then the net work done by these forces are

$$\delta W = \mathbf{f}_{a,b} \cdot \delta\mathbf{r}_a + \mathbf{f}_{b,a} \cdot \delta\mathbf{r}_b$$

$$= \mathbf{f}_{a,b} \cdot (\delta\mathbf{r}_a - \delta\mathbf{r}_b).$$

As argued in Section 17.2, $\mathbf{f}_{a,b} = k(\mathbf{r}_a - \mathbf{r}_b)$, where k is a constant. Therefore

$$\delta W = k(\mathbf{r}_a - \mathbf{r}_b) \cdot (\delta\mathbf{r}_a - \delta\mathbf{r}_b)$$

$$= \frac{k}{2} \delta(\mathbf{r}_a - \mathbf{r}_b)^2.$$

Since the distance between the two points a and b is a constant for a rigid body, $\delta(\mathbf{r}_a - \mathbf{r}_b)^2 = 0$. Hence $\delta W = 0$, or the net work done by the forces $\mathbf{f}_{a,b}$ and $\mathbf{f}_{b,a}$ is zero. By carrying out the sum for all the pairs of internal forces, we can show that the work done by all the internal forces of a rigid body is zero.

When a rigid body rolls without slipping, a frictional force acts on the rigid body. However the rolling frictional force does not do any work because the displacement of the point of contact is zero ($\mathbf{F} \cdot d\mathbf{x} = 0$ because $d\mathbf{x} = 0$). Similarly when a rigid body slides on a *smooth* surface, the normal force is perpendicular to the surface of contact. Since the displacement of the elemental mass at the point of contact is perpendicular to the normal force, the normal force does no work on the elemental mass. Also, the internal forces caused by the normal force do not do any work on the rigid body. Hence, the normal force exerted by a smooth surface does no work on a rigid body. These results help us in applying conservation of energy to rigid bodies in many complex situations.

EXAMPLE 18.4   A plank of length l and mass m is standing vertically on a frictionless surface. The plank starts to fall at $t = 0$. Is the total energy conserved for this system?

SOLUTION   Since the plank slides on a smooth surface, the normal force is perpendicular to the surface. The normal force is also perpendicular to the displacement of the elemental mass that is in contact with the surface. Therefore, the normal force does no work on the plank. Hence, the total energy which is the sum of kinetic energy and gravitational potential energy is conserved.

We could prove the same result by applying equations of dynamics. The arguments are as follows: The free-body diagram of the plank is shown in Fig. 18.3. Note that $y = (l/2)\cos\theta$. The velocity of the CM of the plank is $v = \dot{y} = -l(\sin\theta)\dot{\theta}/2$. Here $v < 0$. We take CM of the plank as the reference point. The equation of motion of the plank is

$$mg - N = -m\dot{v} \tag{18.7}$$

$$N\frac{l}{2}\sin\theta = I\dot{\omega} \tag{18.8}$$

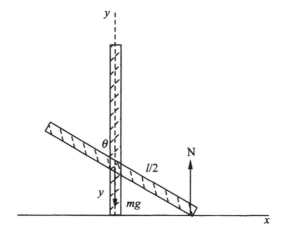

Figure 18.3   *A plank falling under gravity.*

Multiplication of Eq. (18.8) by $\omega = \dot{\theta}$ yields

$$\frac{d}{dt}(\frac{1}{2}I\omega^2) = -Nv \tag{18.9}$$

We multiply Eq. (18.7) by v and apply Eq. (18.9), which yields

$$mgv = Nv - \frac{d}{dt}(\frac{1}{2}mv^2)$$

$$= -\frac{d}{dt}(\frac{1}{2}I\omega^2 + \frac{1}{2}mv^2)$$

Since $v = \dot{y}$, we obtain

$$\frac{1}{2}I\omega^2 + \frac{1}{2}mv^2 + mgy = E$$

which is the statement of conservation of energy. Hence energy conservation holds for the plank.

These relatively simple examples illustrate that if a rigid body is rotating about a fixed axis, we can describe the motion of the rigid body somewhat easily. In the next two sections we will sketch the motion of a wheel and a unicycle. We will assume that these systems also rotate about a single axis. Still there are various other factors which complicate the motion.

18.4   Friction and rotation of wheels

Frictional force acting on a rolling wheel is quite interesting. We will study this system in the following discussion. The rolling condition yields $v_{CM} = \omega R$, where v_{CM} is the velocity of the CM, and ω is the angular velocity of the wheel in the clockwise direction. If the wheel is just rolling, will there be a frictional force? If yes, will it be in the forward direction or in the backward direction?

If there is a frictional force in the backward direction at the point of contact, then the frictional force would decrease the linear velocity of the wheel, while the torque due to the frictional force would increase the angular velocity of the wheel. This is contradictory because $V_{CM} = \omega R$, that is, linear velocity of the CM is proportional to the angular velocity. Similar arguments prove that frictional force in the forward direction is also not possible for a rolling wheel. Hence frictional force is absent altogether on a rolling wheel. A perfect rigid wheel rolls forever with a constant ω.

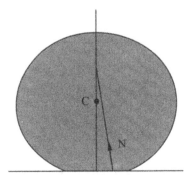

Figure 18.4   *A real rolling wheel gets compressed at the bottom. The normal force on the wheel is shown in the figure.*

We know however that any rolling wheel slows down and finally stops. How does it happen? A real wheel is not completely rigid. It gets compressed at the bottom as shown in

Fig. 18.4. The normal force is not vertical but it is slightly inclined as shown in the figure. This force creates a torque that slows down the wheel. Note that the normal force also decelerates the wheel because it has a component along the $-x$ direction. Thus frictional force slows down both the translational and rotational motion; we need to provide additional force to maintain the motion of the wheel. That is why we need fuel in our vehicles.

18.5   Motion of a unicycle

The dynamics of a bicycle is quite complex. For simplicity, we will focus on a unicycle (one-wheeled cycle) and discuss how the unicycle accelerates and decelerates. In this example we use the clockwise direction as positive as a special case.

To accelerate, a cyclist exerts a torque τ_P on the pedal by applying force on the front of the pedal (see Fig. 18.5 for an illustration).

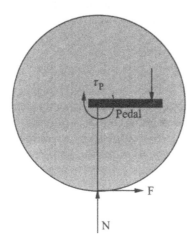

Figure 18.5   *Forces acting on a unicycle.*

Consequently the net torque on the wheel about its centre will be $\tau_P - FR$, where F is the frictional force on the wheel. This torque provides an angular acceleration to the wheel

$$m_W R^2 \alpha = \tau_P - FR,$$

where m_W is the mass of the wheel. The frictional force provides a linear acceleration to the cyclist (mass M) and cycle (mass M_C) as following:

$$(M + M_C)a = F.$$

Using $a = \alpha R$ and the above two equations we obtain

$$a = \frac{\tau_P}{M + M_C + m_W}.$$

This is how the cyclist is able to accelerate. Here both a and α are positive.

To decelerate a unicycle we apply brakes on the wheel. The brakes provide an opposing torque (negative τ_P) on the wheel, that produces frictional force in the opposite direction. The above equations show that both a and α will now become negative, and the unicycle will slow down.

In all the examples discussed so far in this chapter, the angular velocity is along one of the principle axis of the body. In the following sections we will discuss examples in which angular velocity will be along several of the principle axes of the rigid body.

18.6   Rotation about more than one principal axis

The motion of a rigid body becomes quite interesting and nontrivial when angular velocity has components along more than one principal axes. One of the common features in these situations is the precession that has been discussed for torque-free rigid bodies in Section 17.10. In this section we will illustrate an even more complex rotation using a gyroscope.

18.6.1   Gyroscope

The gyroscope, shown in Fig. 18.6(a), is an interesting device having three independent axes of rotation. We strongly urge you to play with a gyroscope in your physics laboratory. A common feature in the gyroscope is that the rotation of the rigid body (the disc in Fig. 18.6(a)) about the vertical axis and the spin axis is torque-free (ignoring friction). We can use this feature to derive an equation for the gyroscope and solve for its motion.

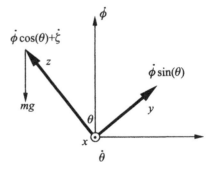

Figure 18.6   *(a) A gyroscope. (b) The components of the angular velocity of the gyroscope along its principal axes.*

Before going ahead with the full-fledged analysis, we will make a remark. If the spin axis of the gyroscope was massless, then the rotation of the gyroscope will be the same as that of a torque-free body discussed in Section 17.10. The disc spins and precesses with a constant polar angle between the spin axis and the vertical axis. The motion gets another dimension when we attach a mass on the spin axis as shown in Fig. 18.6(a). This mass induces a torque about the horizontal axis; this torque induces a variation in the angle θ. In the following we will study this behaviour quantitatively.

Let us focus on the rigid body of Fig. 18.6(a). Its principal axes are shown in Fig. 18.6(b). Since the disc is a symmetric top, $I_{xx} = I_{yy} = I_1$ and $I_{zz} = I_3$. Recall our discussion in Section 17.4 that the orientation of the spin axis is determined using the polar angle θ and the azimuthal angle ϕ, while the spin about the spin-axis is denoted by ζ. Clearly the instantaneous angular velocity about the vertical (Z) axis is $\dot{\phi}$, and the spin angular velocity is $\dot{\zeta}$. Also the instantaneous angular velocity about the horizontal (x) axis is $\dot{\theta}$. We can resolve these angular velocities along the three principal axis—they are $\dot{\theta}$ along x, $\dot{\phi} \sin\theta$ along y, and $\dot{\zeta} + \dot{\phi} \cos\theta$ along z. Hence the instantaneous angular momentum about the three principal axes are

$$L_x = I_1 \dot{\theta}, \tag{18.10}$$

$$L_y = I_1 \dot{\phi} \sin\theta, \tag{18.11}$$

$$L_z = I_3(\dot{\zeta} + \dot{\phi} \cos\theta). \tag{18.12}$$

Since the rotation about the spin axis (z) and the vertical axis (Z) are torque-free, the angular momenta about these axes are conserved. Hence L_z is constant, and

$$L_Z = L_z \cos\theta + L_y \sin\theta$$

$$L_Z = I_3(\dot{\zeta} + \dot{\phi} \cos\theta) \cos\theta + I_1 \dot{\phi} \sin^2\theta = \text{const.} \tag{18.13}$$

If we assume the mass on the spin axis (m) to be concentrated at a distance l from the centre of the disc, the torque due to this mass is $mgl \sin\theta$. The rate of change of the angular momentum about the horizontal axis is due to the variation in $\dot{\theta}$, as well as to the precession of L_x and L_y about the vertical axis, that is

$$\frac{dL_x}{dt} = mgl \sin\theta$$

or, $\quad I_1 \ddot{\theta} + L_z \sin\theta \dot{\phi} - L_y \cos\theta \dot{\phi} = mgl \sin\theta$

or, $\quad I_1 \ddot{\theta} + I_3(\dot{\zeta} + \dot{\phi} \cos\theta) \sin\theta \dot{\phi} - I_1 \sin\theta \cos\theta \dot{\phi}^2 = mgl \sin\theta. \tag{18.14}$

The constancy of L_z and L_Z simplifies Eq. (18.14) into

$$I_1 \ddot{\theta} + \left[L_z \sin\theta - \frac{(L_Z - L_z \cos\theta)}{\sin\theta} \cos\theta \right] \dot{\phi} = mgl \sin\theta, \tag{18.15}$$

where $\quad \dot{\phi} = \dfrac{(L_Z - L_z \cos\theta)}{I_1 \sin^2\theta}.$

Also note that the rotation about the y-axis is not torque free; the side hinges apply a certain torque on the rigid body that can be obtained by computing dL_y/dt.

Equation (18.15) is a second-order differential equation in θ that can in principle be solved given the initial conditions $\theta(t = 0)$ and $\dot{\theta}(t = 0)$. There is however a simpler way to solve for θ. We convert Eq. (18.15) into a first-order differential equation using the energy method. The energy of the rigid body is

$$\frac{L_x^2}{2I_1} + \frac{L_y^2}{2I_1} + \frac{L_z^2}{2I_3} + mgl\cos\theta = E$$

or, $$\frac{1}{2}I_1(\dot{\theta}^2 + \dot{\phi}^2\sin^2\theta) + \frac{L_z^2}{2I_3} + mgl\cos\theta = E$$

or, $$\frac{1}{2}I_1\dot{\theta}^2 + U_{\text{eff}}(\theta) = E' \tag{18.16}$$

where $$U_{\text{eff}}(\theta) = \frac{(L_Z - L_z\cos\theta)^2}{2I_1\sin^2\theta} - mgl(1 - \cos\theta),$$

and $$E' = E - mgl - \frac{L_z^2}{2I_3}$$

Note that E' is a constant. You can also verify that the time derivative of Eq. (18.16) yields Eq. (18.15).

The Eq. (18.16) resembles the equation of a particle in one dimension. We will study this equation for various values of parameters E', L_Z, L_z and mgl. We can non-dimensionalise the above equation by choosing our time scale as $(I_1/2mgl)^{1/2}$ which yields

$$\frac{1}{2}\left(\frac{d\theta}{dt'}\right)^2 + a\left(\frac{1 - b\cos\theta}{\sin\theta}\right)^2 - (1 - \cos\theta) = \bar{E},$$

where

$$a = \frac{L_Z^2}{2I_1},$$

$$b = \frac{L_z}{L_Z},$$

$$\bar{E} = \frac{E'}{mgl},$$

and $t = t'\sqrt{I_1/2mgl}$. Since $\dot{\theta}^2 > 0$, the values of θ are confined between the turning points θ_1 and θ_2 that are given by

$$a\left(\frac{1 - b\cos\theta_{1,2}}{\sin\theta_{1,2}}\right)^2 - (1 - \cos\theta_{1,2}) = \bar{E},$$

The variable θ performs periodic motion between these two values (recall our discussion on 1D motion in a confining potential). The motion of the tip of the gyroscope axis is shown in Fig. 18.7. The spin axis falls down due to gravity, but climbs up again because of the angular momentum. This type of motion is called *nutation*. In Fig. 18.7 we illustrate two different type of nutation. In case (a), $\dot{\phi}$ is always positive, while in case (b), $\dot{\phi}$ becomes zero for a certain $\theta_c = \cos^{-1}(1/b)$ lying between θ_1 and θ_2. We leave the solving of the above problem using a computer as an exercise for interested students.

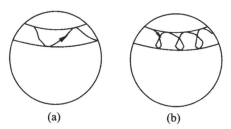

(a)   (b)

Figure 18.7   *The motion of the spin axis of the gyroscope. $\dot{\phi}$ is always positive in case (a), while it changes sign in case (b).*

In summary, the motion of a gyroscope consists of a spin about the spin axis (z-axis), precession (Z-axis), and nutation (x-axis). In the absence of frictional force, θ variation is periodic. However these motion get damped in the actual gyroscope due to the frictional torque. The frictional torque on the horizontal axis induces a term of the type $\gamma\dot{\theta}$ in Eq. (18.14) that damps the oscillations in θ. The frictional torque along the vertical axis and the spin axis damps $\dot{\phi}$ and $\dot{\zeta}$. Consequently the gyroscope loses its angular momentum due to these frictional torques, and asymptotically the gyroscope reaches its static equilibrium (the axis along the vertical).

A major application of the gyroscope is in navigation. Note that a torque-free gyroscope (set $m = 0$ in the above example) whose angular momentum and angular velocity are aligned will maintain its direction of rotation irrespective of the orientations or position of its base. Hence the direction of the spin can be used as a reference direction or initial direction for navigation.

There are various types of gyroscopes. We will briefly study them in the next section. Also refer to the website http://www.gyroscope.org for illustrations of different kinds of gyroscopes.

18.6.2   Other types of gyroscopes: Top and bicycle wheel

A top or a bicycle wheel whose axis is attached to a pivot are some other examples of gyroscopes. These systems are very similar to the disc gyroscope discussed in the earlier section, the only difference being that the centre of mass of the top or the bicycle wheel is not at the pivot unlike the disc gyroscope where the disc's centre of mass is at the pivot. Hence, for the top or the bicycle wheel, the moment of inertia I_1 has to be replaced by $I_1 + ml^2$, where m is the mass of the top or the wheel, and l is the distance of the centre of mass from the pivot. The analysis for the top and the bicycle wheel is very similar to that for the disc gyroscope discussed in the previous section.

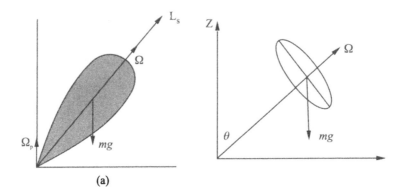

(a)

Figure 18.8   (a) Motion of a top in a gravitational field (b) Motion of a bicycle wheel in a gravitational field. Both these objects are hinged at the origin.

As a special case, let us consider the motion of a fast rotating top in the gravitational field. Under this approximation, the most dominant terms of Eq. (18.14) are $L_z \sin\theta\dot{\phi}$ and $mgl \sin\theta$. Equating these two terms we obtain

$$\dot{\phi} = \Omega_P = \frac{mgl}{L_z}, \tag{18.17}$$

which is the precession angular velocity of the top. The angular momentum due to the precession has been ignored in deriving the above equation. This approximation is valid if the ignored angular momentum is much smaller than the spin angular momentum $\mathbf{L}_z$ or $\mathbf{L}_S$, which would be the case for a fast spinning heavy top.

18.7   Euler's equation

When a rigid body rotates about a fixed axis, we compute the time evolution of only one angle. However when the spin axis itself rotates, then the rigid body is specified by Euler angles discussed in an earlier chapter. In this section we will derive a set of dynamical equations for these angles.

While discussing rotating frames, we had found a relationship between rate of change of a vector in an inertial reference frame, and in another reference frame that is rotating with respect to the inertial frame. In this section we will write down $d\mathbf{L}/dt$ in these two reference frames.

Consider a rigid body whose moments of inertia about the three principal axes are I_{xx}, I_{yy} and I_{zz}, and their directions at a given instant are along $\hat{\mathbf{x}}$, $\hat{\mathbf{y}}$, and $\hat{\mathbf{z}}$ respectively in the body frame. If the rigid body has angular velocity

$$\mathbf{\Omega} = \Omega_x \hat{\mathbf{x}} + \Omega_y \hat{\mathbf{y}} + \Omega_z \hat{\mathbf{z}},$$

then $\mathbf{L} = I_{xx}\Omega_x \hat{\mathbf{x}} + I_{yy}\Omega_y \hat{\mathbf{y}} + I_{zz}\Omega_z \hat{\mathbf{z}}.$

The reference frame attached to the rigid body is the rotating frame. The rate of change of angular momentum $\mathbf{L}$ in the laboratory and rotating frames are related as

$$\left[\frac{d\mathbf{L}}{dt}\right]_{\text{inertial}} = \left[\frac{d'\mathbf{L}}{dt}\right]_{\text{rotating}} + \boldsymbol{\Omega} \times \mathbf{L} = \mathbf{N},$$

where $\mathbf{N}$ is the net torque acting on the body. Using the definition of angular momentum, we obtain

$$I_{xx}\frac{d\Omega_x}{dt} + (I_{zz} - I_{yy})\Omega_y\Omega_z = N_x,$$

$$I_{yy}\frac{d\Omega_y}{dt} + (I_{xx} - I_{zz})\Omega_z\Omega_x = N_y,$$

$$I_{zz}\frac{d\Omega_z}{dt} + (I_{yy} - I_{xx})\Omega_x\Omega_y = N_z.$$

Here the time derivatives are measured in the rotating frame. The above equations are called *Euler's equations*. For torque-free rotation, $\mathbf{N} = 0$, so Euler's equations become

$$I_{xx}\frac{d\Omega_x}{dt} + (I_{zz} - I_{yy})\Omega_y\Omega_z = 0, \tag{18.18}$$

$$I_{yy}\frac{d\Omega_y}{dt} + (I_{xx} - I_{zz})\Omega_z\Omega_x = 0, \tag{18.19}$$

$$I_{zz}\frac{d\Omega_z}{dt} + (I_{yy} - I_{xx})\Omega_x\Omega_y = 0. \tag{18.20}$$

The solution of Euler's equations for a general situation could be quite complex. In the following examples we will apply Euler's equations to relatively simpler examples. However we will make a remark before we move on to apply the above equations to certain problems. It is quite confusing to imagine $d\Omega_i/dt$ measured in the rotating frame; one may think that $d\Omega_i/dt$ is zero in the rotating frame. However this is not so. The vector Ω is the same in both laboratory and the rotating frame. Also, for a disc spinning about the z-axis, $d\Omega_z/dt$ is the same in both the laboratory frame and in the body frame. However the difference appears when the spinning axis itself starts rotating about some other axis; the term $\boldsymbol{\Omega} \times \mathbf{A}$ becomes important under these situations.

In the following two sections we will apply Euler's equations to several problems.

18.7.1   Dynamical stability

Recall the definition of stability introduced in Chapter 14 for static objects. Stable systems return to their equilibrium position, while unstable ones move further away from their equilibrium positions. We can generalise this concept to objects in motion. When we apply a small disturbance to a moving body, and the body comes back to its original configuration, then it is said to be *dynamically stable*. However if the system moves further away from its original configuration, then it is said to be *dynamically unstable*.

When a mechanical system has certain angular momentum, we require a torque to change the angular momentum of the system. Systems with large angular momentum require larger torque for a small change in the angular momentum. Hence systems with large angular momentum have a stronger dynamic stability. This is the reason why a fast-moving bicycle is easier to balance than a slow-moving one. Also, a spinning bullet is more stable than the a non-spinning bullet. Hence, angular momentum brings stability to many systems.

We will demonstrate the above results quantitatively using Euler's equation by taking a rigid body whose three moments of inertia are unequal (Fig. 18.9). Suppose the body is rotating about the z-axis with angular frequency Ω_z. If $\Omega_x = \Omega_y = 0$, then the body keeps rotating along the z-axis with angular frequency Ω_z. We would like to examine the dynamics when we introduce a small Ω_x and/or Ω_y on the body. These perturbations could be present in the initial conditions. Since gravity passes through the CM of the body, the net torque on the body is zero. We can ignore the torque induced by air friction or other small external forces. An application of Euler's equation to the rigid body yields the following equations:

$$I_{xx}\dot{\Omega}_x = (I_{yy} - I_{zz})\Omega_z\Omega_y \tag{18.21}$$

$$I_{yy}\dot{\Omega}_y = (I_{zz} - I_{xx})\Omega_z\Omega_x. \tag{18.22}$$

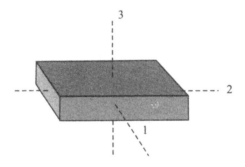

Figure 18.9   *Stability of a rotating rigid body.*

Taking another time derivative of Eqs. (18.21) and (18.22), and using $\dot{\Omega}_z = 0$, we obtain

$$\ddot{\Omega}_{x,y} = -\left[\frac{(I_{zz} - I_{yy})(I_{zz} - I_{xx})}{I_{xx}I_{yy}}\Omega_z^2\right]\Omega_{x,y}.$$

The bracketed term is positive when $I_{zz} < I_{xx}, I_{yy}$ or $I_{zz} > I_{xx}, I_{yy}$; for these cases, the initial fluctuations oscillates (not grow). In the presence of small friction, the fluctuations will die down. On the other hand when I_{zz} lies between I_{xx} and I_{yy}, the bracketed term is negative, and the fluctuations will grow in time, and the body is dynamically unstable. The rigid body shown in Fig. 18.9 is dynamically stable when rotated about axes 2 and 3, but is dynamically unstable when rotated about axis 1 because $I_2 < I_1 < I_3$. You can test this theoretical result by spinning your eraser about three axis and testing its stability. Give a spin to the eraser about different axis and throw it. The eraser is dynamically stable when it is spun about its axis with maximum or minimum moment of inertia.

In Section 17.10 we studied torque-free rotation of a rigid body using rotational kinematics. In the following section we will study this system using Euler's equation (rotational dynamics).

18.7.2   Torque-free rotation

Let us apply Euler's equations to the torque-free rotation of a symmetric top discussed in Section 17.10. For these bodies $I_{xx} = I_{yy} = I \neq I_{zz}$. An application of Eq. (18.20) of Euler equation yields $\dot{\Omega}_z = 0$, hence $\Omega_z = $ const. When we use complex velocity variable $\zeta = \Omega_x + i\Omega_y$, the other two Euler equations (Eqs. (18.18) and (18.19)) yield

$$I\frac{d\zeta}{dt} - i(I_{zz} - I)\Omega_z\zeta = 0.$$

The solution of the above equation is

$$\zeta(t) = \zeta(0)\exp\left(i\Omega_z\frac{I_{zz} - I}{I}t\right).$$

For the initial conditions $\Omega_x(0) = A$ and $\Omega_y(0) = 0$, the solution is

$$\Omega_x(t) = A\cos(\omega t), \tag{18.23}$$

$$\Omega_y(t) = A\sin(\omega t), \tag{18.24}$$

where $\omega = \Omega_z\left(\dfrac{I_{zz} - I}{I}\right).$ \hfill (18.25)

Hence, $\Omega_\rho = \Omega_x\hat{x} + \Omega_y\hat{y}$ rotates around the z-axis with angular velocity ω as shown in Figs. 18.10 and Fig. 18.11.

We will compare the above results with those obtained using kinematic calculations in Section 17.10. As shown in Fig. 18.10, the angular velocity can be written as $\Omega_z\hat{z} + \Omega_\rho\hat{\rho}$ or $\boldsymbol{\Omega}_S + \boldsymbol{\Omega}_P$, where $\boldsymbol{\Omega}_S$ and $\boldsymbol{\Omega}_P$ represent spin and precession angular velocities.

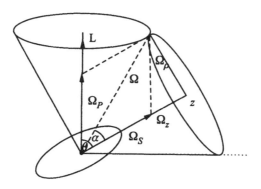

Figure 18.10   *The angular velocity of a torque-free top is $\boldsymbol{\Omega}_S + \boldsymbol{\Omega}_P = \Omega_z\hat{z} + \Omega_\rho\hat{\rho}$. Here $I_{zz} < I$.*

The relationship between these two forms is

$$\mathbf{\Omega} = \mathbf{\Omega}_S + \mathbf{\Omega}_P$$

$$= (\Omega_S + \Omega_P \cos\theta)\hat{\mathbf{z}} + \Omega_P \sin\theta\hat{\rho}.$$

Hence $\Omega_z = \Omega_S + \Omega_P \cos\theta$ and $\Omega_\rho = \Omega_P \sin\theta$ are the components of angular velocity in the body frame. Note that $\Omega_S \neq \Omega_z$.

In the laboratory reference frame, $\mathbf{\Omega}$ precesses around the z-axis with precession angular velocity Ω_P. The motion is depicted in Fig. 18.10. In the body reference frame, Ω_ρ rotates around the body-z axis with angular velocity ω, and the body-z axis itself rotates around the space-z axis (see Eqs. (18.23) and (18.24)). This motion also forms a cone as shown in the figure. Since the vector $\mathbf{\Omega}$ is common to both the cones, the distance covered by $\mathbf{\Omega}$ in both the cones in time dt must be the same. By equating these distances, we obtain

$$\Omega_\rho \omega dt = -\Omega \sin(\theta - \alpha)\Omega_P dt,$$

where α is the half-angle of the body cone. From the above equation we can compute ω as

$$\omega = -\frac{\Omega \sin(\theta - \alpha)}{\sin\theta}$$

$$= -\Omega \cos\alpha + \Omega \cot\theta \sin\alpha$$

$$= -\Omega_z + \Omega_\rho \cot\theta$$

$$= -\Omega_z + \Omega_P \cos\theta$$

$$= -\Omega_S$$

As expected, the observer in the body frame finds that $\hat{\rho}$ rotates with angular velocity $-\Omega_S$, which is exactly opposite to the observations of the laboratory observer who finds that the body spins with angular velocity Ω_S.

Using Eq. (17.13) we obtain

$$\omega = -\Omega_S = \Omega_z \left(\frac{I_{zz} - I}{I}\right). \tag{18.26}$$

which is the same as what we had calculated earlier using Euler's equations (see Eq. (18.25)). For bodies with $I_{zz} > I$, ω (or $-\Omega_S$) and Ω_z are in the same direction as shown in Fig. 18.11. However, ω and Ω_z are in the opposite directions for bodies with $I_{zz} < I$ as shown in Fig. 18.10.

We will illustrate this phenomena using the following examples.

1. **Motion of a frisbee**

    For a frisbee $I_{zz} = 2I_{xx} = 2I$. Hence $\omega = \Omega_z$. In the coordinate system attached to the frisbee, the principal axes y and z rotate with angular velocity ω, which is equal to the angular velocity of the rigid body around the z-axis.

2. **Earth**

The spin axis and the precession axis of the Earth make an angle of the order of a few tenths of a second. For the Earth $I_{xx} = I_{yy} = 0.329591MR^2$, and $I_{zz} = 0.330675MR^2$. Therefore $\omega/\Omega_z = (I_{zz} - I_{xx})/I_{xx} \approx 1/304$, or the time period corresponding to ω is approximately 304 days. Hence we expect the wobble axis (ω_ρ) to rotate around the spin axis with the above time period. Chandler studied this phenomena and found that the Earth's axis of rotation is precessing with a period of about 435 days. The difference in time period is believed to be due to the fact that the Earth is not rigid. For details refer to Goldstein et al. (2002).[2]

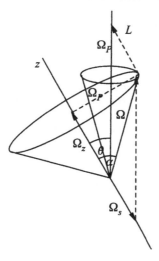

Figure 18.11   *The angular velocity vectors for rigid bodies with $I_{zz} > I_{xx} = I_{yy}$.*

In this chapter we discussed rigid body dynamics. Rotation of a rigid body about a single axis is simple. However, the motion becomes complex when the angular velocity of the rigid body has components along more than one principal axes. In this chapter we looked at some simple examples. You can refer to advanced books for discussions on complex rotational dynamics.

Exercises

1. A cylinder and a sphere of the same mass and radius (but different densities) are rolled down an incline. Which of the two will come down faster?

---

[2]The torque-free precession of the Earth is different from another precession due to the tidal effects. Because of the oblate nature of the Earth, the Sun and the moon cause a torque on the Earth resulting in a precession of the spin axis in a cone whose half-angle is 23.5 degrees. The time period of this precession is around 26000 years. At present Earth's axis points towards the pole star, however it will point to a different direction at a later time. See website http://www.opencourse.info/astronomy/introduction/03.motion_earth/ for more details

2. A hollow cylinder and a solid cylinder of same mass and size are made of unknown material of different densities? Is it possible to identify the cylinders by rigid-body dynamics? If yes, how?

3. Two masses m_1 and m_2 are hanging on the two sides of a pulley that has moment of inertia I. Assume the string to be massless and inextensible. Compute

   (a) the acceleration of the masses

   (b) the tension of the string.

4. A thin uniform rod of length l and mass M is revolving with a constant angular velocity ω about point O from the ceiling. It describes a conical surface. Compute the angle of deviation of the rod from the vertical, and the reaction force at point O.

5. A thin uniform rod of length l and mass M is hinged to the ceiling. Show that the rod performs an SHM for small displacements. Compute the frequency of oscillation.

6. A waterfall is falling from a height 10 metres and it brings down 1000 litres of water per second. Design a small generator for a village community. How much power can be generated from this setup?

7. A uniform thin rod of mass M and length L can rotate freely on a horizontal plane about the hinge at the mid-point O as shown in Fig. 18.12(a).

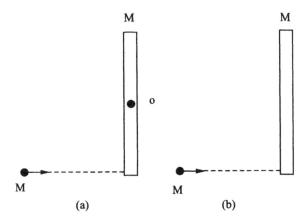

Figure 18.12   *Exercise 7; Exercise 9.*

A point particle of mass M collides inelastically with velocity v at the bottom of the rod. Assume the duration of the collision to be very short (collision approximation). Compute the angular velocity of the rod after the collision. The hinge experiences an impulse. Compute the impulse in the direction perpendicular to the rod.

8. Consider the setup of Problem 7.

(a) Where should be the position of the hinge so that the reaction force on the hinge is zero?

(b) Suppose the mass M collides elastically, but the direction of impact is still perpendicular to the rod. Will the position of the hinge change compared to part (a)?

9. A uniform thin rod of mass M and length L is lying on a horizontal frictionless table. A point particle of mass M collides inelastically with velocity v at the bottom of the rod as shown in Fig. 18.12(b). Assume collision approximation.

(a) Compute the velocity of CM before and after the collision.

(b) Compute the angular momentum of the CM before and after the collision.

(c) Compute the angular velocity of the rod.

(d) Describe the motion of the rod.

10. In cricket, a batsman always likes to hit the ball such that the reaction force on the batsman's hand is as small as possible. Where should the point of impact of the bat and ball be to achieve this objective?

11. The upper end of a ladder is resting against a frictionless wall, while its bottom end stands on a rough floor. The coefficient of friction between the ladder and the floor is k. Determine the angle between the ladder and the wall at which the ladder will be in equilibrium.

12. A ladder is leaning against a frictionless wall and the ground, which is also frictionless. The ladder starts to slip downward.

(a) Obtain an expression for the angular velocity of the ladder as a function of time (HINT: Constraint forces do not do any work on the ladder).

(b) Show that the top of the plank loses contact with the wall when it is at two-thirds of its initial height.

13. Refer to Fig. 18.13. The middle rod of length $2L$ has mass density ρ per unit length, while the connecting rods of length l each are massless. The bobs have mass M each. The setup is hinged at point A. The angles between the middle rod and the connecting rods is 30 degrees each.

(a) For what values of l will the system be stable?

(b) Under stable regime, compute the frequency of oscillation.

14. A yo-yo of mass M has an axle of radius b and spool of radius R, and its moment of inertia is approximately $MR^2/2$. The yo-yo rests on a horizontal table. At $t = 0$, the yo-yo is pulled horizontally with a constant force T using the string.

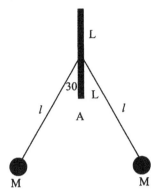

Figure 18.13   *Exercise 13.*

(a) The yo-yo slides without rolling for T beyond T_{critical}. Derive T_{critical}.

(b) Under what conditions does the yo-yo roll without slipping?

15. The yo-yo described in Exercise 14 is pulled so that the string makes an angle θ with the horizontal. Describe the motion of the yo-yo for different values of θ.

16. The yo-yo described in Exercise 14 is held vertically in air, and released from rest. What is the tension in the cord as the yo-yo descends and as it ascends?

17. The yo-yo described in Exercise 14 is held vertically in air, and released from rest. While the yo-yo is descending, the vertical support of the yo-yo is pulled up with an acceleration A. Describe the motion of the yo-yo.

18. A cylinder of radius R and moment of inertia I is let go on an incline that makes an angle α with the horizontal. The coefficient of friction between the ball and the plane is μ. Derive the conditions under which the cylinder rolls and slides. Describe the motion in each case.

19. The sides of an eraser are 1, 2, and 3 cms. The density of the material is 5 gm/cc. Analyse the stability of the eraser when it is rotated by axes 1, 2, and 3.

20. Compute the torque on the gyroscope studied in this chapter.

Projects

1. Compute the torque on the Earth due to its oblate shape. What is the consequence of this torque?

2. Visit the website http://www.gyroscopes.org and study various types of gyroscopes. Also study the application of a gyroscope in navigation.

3. Solve the motion of a gyroscope on a computer.

4. Study about a yo-yo.

19

Special Theory of Relativity: Kinematics

In previous chapters we studied mechanical systems in the Newtonian framework that assumes absolute space and absolute time. It is found that the Newtonian framework is not valid for particles moving with speeds close to the speed of light. Einstein formulated a more refined theory of mechanics called the *theory of relativity*. This theory is quite profound, and it provided a new paradigm in physics. A thorough discussion of this topic is impossible in just a few pages. But we will try to highlight the basics of this theory. Einstein gave two versions of the relativity theory. First, the *special theory of relativity* (SR) which deals with inertial frames, and second, the *general theory of relativity* (GR), which deals with accelerating frames of reference. In the present and the next chapter we will focus on the special theory of relativity. The general theory of relativity is quite advanced, and cannot be covered in introductory mechanics book.

The spacetime of SR is very different from the spacetime of Newtonian mechanics. We will discuss this in the following section.

19.1 Spacetime

We illustrate the structure of spacetime in SR by making an analogy with the three-dimensional space you have studied earlier in Newtonian dynamics. In Newtonian space, the distance between two points is the same in all references frames. In Section 7.3 we considered two reference frames that make an angle θ with respect to each other. Imagine a vector OP, where O is the origin of the coordinate frames and P is any point. The components of OP in the two frames are $(\Delta x, \Delta y, \Delta z)$ and $(\Delta x', \Delta y', \Delta z')$ respectively. The quantities $\Delta x'$ and $\Delta y'$ are related to Δx and Δy as follows:

$$\Delta x' = \Delta x \cos \theta + \Delta y \sin \theta,$$

$$\Delta y' = -\Delta x \sin \theta + \Delta y \cos \theta,$$

$$\Delta z' = \Delta z$$

It is easy to see that the distance OP measured in the two coordinate frames are the same

$$(\sqrt{(\Delta x)^2 + (\Delta y)^2 + (\Delta z)^2} = \sqrt{(\Delta x')^2 + (\Delta y')^2 + (\Delta z')^2})$$

even though the components are different. The above type of space is called Euclidean space in honour of the Greek geometer Euclid.

Recall our discussion in Chapter 7 where we classified physical quantities as vectors, scalars etc. The quantities that remain unchanged under rotation are called scalars, and the ones that transform as coordinates (x, y, z) are called vectors.

Now we will discuss the spacetime structure in SR. We have two reference frames: (i.) the laboratory frame S, and (ii.) the rocket frame S' that is moving along the x-axis of the laboratory frame with a constant velocity V_r. At $t = 0$, the origins of both the systems coincide with each other. Observers in the two reference frames record events. Let us focus on two events (say bursts of firecrackers): the coordinates of the first event in S and S' are $(\mathbf{r} = \mathbf{r}' = 0,\ t = t' = 0)$, and that of the second event are (x, y, z, t) and (x', y', z', t') respectively.

In the Galilean transform discussed earlier, the time interval and the physical distance between the two events are the same for both the observers. This result is valid for $V_r \ll c$, where c is the velocity of light. Experiments show that when V_r is comparable to c, $\Delta t' \neq \Delta t$. Surprising! In addition, experiments show that

$$(c\Delta t)^2 - (\Delta x)^2 - (\Delta y)^2 - (\Delta z)^2 = (c\Delta t')^2 - (\Delta x')^2 - (\Delta y')^2 - (\Delta z')^2. \qquad (19.1)$$

The above quantity is called the *spacetime interval* which is the same for both the observers. Making an analogy with the transformation under rotation, we interpret the above result as follows: The events are points in 4-dimensional *spacetime*. Relative to a reference event, observers find different space and time coordinates for the events. However the interval between an event and a reference event calculated using Eq. (19.1) is the same for all observers. The invariance of the interval, i.e. its independence from the choice of the reference frame, induces one to recognise that time and space cannot be separated from each other. Space and time are part of the single entity called *spacetime*.

The above type of spacetime is called *Lorentzian*. Note that our spacetime is four dimensional, and is different in character from Euclidean space. In Euclidean space the distance between two points is $\sqrt{\sum_i (\Delta x_i)^2}$, but in Lorentzian space, the distance is

$$\sqrt{(\Delta x_0)^2 - (\Delta x_1)^2 - (\Delta x_2)^2 - (\Delta x_3)^2}$$

with $\Delta x_0 = c\Delta t$. The distance between any two points in Euclidean space is always positive, but it could be zero or imaginary in Lorentzian space.

Einstein's special theory of relativity is the theory of kinematics and dynamics in Lorentzian spacetime. We will start our discussion with the postulates of SR.

19.2   The inertial reference frame; Observers in relativity

Inertial reference frames are fundamental in SR. In fact, SR gives a prescription to identify an inertial reference frame. Recall that in Newtonian framework, an inertial reference frame is one in which an isolated particle continues to move with its initial velocity; if the particle is at rest, it will remain at rest forever. A rocket moving with a constant velocity is an example of an inertial reference frame in both special relativity and Newtonian mechanics.

As discussed in Section 10.5, according to the *principle of equivalence* a freely falling elevator under gravity is also an inertial frame. Gravity acts on both the elevator and the particle, so if no other force (electrical for example) acts on the particle there is no relative acceleration between the observer and the particle. The distance between two particles will remain the same in the elevator as long as the distance between the particles is small (see Section 10.5). When the distance between the particles are large, tidal forces come into play. We can detect the external gravitational force using the tidal forces. In fact, this observation is used to identify a reference frame. If the distance between two *free* particles does not exceed a certain small value in a unit time, then we can call that reference frame to be inertial. Of course, this definition is based on the permissible error. If the error limit is decreased, then some frames which were considered inertial will no longer be inertial.

In conclusion, a freely falling reference frame is an inertial reference frame in SR. It is a local reference frame because the law of inertia is valid only in a small elevator. Tidal forces spoil the law of inertia in large elevators.

How do we make measurements in spacetime? We form a grid of sticks of unit length. At each joint we place synchronised clocks. When an event takes place at a point, the observer at that point records the time. This way observers record position as well as time of an event. Note that observers in different reference frames may record different values for position and time of the same event.

The units of length and time are metre and second respectively. It is however convenient to specify time in the same unit as length. This can be achieved by multiplying time by c, i.e., $ct \rightarrow t$. In this system, velocity will be dimensionless : $v \rightarrow v/c = \beta$, and Eq. (19.1) becomes

$$(\Delta t)^2 - (\Delta x)^2 - (\Delta y)^2 - (\Delta z)^2 = (\Delta t')^2 - (\Delta x')^2 - (\Delta y')^2 - (\Delta z')^2. \tag{19.2}$$

After this definition of an inertial frame, we will state the postulate of SR.

19.3   The principle of relativity

Einstein's principle of relativity states that

All the laws of physics are the same in every inertial reference frame.

Usually the laws of physics are expressed in terms of equations. According to the above postulate, the same equation is valid in all inertial reference frames. These equations contain physical constants (called fundamental constants) like speed of light, universal gravitational constant, etc. The above postulate also implies that the fundamental constants must be the same in all inertial reference frames.

Since the laws of physics are the same in all inertial reference frames, all inertial frames are equivalent, and no inertial frame is special. The laws of physics cannot provide a way to distinguish one inertial frame from another. Imagine a train that is completely sealed from the outside world; we cannot see anything outside, and no external field can penetrate inside the train. According to the postulate of relativity, no experiment can help us determine whether the train is moving with a constant velocity, or if it is stationary.

The laws of electromagnetism hold true in all inertial reference frames. The equations of electrodynamics (called Maxwell's equation) contains speed of light c as a constant ($c = 2.997925... \times 10^8$ metre per second). According to the principle of relativity, this experimental value must be the same in all inertial frames. That is, the observed value of c is the same irrespective of the relative motion of the source or the observer.[1] This is very surprising indeed.

The constancy of the speed of light is contradictory to the Galilean transformation of velocity, which is based on the assumptions of absolute space and time. In a Newtonian framework c should have increased (decreased) if the observer moves towards (away from) the source. The constancy of the speed of light in all frames of reference was a major shock to the physicists at the turn of the twentieth century. Many physicists constructed theoretical models to explain the constancy of c in the framework of absolute space and absolute time. It was Einstein who abandoned the idea of absolute space and absolute time and constructed a fundamentally new framework called special theory of relativity. Historically, the constancy of c was first observed experimentally by Michelson and Morley around 1887. Many more experiments afterwards have observed the constancy of the speed of light with better accuracy. Note that the accuracy of the best experimental result is three parts in a million, i.e. it is good up 2 metre per second (Kennedy-Thorndike 1932).

Einstein's principle of relativity is very simple to state; but it is very powerful. There are amazing predictions of this theory. We will discuss some of them in the present chapter and the next one. But first we will start by deriving the transformation rules between coordinates and time in two inertial reference frames.

19.4 Lorentz transformation

In Section 19.1 we introduced the concepts of events and intervals between two events. Relative to a reference event, the position and time coordinates of an event could be different in two different inertial frames. However the Lorentzian distance (Eq. (19.2)) of the event from the reference event remains the same. In the present section we will derive the transformation rule between the position and time coordinates in Lorentzian spacetime. This transformation rule is called the *Lorentz transformation*.

Consider two reference frames: (i.) the laboratory frame S, and (ii.) the rocket frame S' that is moving along the x-axis of the laboratory frame with a constant velocity $\beta_r = V_r/c$. We assume that the x-axis of both the frames coincide, and the origin of both frames coincide at $t = t' = 0$. We will derive the transformation rules in several steps.

1. First we will consider the transformation rules for vertical coordinates. Both the frames have a one metre stick each vertically placed. When they cross each other, will the tip of the stick coincide or not? If the rocket's stick is below the lab's stick, we will have to conclude that the moving stick is shrunk in the vertical direction.

---

[1] Contrast this with the propagation of sound. For a stationary observer in still air, the speed of sound propagation is C_s. However for an observer moving towards the source, the speed becomes $C_s + V_r$, where V_r is the relative speed of the observer. As described above, light behaves very differently from sound.

However, the lab frame is also moving leftward relative to the rocket frame, hence it should shrink relative to the rocket's stick, which is not the case. This inconsistency can be removed only when the tips of both the sticks coincide. Note that we have used left–right symmetry in our argument, which is natural considering the isotropic nature of space. Hence,

$$y' = y,$$

$$z' = z.$$

The above result could also be obtained by using the following arguments. Consider the two events when the sticks coincide: one occurring at the bottom of the stick, and the other at the top of the stick. Since $\Delta x = \Delta t = \Delta x' = \Delta t' = 0$, we obtain $\Delta y = \Delta y'$ from Eq. (19.1).

2. Consider two events that takes place at the origin of the rocket frame at time $t' = 0$ (reference event) and $t' = \tau$. Hence, $\Delta x' = 0$ and $\Delta t' = \tau$. In the lab frame $\Delta x = \beta_r \Delta t$. What is the time interval Δt measured in the laboratory frame? We can obtain $\Delta t'$ using invariance of interval (Eq. (19.2)):

$$(\Delta t)^2 - (\Delta x)^2 = (\Delta t')^2 - (\Delta x')^2$$

or

$$\Delta t = \frac{\tau}{\sqrt{1 - \beta_r^2}} = \gamma(\Delta t'), \tag{19.3}$$

where $\gamma = 1/\sqrt{1 - \beta_r^2}$. So the time interval is not the same in both the frames. In the lab frame, the time interval is larger than that in the rocket frame. This phenomena is called *time dilation*. Also

$$\Delta x = \beta_r \gamma(\Delta t').$$

We can illustrate the above result using an experiment in which a laser light goes up and comes down vertically in the rocket frame (Exercise 3).

3. In the previous example, we considered events that occur at the same spot in the rocket frame. Now consider a general case when $x' \neq 0$. Note that the reference event takes place at the origin in both S and S'. The relationship between the space and time intervals are linear as given below:

$$x = ax' + bt', \tag{19.4}$$

$$t = cx' + dt'. \tag{19.5}$$

where $a, b, c,$ and d are constants which depend only on β_r. Why should we choose a linear relationship? Imagine $x' = 0$, and $t = K(t')^2$, a nonlinear relationship. For events occurring at $t' = 1, 2, 3$, $t = K, 4K, 9K$. The gap between the 1st and 2nd events is $3K$, and the gap between the 2nd and 3rd event is $5K$. Now let us shift the clock, and imagine that the events occur at $t' = 0, 1, 2$ units. For this choice of reference event, the corresponding gaps in the lab frame will be K and $3K$, which is absurd. The gaps must be the same for both the choices of reference for the time. This contradiction can be resolved only if t and t' are related by a linear relationship. Similar arguments can be used to show that x and x' are also related by a linear relationship.

The Eqs. (19.4) and (19.5) are valid for all x' and t'. By choosing $x' = 0$, we obtain $b = \beta_r \gamma$ and $d = \gamma$ (proved in item 2). The other two constants a and c can now be obtained by using the invariance of interval ($t^2 - x^2 = t'^2 - x'^2$) that yields

$$t^2 - x^2 = (d^2 - b^2)t'^2 - (a^2 - c^2)x'^2 + 2(cd - ab)x't'.$$

The above relationship is valid for all x and t. Hence,

$$d^2 - b^2 = 1, \tag{19.6}$$

$$a^2 - c^2 = 1, \tag{19.7}$$

$$cd - ab = 0. \tag{19.8}$$

Equation (19.6) is automatically satisfied. Equation (19.8) implies that $c = a\beta_r$, substitution of which in Eq. (19.7) yields

$$a = \gamma, \quad c = \beta_r \gamma.$$

Hence,

$$x = \gamma(x' + \beta_r t')$$

$$t = \gamma(t' + \beta_r x')$$

$$y = y'$$

$$z = z'. \tag{19.9}$$

This set of transformations are called *Lorentz transformations*. In systems of units in which t is in seconds, the above equations are

$$x = \gamma(x' + V_r t')$$

$$t = \gamma(t' + \frac{V_r x'}{c^2})$$

$$y = y'$$

$$z = z'. \tag{19.10}$$

4. An inversion of the above transformation yields

$$x' = \gamma(x - \beta_r t)$$

$$t' = \gamma(t - \beta_r x)$$

$$y' = y$$

$$z = z. \tag{19.11}$$

The minus sign indicates that the unprimed reference frame is moving leftward with velocity β_r relative to the primed reference frame. This is consistent because the lab frame indeed is moving leftward with velocity β_r relative to the rocket frame.

5. We can easily verify from the transformation equations that $t^2 - x^2 - y^2 - z^2 = t'^2 - x'^2 - y'^2 - z'^2$. This has to be case since Lorentz transformation was derived using this invariance property.

6. Lorentz transformation involves a factor $\gamma = c/\sqrt{c^2 - V_r^2}$ which is meaningful only for $V_r < c$. This observation indicates that the relative velocity of the reference frames must always be less than c. In subsequent discussions we will show that no material particle (a particle with nonzero mass) can move with velocity equal to or greater than c.

The Eqs. (19.9) is the most general transformation from primed coordinates to unprimed coordinates. When $\beta_r \ll 1$ (non-relativistic limits), $\gamma \to 1$ and we obtain

$$x = (x' + \beta_r t')$$

$$t = t'$$

$$y = y'$$

$$z = z'. \tag{19.12}$$

which is Galilean transformation. Hence Lorenz transformation is consistent with Galilean transformation at small speeds.

Consider a stick of length L in the rocket frame. We want to make a length measurement of this stick in the laboratory frame. To do this we should make measurements of both the ends at the same time in the lab frame, i.e., $\Delta t = 0$ and $\Delta x = L_{\text{apparent}}$. Note that for the moving frame, $\Delta x' = L$. Substitution of these expressions in the equation $\Delta x' = \gamma(\Delta x - \beta_r \Delta t)$ yields

$$L_{\text{apparent}} = \frac{L}{\gamma}. \tag{19.13}$$

Hence, the apparent length of the rod in the moving frame is shrunk by a factor $\sqrt{1 - \beta_r^2}$. This relativistic effect is called *length contraction*.

The spacetime of special relativity is very different from that of Newtonian dynamics. As mentioned earlier, the time interval between two events is not the same for the laboratory and the train observer. In fact, two simultaneous events occurring at two different points in the train's reference frame ($\Delta t' = 0, \Delta x' \neq 0$) will not be simultaneous in the laboratory frame ($\Delta t' \neq 0$). In the same way, the length of an object may not measure the same in the train and laboratory frame. This is not because of error in measurement or due to the limitation of the observer. The nature of spacetime is such that the time interval and space separation between two events are different for the two observers. The quantity that is measured to be the same by both the observers is $\sqrt{(\Delta t)^2 - (\Delta x)^2 - (\Delta y)^2 - (\Delta z)^2}$. Clearly the nature of spacetime for special relativity is fundamentally very different from that of Newton's postulated spacetime. Newton does go wrong here.

EXAMPLE 19.1   In a laboratory frame of reference two events take place at the same place, but are separated in time by 12 seconds. (a) What is the spatial distance between these two events in a rocket frame in which the events are separated in time by 13 seconds? (b) What is the relative speed between the rocket and the laboratory frame?

SOLUTION   We use the invariance of the interval. The time interval in the lab (unprimed) and rocket (primed) frames are 12c and 13c metres respectively. $\Delta x = 0$. Using

$$(\Delta t')^2 - (\Delta x')^2 = (\Delta t)^2 - (\Delta x)^2$$

we obtain

$$(\Delta x')^2 = (13c)^2 - (12c)^2 = (5c)^2.$$

Hence $\Delta x' = 5c$ metres.

For an observer in the rocket, the origin of the lab frame moves $5c$ metres in 13 sec. Therefore, the velocity of the lab frame relative to the rocket frame is $5c/13$ metres/sec, or $\beta_r = 5/13$.

EXAMPLE 19.2   Many elementary particles decay into other lighter particles. These particles have a certain statistical mean lifetime in a reference frame moving with the particle. This time is called the proper mean lifetime of the particle. According to Eq. (19.3), the lifetime measured in other reference frames would be different. The proper mean lifetime of π^+ meson is 2.5×10^{-8} sec.

1. What is the mean life measured in the laboratory frame for a burst of π^+ travelling with $\beta = 0.80$?

2. What distance is the average distance travelled by the particles in the laboratory frame during the mean lifetime?

3. What would be distance travelled without relativistic effects?

4. Do the above for $\beta = 0.99$.

SOLUTION   We use Lorentz transformation to solve this problem.

1. In the rest frame, $\Delta x' = 0$ and $\Delta t' = 2.5 \times 10^{-8}$sec or 7.5 metres. In the lab frame, $\Delta x = \beta_r \Delta t = 0.8\Delta t$. Using

$$(\Delta t')^2 - (\Delta x')^2 = (\Delta t)^2 - (\Delta x)^2,$$

we obtain

$$\Delta t = \frac{\Delta t'}{\sqrt{1 - \beta_r^2}} = \frac{7.5}{3/5} = 12.5\,\text{m} = 4.17 \times 10^{-8}\,\text{sec.}$$

2. The distance travelled by the particle is $\beta\Delta t = 0.8 \times 12.5 = 10$ metres.

3. Without relativistic effect, $\Delta t = \Delta t'$, and the distance travelled by the particle is $\beta\Delta t' = 6$ metres.

4. With $\beta_r = 0.99$, the mean lifetime in the lab frame is

$$\Delta t = \frac{7.5}{\sqrt{1 - 0.99^2}} = 53.2\,\text{m} = 17.7 \times 10^{-8}\,\text{sec,}$$

and the distance travelled by the particle will be $\beta_r \Delta t = 0.99 \times 53.2 = 52.6$ m.

19.5   Geometric interpretation

The Lorenz transformation has a very interesting geometric interpretation that we will discuss now. Events are represented by points in the xt plane. Suppose an event has coordinates (x, t) in the lab frame, and (x', t') in the moving frame. We draw x' and t' axes in the xt-plane. The t'-axis is given by $x' = 0$ or $(x = \gamma\beta_r t', t = \gamma t')$, hence it satisfies $t/x = 1/\beta_r$. Similarly, x'-axis is given by $t' = 0$ or $(x = \gamma x', t = \gamma\beta_r x')$ that satisfies $t/x = \beta_r$. These axes are plotted in Fig. 19.1.[2]

To gain a geometrical insight of xt space we consider two events that takes place at the origin of the rocket frame at time $t' = 0$ and $t' = \tau$ marked as points O and P in Fig. 19.1. We treat the event at $t' = 0$ as the reference event. Hence, $\Delta x' = 0$ and $\Delta t' = \tau$. The coordinates of the second event in the lab frame are $x = \gamma V_r \tau$ and $t = \gamma\tau$. The equation satisfying x, t is a hyperbola:

$$t^2 - x^2 = \tau^2 = \text{const}$$

as shown in Fig. 19.1.

The coordinates x and t are the base and the perpendicular of a right-angled triangle shown in Fig.19.1. The hypotenuse of the triangle is given by $\sqrt{t^2 - x^2} = \tau$. This is a

---

[2]Contrast the above transformation with Euclidean rotation discussed in Section 7.3 (Fig. 7.2). Also contrast the above axes with (x, t) diagram of Fig. 2.3.

property of a Lorentzian right-angled triangle. Contrast this with an Euclidean right-angled triangle where the hypotenuse is $\sqrt{t^2 + x^2}$.

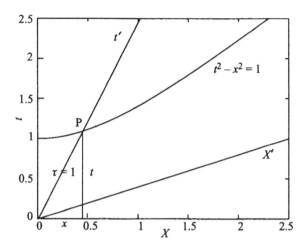

Figure 19.1   *Representation of events in xt space.*

It is also interesting to interpret $t = \tau \cosh \theta$ and $x = \tau \sinh \theta$, with $\cosh \theta = \gamma$ and $\sinh \theta = \gamma \beta_r$, and $\theta = (0, \infty)$.[3] We can easily write down the Lorentz transformation in terms of θ as

$$x = x' \cosh \theta + t' \sinh \theta,$$

$$t = x' \sinh \theta + t' \cosh \theta.$$

Note that the Lorentz transformation appears symmetric in x and t, which is not the case for Galilean transformation.

Let us consider (t, x, y) spacetime. The surface $t = \sqrt{x^2 + y^2}$ has an important property. For any point on this surface, the Lorentzian interval $\sqrt{t^2 - x^2 - y^2}$ is zero. Physically, a flash of light which starts from the origin at $t = 0$ in the direction of (x, y) will reach (x, y) at time t. The arrival of light at (x, y) is the event (t, x, y) in spacetime. Every point on this cone corresponds to the arrival of light from the origin, hence this cone is called a *light cone*. A light cone is the surface of the cone shown in Fig. 19.2.

All material particles (particles with nonzero mass) move with speed less than c. Therefore, if a free particle starts from the origin at $t = 0$, it will always fall short of reaching the light cone. At time t, the particle will reach some point inside the light cone. The spacetime region inside the light cone is called the *time-like region*. Each point in this zone correspond to the arrival of some material particle that starts from the origin at $t = 0$.

---

[3]The rotated axes under the Euclidean rotation are shown in Fig. 7.2. Under this rotation $t^2 + x^2 = R^2 = $ const, which is the equation of a circle of radius R. (t, x, R) form base, perpendicular, and hypotenuse of a right-angled triangle with $R = \sqrt{t^2 + x^2}$.

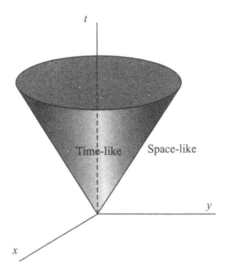

Figure 19.2   *Division of spacetime into time-like, light-like, and space-like regions*

A particle or light starting from the origin at $t = 0$ cannot reach any point whose distance is more than t. Hence all the points outside the light cone are not physically realisable events. The region outside the light cone is called the *space-like region*. Thus we can classify the zones of spacetime into time-like, light-like, and space-like regions. See Fig. 19.2 for an illustration.

There are many curious things about the spacetime of special relativity. Consider three events A,B,C shown in Fig. 19.3: event A occurring at the origin at $t = t' = 0$; event B occurring inside the time-like region with $t_B > 0$; and event C occurring inside the space-like region with $t_C > 0$. Clearly, for the laboratory observer, the events B and C occur after the event A. What does the moving observer see? As shown in Fig. 19.3, for the moving observer, the event B occurs at time $t'_B > 0$, but the event C occurs at time $t'_C < 0$. Hence the moving observer reports that the event B occurs after the event A, but the event C occurs before the event A. Sounds strange! It can be shown quite rigorously that the time ordering of the event is preserved in the time-like region. However it is not preserved for the events in the space-like region (see Exercise 12).

Does the above result violate causality? Let us take an example. An observer reports that a dynamite exploded near a mountain which caused a landslide. For this observer, the landslide occurs after the explosion and is caused by the explosion. If the above cause and effect is correct, then we would like that all the observers (moving or stationary) should agree with the above cause and effect sequence. This assumption of physics is called *causality*.

The events A, B, and C discussed above appear to violate causality, but is not so for the following reasons. Any wave or signal generated by the event A cannot reach the location of the event C before the event C occurs because the distance is more than ct_C. We call the events A and C *causally unconnected*. Hence, the time-ordering of the events A and C need not be preserved to maintain causality. On the contrary, the waves or signals generated by the event A can reach the location of the event B before the event B occurs. Hence the events A and B can be *causally connected*. Fortunately, the time-ordering is preserved for the events A and B as well as any events inside the time-like region. Hence causality is preserved in special relativity.

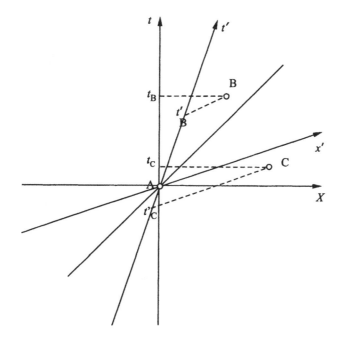

Figure 19.3   *Event A occurs at the $(x = 0, t = 0)$. Event B is in the time-like region, and event C occurs in the space-like region. The event B(C) occurs at $t_B(t_C)$ in the laboratory frame and at $t'_B(t'_C)$ in the moving frame.*

EXAMPLE 19.3   A runner carrying a 20 m stick is running with velocity $(\sqrt{3}/2)c$. She runs into a barn of size 10 m. The attendant of the barn sees the stick to be shrunk to 10 m. As soon as both the ends of the stick just touch both the end walls of the barn, the attendant of the barn shuts the door and arrests the stick. Is there any fallacy in this argument?

SOLUTION   Let us mark the events in the xt-plane. Point A denotes the event when the attendant observes that the right end of the stick touches the right wall. Point B denotes the event when the attendant finds that the left end of the stick coincides with the left wall (see Fig. 19.4). These events are shown in the (xt) diagram. Both the events occur at the same time for the attendant, but the events are not simultaneous for the runner. In fact, for the runner, event A occurs at $\Delta t'$ corresponding to A''

$$\Delta t' = -\gamma \beta_r \Delta x = -2 \times \frac{\sqrt{3}}{2} \times 10 \text{ m} \approx -17.3 \text{ m}.$$

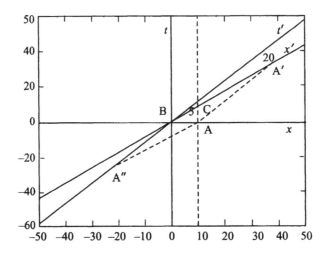

Figure 19.4 *The barn attendant measures the length of the stick using events A and B. The runner measures the size of the barn using events B and C.*

Hence according to the runner, event A occurs before event B. For her, by the time the left end of the stick reaches the left end of the barn, the right end would have gone past the right wall. Hence the two observers do not agree that the stick is inside the barn for a certain time interval. Hence the stick cannot be trapped inside the barn.

Also note that the runner would find that the barn has shrunk to 5 m. Hence she would find it impossible to fit the stick inside the barn. The runner performs the length measurement using events B and C shown in Fig. 19.4. These events yield $x' = 5$ m and $t = \gamma\beta_r x' = 5\sqrt{3}$ m. We have to be cautious in our interpretation of the triangles etc. since the space is not Euclidean. For example Pythagoras theorem does not work in this space.

The xt diagram also contains other information. The distance BA$' = 20$ m is the rest-length of the rod. The reader is advised to mark all the distances and time in the figure (e.g., AC).

We remark that the misconception in the problem arises because of the wrong assumption that the ends of the stick occupy the same space points in absolute space. The confusion is due to our hang up with absolute space.

We can use Lorentz transformation to derive the relationship between velocity of a particle measured in the laboratory frame and the rocket frame.

19.6 Velocity transformation

A particle is being observed in the laboratory and the rocket frames. Its position in these frames are denoted by x and x' respectively. Therefore, the velocities of the particle in the laboratory and rocket frames are dx/dt and dx'/dt'. The relationship between the two velocities is

$$\frac{dx}{dt} = \frac{\gamma_r(dx' + \beta_r dt')}{\gamma_r(dt' + \beta_r dx')}$$

or

$$\beta_x = \frac{\beta_x' + \beta_r}{1 + \beta_x'\beta_r}. \tag{19.14}$$

Here $\gamma_r = (1 - \beta_r^2)^{-1/2}$. We can invert the above equation to compute velocity of the particle in rocket frame

$$\beta_x' = \frac{\beta_x - \beta_r}{1 - \beta_x\beta_r}. \tag{19.15}$$

Under non-relativistic limit, β_x' and β_r are much less than one. In this limit, $\beta_x\beta_r \to 0$, and the transformation rule is $\beta_x = \beta_x' + \beta_r$ or $v_x = v_x' + V_r$, which is the rule for velocity addition in the Galilean transformation. Hence the velocity transformation rule of relativity is consistent with the Galilean transformation at slow speeds.

For light, $\beta_x' = 1$. When we substitute this value in the above expression we obtain $\beta_x = 1$. Hence, the speed of light is the same in both the frames of reference consistent with the principles of relativity.

Using similar algebra we can derive the velocity of the particle in y and z directions:

$$\frac{dy}{dt} = \frac{dy'}{\gamma_r(dt' + \beta_r dx')}$$

$$\beta_y = \frac{\beta_y'}{\gamma_r(1 + \beta_r\beta_x')}$$

and $\beta_z = \dfrac{\beta_z'}{\gamma_r(1 + \beta_r\beta_x')}$

These transformation rules are very useful in studying kinematics and dynamics of relativistic particles. We illustrate the application of the above formulas using several examples.

EXAMPLE 19.3   We observe a galaxy receding in a particular direction at a speed $\beta_1 = 0.3$, and another galaxy receding in the opposite direction with the same speed. What is the speed of recession for an observer in one of galaxies?

SOLUTION   We use Eq. (19.15) to compute the velocity of a galaxy wrt the other galaxy. We are given $\beta = 0.3$, and $\beta_r = -0.3$. Substitution of these values in the equation yields

$$\beta' = \frac{0.3 - (-0.3)}{1 - (0.3)(-0.3)} = \frac{0.6}{1 + 0.09} = 0.5505.$$

Hence an observer in one galaxy observes the other galaxy recede with speed $0.5505c$. Note that Galilean transformation would yield $v' = 0.6c$.

EXAMPLE 19.4   A flash of light is emitted in a rocket at an angle ϕ' with respect to the x'-axis. Show that the angle ϕ that the flash light makes with respect to the x-axis of the lab frame is given by the equation

$$\cos \phi = \frac{\cos \phi' + \beta_r}{1 + \beta_r \cos \phi'}.$$

SOLUTION   The addition of velocity along the x-axis is given by

$$\beta_x = \frac{\beta'_x + \beta_r}{1 + \beta'_x \beta_r}.$$

For the light, $\beta = \beta' = 1$, and $\beta_x = \cos \phi$ and $\beta'_x = \cos \phi'$. The substitution of these values immediately gives us the desired result.

We end this chapter with a brief discussion on Lorentz scalars and Lorentz vectors.

19.7   Lorentz scalars and Lorentz vectors

In Section 7.3 we defined vectors and scalars in the Newtonian framework using rotation as an operation. Vectors are the quantities whose components transform in the same way as (x, y, z) under rotation. Quantities that remain unchanged under rotation are called scalars. We can generalise the above ideas of vectors and scalars to spacetime in SR. The corresponding operation is Lorentz transformation.

The physical quantities whose components transform as (t, x, y, z) under Lorentz transformation are called Lorentz vectors. The quantities that remain unchanged under Lorentz transformation are called Lorentz scalars.

We provide several examples of Lorentz vectors and scalars.

1. The distance in Lorentz space $\sqrt{t^2 - x^2 - y^2 - z^2}$ is invariant under Lorentz transformation, hence $\sqrt{t^2 - x^2 - y^2 - z^2}$ is a Lorentz scalar.

2. When a particle moves by a small distance $d\mathbf{r}$ in time dt, then by considering the starting points and end points as events, we deduce that $d\tau = \sqrt{(dt)^2 - (d\mathbf{r})^2}$ is a Lorentz scalar.

3. The quantity $\left(\dfrac{dt}{d\tau}, \dfrac{d\mathbf{r}}{d\tau} \right)$ is a Lorentz vector.

4. The quantity $\left(\dfrac{dt}{dt}, \dfrac{d\mathbf{r}}{dt} \right)$ is not a Lorentz vector. That is, particle velocity is not a component of a Lorentz vector.

5. Acceleration of a particle and force acting on a particle are not components of Lorentz vectors.

6. The phase of light is a Lorentz scalar. Recall that the phase of wave is given by $\omega t - \mathbf{k} \cdot \mathbf{x}$. Since (t, x, y, z) form a Lorentz vector, (ω, k_x, k_y, k_z) must transform in the following manner to keep the phase as a Lorentz scalar:

$$
\begin{aligned}
\omega' &= \gamma(\omega - \beta_r k_x) \\
k'_x &= \gamma(k_x - \beta_r \omega), \\
k'_y &= k_y \\
k'_z &= k_z,
\end{aligned}
$$

where $\beta_r = v_r/c$ is the non-dimensional relative velocity between the two reference frames. It is easy to verify that $\omega t - \mathbf{k} \cdot \mathbf{x} = \omega' t' - \mathbf{k}' \cdot \mathbf{x}'$.

In the next chapter we will derive linear momentum and energy of a particle in SR using the above properties.

Exercises

1. Write down all the equations discussed in the present chapter with time in the units of second.

2. A young woman voyages to the one of the nearest star 5 light-years away. She travels in a spaceship at a velocity of $c/2$. When she returns to Earth, how much younger would she be than her twin brother who stayed home?

3. Time dilation in relativity can be derived by the following simple argument. Consider a moving train inside which a laser beam is fired vertically upward. The laser beam gets reflected from the roof of the train, and received at the starting point. What will be the motion of the laser beam for a stationary observer outside the train? Using the constancy of the speed of light, show that time is dilated for the outside observer.

4. Elementary particle pion is unstable and decays with a proper half-life of 1.8×10^{-8}sec. What is the pion's half-life measured in a frame S that moves with (a) 100 m/s; (b) $0.9c$; (c) $0.99c$.

5. A rod of length L_0 is placed inside a moving train such that it makes angle θ_0 with the horizontal. The train is moving with velocity $0.9c$ with respect to observers on the railway platform. What would be the measured length and angle θ for the observers on the platform?

6. Derive the transformation rule for the acceleration of a particle in the laboratory and rocket frame of reference. Argue that acceleration is not a component of Lorentz vectors.

7. A relativistic rocket emits light with wavelength λ from its headlight in its rest frame. What would be the measured wavelength of the light for a laboratory observer who measures that the rocket is approaching her with speed $0.5c$.

8. A bright star in a galaxy emits light with wavelength $\lambda = 6000$ Å. The above galaxy is receding from our galaxy with $v = 0.5c$. What would be the wavelength of the emitted light as observed in the reference frame of the Earth.

9. An experimenter inside a train shines a laser torch directly upward. The speed of the train is $0.5c$. Compute the components of velocity of the laser in the laboratory frame. What is the speed of light in the laboratory frame?

10. The above experimenter (Exercise 9) shoots an electron vertically upward with speed $0.9c$. What would be the measured speed in the laboratory frame? What angle will the electron make relative to the horizontal in the lab frame of reference?

11. All material particles move with speed less than the speed of light. Show that if the speed of a particle is less than c in a reference frame, then this property would be seen in all other inertial reference frames.

12. Consider the events A,B,C discussed in Section 19.5 (Fig. 19.3). The event A occurs at the origin in both the laboratory and the moving reference frame. Suppose the coordinates of the events of B and C are (x_B, t_B) and (x_C, t_C) respectively in the laboratory frame. Show that for the moving observer $t_B' > 0$, but t_C' could be either positive or negative depending on the velocity of the moving reference frame.

20

Relativistic Dynamics

In the last chapter we studied the structure of spacetime. In the present chapter we will look at dynamical quantities like momentum and energy in special relativity.

20.1 Relativistic linear momentum

In Newtonian dynamics, linear momentum is given by

$$\mathbf{p} = m\mathbf{v} = m\frac{d\mathbf{r}}{dt} \tag{20.1}$$

If we use the above formula, we find that the momentum is not conserved in special relativity (SR). We will illustrate this result in the following discussion using a two-body collision experiment. The non-confirmation of the conservation law is not a welcome sign. Luckily, if we modify the formula for the linear momentum to

$$\mathbf{p} = \frac{m\mathbf{v}}{\sqrt{1 - \beta^2}}, \tag{20.2}$$

then the conservation of linear momentum holds in special relativity. We will illustrate the above formula using a collision experiment described below.

Two particles a and b of equal masses m collide with each other in the lab frame as shown in Fig. 20.1(a). The particles move with speed v, and make an angle of 45 degrees with the x-axis. According to both the above formulas, in the lab frame, the linear momentum are equal and opposite. Hence the principle of conservation of linear momentum holds in the laboratory frame. We observe the same experiment in another reference frame S that is moving along the x-axis with velocity $v/\sqrt{2}$. In S, the trajectories of the particles appear as shown in Fig. 20.1(b).

Let us compute the linear momentum of the particle along the y-axis. Using relativistic-velocity transformation rules, we find that the vertical velocities ($\bar{v}_y$) of the particles a and b before the collision in S are

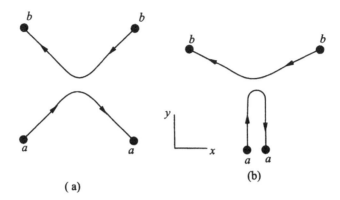

Figure 20.1   *Two particles a and b of equal mass m collide with each other. These particles make an angle of 45 degrees with the x-axis before and after the collision. Particle trajectories observed in (a) the laboratory reference frame, (b) in the reference frame S moving along the x-axis with velocity $v/\sqrt{2}$.*

$$\bar{\beta}_{ay} = \frac{\beta/\sqrt{2}}{\gamma(1 - \beta^2/2)} \tag{20.3}$$

$$\bar{\beta}_{by} = \frac{-\beta/\sqrt{2}}{\gamma(1 + \beta^2/2)}$$

where $\beta = v/c$, $\bar{\beta}_{a,b} = \bar{v}_{a,b}/c$, and $\gamma = 1/\sqrt{1 - \beta_r^2} = 1/\sqrt{1 - \beta^2/2}$ ($\bar{f}$ denotes the value of the quantity in the moving frame). After the collision, the vertical velocities are

$$\bar{\beta}'_{ay} = \frac{-\beta/\sqrt{2}}{\gamma(1 - \beta^2/2)}$$

$$\bar{\beta}'_{by} = \frac{\beta/\sqrt{2}}{\gamma(1 + \beta^2/2)}$$

From the above calculation, we conclude that

$$m(\bar{\beta}_{ay} + \bar{\beta}_{by}) \neq m(\bar{\beta}'_{ay} + \bar{\beta}'_{by}).$$

Hence linear momentum along the vertical direction is not conserved if we take mv as the formula for linear momentum. Let us try to discover an appropriate formula for linear momentum which preserves the conservation law.

According to Eq. (20.1), the vertical velocities of particle a in the lab and moving frame are dy/dt and dy'/dt' respectively with $dy = dy'$. Hence the vertical velocities of the particle is seen differently in the two frames of reference. However, the value of the quantity $dy/d\tau$ ($d\tau = \sqrt{(dt)^2 - (d\mathbf{r})^2}$) is the same in both the reference frames. Therefore, $m\,dy/d\tau$ is a good candidate for the vertical momentum that has the same value in both lab and moving frames. Since, the vertical momentum $m\,dy/d\tau$ is conserved in the lab frame, it will also be conserved in the moving frame. Thus

$$p_y = \frac{mdy}{d\tau} = \frac{mv_y}{\sqrt{1-\beta^2}}$$

is the vertical linear momentum of the particle. A generalisation of the above formula yields Eq. (20.2) as the relativistic linear momentum of a particle of mass m that is moving with velocity $\mathbf{v}$.

In the following section we will derive the formula for the linear momentum using symmetry arguments.

20.2   Derivation of linear momentum and energy using symmetry arguments

Consider a free particle of mass m moving with a constant velocity $\mathbf{v}$. From our earlier discussion on Lorentz transformations, we know that velocity of a particle is not a Lorentz vector; neither is it a component of a Lorentz vector. Hence $\mathbf{p} = m\mathbf{v}$ is not a Lorentz vector. However the quantity $m(\frac{dx}{d\tau}, \frac{dy}{d\tau}, \frac{dz}{d\tau}, \frac{cdt}{d\tau})$, where $d\tau = \sqrt{(dt)^2 - \mathbf{dr} \cdot \mathbf{dr}/c^2}$, is a Lorentz vector. The first three components of the above quantity are

$$m\left(\frac{dx}{dt}, \frac{dy}{dt}, \frac{dz}{dt}\right)\gamma = \frac{m\mathbf{v}}{\sqrt{1-\beta^2}} = \mathbf{p}. \tag{20.4}$$

This quantity is the momentum of the particle in special relativity consistent with the discussion in the earlier section.

The fourth component of $m(\frac{dx}{d\tau}, \frac{dy}{d\tau}, \frac{dz}{d\tau}, \frac{cdt}{d\tau})$ is $mdt/d\tau$, the energy of the particle. Let us see how. The quantity $mcdt/d\tau$, denoted by p_0, is

$$p_0 = m\frac{cdt}{d\tau} = \frac{mc}{\sqrt{1-\beta^2}},$$

for small β, p_0c is

$$p_0c = \frac{mc^2}{\sqrt{1-\beta^2}} \approx mc^2 + \frac{1}{2}mv^2.$$

The second term is the kinetic energy, and the first term is called *rest mass energy*. When the particle is at rest, it has energy mc^2. This observation induces us to believe that p_0c is the total energy of a free particle. It turns out that the above interpretation is consistent with experiments, and p_0c is the formula for the energy of a relativistic particle.

So the total energy is identified by

$$E = p_0c = mc^2 + T$$

where T is the kinetic energy of the particle. The energy can also be written as

$$E = \frac{m}{\sqrt{1-(v/c)^2}}c^2, \tag{20.5}$$

and $m/\sqrt{1-(v/c)^2}$ is identified as the *effective mass*. Equation (20.5) is called *mass energy equivalence*. A system with certain mass contains the energy given by Eq. (20.5). Similarly a system having certain energy has inertia. We can convert mass into energy and vice versa. For example, when an electron meets its anti-particle (called positron), the mass is converted into electromagnetic energy. Conversion of mass into energy forms the basis for nuclear reactors and nuclear bombs.

The above result clearly shows that the mass of a system is not conserved. This observation is in direct contradiction with Newtonian dynamics where total mass of a system is conserved.

We also observe that the effective mass of the particle increases with the increase of its velocity. When $v \to c$, the effective mass approaches infinity. To reach $v = c$ we need infinite energy; this is the reason why a particle with finite mass cannot attain the speed of light. Exceptions to this rule are photons, quanta of light, that always move with the speed of light. The mass of the photons is zero, hence there is no contradiction. By dividing Eq. (20.5) by absolute value of $\mathbf{p}$ (from Eq. (20.4)) we obtain

$$\frac{E}{pc} = \frac{c}{v}.$$

Since $v = c$ for photons, we obtain

$$E = pc$$

for photons.

From the above definitions

$$(E/c, p_x, p_y, p_z) = m \left(\frac{cdt}{d\tau}, \frac{dx}{d\tau}, \frac{dy}{d\tau}, \frac{dz}{d\tau} \right).$$

Since $\left(\frac{dt}{d\tau}, \frac{dx}{d\tau}, \frac{dy}{d\tau}, \frac{cdt}{d\tau}\right)$ is a Lorentz vector, $(E/c, p_x, p_y, p_z)$ is also a Lorentz vector. A simple algebra shows that

$$\frac{E^2}{c^2} - p_x^2 - p_y^2 - p_z^2 = (mc)^2$$

is a Lorentz scalar.

In the dimensions of mass, momentum and energy are given by

$$\mathbf{p} = \frac{m\beta}{\sqrt{1-\beta^2}},$$

$$E = \frac{m}{\sqrt{1-\beta^2}}$$

and

$$E^2 - p^2 = m^2.$$

Using the transformation rules for $(d\mathbf{r}, dt)$, we can easily derive the transformation rules for the energy and linear momentum from one reference frame to another. The relationships are

$$E' = \gamma(E - \beta_r p_x).\tag{20.6}$$

$$p'_x = \gamma(p_x - \beta_r E),\tag{20.7}$$

$$p'_y = p_y,\tag{20.8}$$

$$p'_z = p_z,\tag{20.9}$$

where β_r is the relative velocity of the primed reference frame with respect to the laboratory frame. The Eqs. (20.6–20.9) are very interesting. Clearly, if for a system of particles $\sum_a E_a$ and $\sum_a \mathbf{p}_a$ (a is the particle label) are conserved in the laboratory reference frame (i.e., they have the same value before and after the process), then the above transformation rules imply that $\sum_a E'_a$ and $\sum_a \mathbf{p}'_a$ would also be conserved in the moving reference frame. Note however that the values of energy and linear momentum would be different in the two reference frames.

Equations (20.6–20.9) are very useful in analysing relativistic collisions and other relativistic processes when we need to go from one frame of reference to another (e.g., centre of mass frame to the laboratory frame). We cannot delve into these problems due to lack of space and time.

EXAMPLE 20.1   What is the kinetic energy of an electron that is moving with $v = 0.99c$?

SOLUTION   Energy of the electron is

$$E = m + T = \frac{m}{\sqrt{1 - \beta^2}}$$

using $\beta = 0.99$, we get $E = 7.09m$ and kinetic energy $T = E - m = 6.09m = 55.42 \times 10^{-31}$ kg. In units of joule we get

$$T = 55.42 \times 10^{-31} \times c^2 = 4.98 \times 10^{-13} \text{ J.}$$

There is another popular unit called *electron-volt (eV)*. Using the conversion 1 eV $= 1.6 \times 10^{-19}$ J, we obtain $T = 3.11 \times 10^6$ eV.

Note that the Newtonian mechanics would provide $T = mv^2/2 \approx 8 \times 10^{-14}$ joules, which is around 0.16 times lower than the relativistic calculation.

EXAMPLE 20.2   What is the linear momentum of an electron that is moving with $v = 0.99c$?

SOLUTION   The linear momentum of the electron is

$$p_0 = \frac{mv}{\sqrt{1 - \beta_r^2}} = \frac{m \times 0.99c}{\sqrt{1 - 0.99^2}} = 1.9 \times 10^{-21} \text{ kg m/sec.}$$

According to Newtonian mechanics, p_0 would have been $0.99mc = 2.7 \times 10^{-22}$ kg m/sec that is order of magnitude lower than the relativistic value.

EXAMPLE 20.3   Two identical particles 1 and 2 collide and stick together to make a third particle 3. What is the mass of particle 3 in the reference frame in which the CM is at rest? Take the velocity of particles 1 and 2 to be v and $-v$ respectively. Work out the numbers for $m_1 = m_2 = 1$ gm, and $v = 10^3$ m/s. The change in mass from 2 gm is the relativistic effect. Work out a similar number for two protons that are moving towards each other with $0.99c$.

SOLUTION   In the CM frame, the net momentum is zero. The total energy is

$$E = \frac{2m}{\sqrt{1 - \beta^2}},$$

where $\beta = v/c$. After the collision, the new particle has zero velocity, so the mass of the particle is (using $E^2 - p^2 = m^2$)

$$m_3 = E = \frac{2m}{\sqrt{1 - \beta^2}}.$$

For $m = 1$ gm and $\beta = 10^3/(3 \times 10^8)$, we obtain

$$m_3 = 2(1 + \frac{1}{2}\beta^2) \approx (2 + 10^{-11})\text{gm}.$$

If we consider the two colliding particles to be protons moving towards each other with $v = 0.99c$ then

$$m_3 = E = \frac{2m}{\sqrt{1 - \beta^2}} = 7.088 \times 2m.$$

Hence the mass of the third particle is 7.088 times larger than the original mass. Since $m = 1.6 \times 10^{-27}$ kg, the numerical value of m_3 is 2.26×10^{-26} kg. Clearly mass is not conserved in special relativity.

With this discussion we close our discussion on relativistic dynamics.

20.3   Brief prelude to general relativity

Einstein did not stop at special relativity. He postulated that accelerating reference frames are equivalent to inertial frames in gravity (principle of equivalence, see Chapter 10). Using this idea he could construct a modern theory of gravity. This theory is called the *general theory of relativity*. It is considered to be the most beautiful theory of physics. We will not be able to discuss it here. We just state one result: In a gravitational field, clocks at higher altitude run slower than those at lower altitudes. We can appreciate this result using the following thought experiment.

Imagine that a particle of mass m falls by a distance h (see Fig. 20.2).

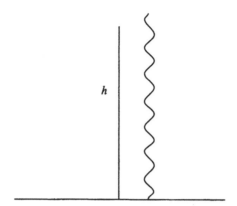

Figure 20.2   *A photon rising up in a gravitational field. Its frequency at higher altitudes is lower than that at lower altitudes.*

At the bottom, the particle has kinetic energy $E = mc^2 + mv^2/2 = mc^2 + mgh$. Now convert the whole mass into a photon with frequency $\nu = E/h$, and shoot it upward. When it reaches height h, its energy must be $E' = mc^2$, otherwise we can run a perpetual machine. Therefore,

$$\frac{E'}{E} = \frac{h\nu'}{h\nu} = \frac{mc^2}{mc^2 + mgh} \approx 1 - \frac{gh}{c^2}.$$

That is, $\nu' < \nu$ or $\Delta t' > \Delta t$. Hence clocks runs slower at higher altitude than those at lower altitudes. A photon climbing up Earth's gravitational field loses energy, and gets redshifted. With this discussion we close our discussion on relativity.

Exercises

1. Write down all the equations discussed in the present chapter with proper dimensions.

2. Expand the expression for relativistic energy up to an order of v^4.

3. A neutral π meson, whose rest mass is 135 MeV, decays symmetrically into two photons while moving with high speed. The energy of each photon in the lab frame is 50 MeV.

   (a) Find the speed of the meson before its decay.

   (b) Find angle θ in the laboratory system between the momentum of each photon and the initial line of motion.

4. The rest mass of proton is 931 MeV/c^2. What is the energy and momentum of the proton whose speed is $0.999c$?

5. In the present day particle accelerators, protons can be accelerated up to an energy of 10^{13}eV. What is the speed of the accelerating proton?

6. A particle of rest mass M spontaneously decays from rest into two particles with rest masses m_1 and m_2.

   (a) Find the energy and momentum of the daughter particles in the rest frame of the original particle.

   (b) If the original particle was travelling with $0.5c$ in the laboratory frame, what would be the velocities of the daughter particles in the laboratory frame?

7. Consider the collision experiment described in Section 20.1.

   (a) Compute the relativistic momentum of the particles in the laboratory frame before and after the collision.

   (b) Compute the relativistic momentum of the particles in the reference frame S before and after the collision.

   (c) Verify the conservation of linear momentum in both the reference frames.

   (d) Compute the relativistic energy of the particles in both the reference frames before and after the collision.

   (e) Verify the conservation of energy in both the reference frames.

   (f) Are these quantities consistent with Eqs. (20.6–20.9)?

8. Estimate the energy released from a nuclear reaction $4H^1 \rightarrow He^4$. The mass of proton and He^4 nucleus are 1.6736×10^{-24} gm and 6.6477×10^{-24} gm respectively. How much energy will be released by 1 kg of proton nucleus? Compare it with the energy released from 1 kg of petrol (40 Mega Joule/kg).

9. [COMPTON SCATTERING]   A photon with wavelength λ gets scattered by a stationary free electron. What would be the wavelength of the photon if its scattering angle is θ,

10. A particle of mass m_a, energy E_a, and velocity v_a collides with an stationary particle of mass m_b. The two particles get fused during the collision? Compute the final mass, energy, and velocity of the final particle.

11. The average power incident at the top of the Earth's atmosphere from the Sun is approximately 1.4 kW/m^2. If the energy in the Sun is produced by burning hydrogen into helium, how much hydrogen is burnt in the Sun every second?

12. The projected energy requirement in India by 2030 is 0.4 million MW. If we were to obtain this energy by burning hydrogen to helium, how much hydrogen would be required?

13. In the laboratory frame, a particle of mass 10^{-25} gm is moving with velocity $(c/2)(\hat{\mathbf{x}} + \hat{\mathbf{y}})$.

   (a) Compute its linear momentum and energy,

(b) What would be the linear momentum and energy of this particle in the reference frame that is moving with velocity $(c/4)\hat{x}$ wrt the laboratory frame.

14. A neutron decays into a proton, electron, and neutrino. But a proton does not decay into a neutron, positron, and a neutrino. Explain why?

15. A free electron does not emit photons. Why not?

Epilogue

We have come to the end of our discussion on mechanics. We started with Newton's laws and its assumptions. One of the key assumptions was that of absolute space and absolute time, whose consequence is that time flows uniformly for all observers in Newtonian mechanics.

Newton's laws of motion had a revolutionary impact on science. For the first time in the history of science, a variety of phenomena could be explained quantitatively using a simple set of physical laws. Concepts of energy, angular momentum etc. were introduced that helped in solving scientific and engineering problems. All these ideas as well as modern topics like symmetries, vectors, phase space etc. have been discussed in our book.

But when some experimentalists found that the speed of light is the same for all observers independent of the relative motion of the source and the observer, problems arose. This constant speed is in direct conflict with Newton's assumption of absolute space and absolute time. Einstein constructed a new formulation called the theory of relativity that is based on relative spacetime rather than absolute space and absolute time. Einstein's relativity theory is consistent with the constancy of speed of light. The theory itself is not an extension of newton's laws, but a completely new formulation of spacetime with a new set of startling predictions.

Mechanics is the starting course in the study of physics. Many concepts introduced here like potential, energy, inertial frame, linear and angular momentum etc. are encountered in all areas of physics. I hope you had fun learning these concepts.

Appendix A

Present Paradigm of Physics and Science

In ancient times, the description of nature used to be mystical and philosophical. It was only around 500 years back that a new approach called the scientific method was born. In this method, it is believed that the *real world* is governed by certain laws that we can discover. We can contrast this with ancient ideas that nature is under the control of gods. In the following discussion we will briefly describe the basic tenets of the modern scientific method.

A.1 Scientific method

To understand the real world we take recourse to observations. We construct a physical theory consistent with these observations. The theory should provide certain predictions which must be verifiable through experiments. If the predictions of the theory match with experimental results, then the theory is said to have survived the test. Clearly no physical theory can ever be proved. It is valid as long as the results of all the experiments done so far are consistent with the predictions of the theory; a single experiment whose results differ from the theoretical prediction can destroy the theory. Once we discover an inconsistency between theory and experiment, we either (a) modify the existing theory to make it consistent with the results of the experiment, or (b) construct a new theory altogether.

Let us consider some example discussed in this book. The observed trajectories of outer planets were inconsistent with the predictions of earliest geocentric theories. Ptolemy managed to achieve consistency by adding more spheres for each planet (revision of type (a)). Even finer observations of planetary orbits were found to inconsistent with Ptolemy's revised theory, so a completely new theory, a heliocentric model was constructed that was found to be consistent with the observations; this is revision of type (b) discussed above. Other examples of type (b) theories is Newton's laws of motion over Aristotle's theory. Later, Newton's theory itself was replaced by Einstein's relativity theory, which is another major revision of type (b).

This general scheme of scientific method is illustrated in the flow diagram shown in Fig. A.1.

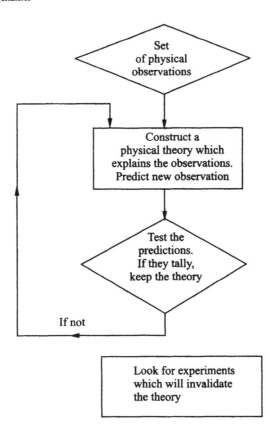

Figure A.1   Basic paradigm of modern science

There could be several models to explain a physical phenomenon. How do we choose which model to use for a given situation? When we are searching for a theory, what are the guidelines we must follow? In the following section we will study some generally accepted properties of physical theories.

A.2   Nature of physical theories

A.2.1   Simplicity

It may be possible that several theories could explain certain observations. Which of these theories should be chosen as a physical theory? According to *Occam's Razor, when you have two competing theories which make exactly the same predictions, the one that is simpler is better.* Similarly, a theory with a minimum set of axioms is preferred over one with more axioms. If we use this criteria, the heliocentric theory is a clear winner over Aristotle and Ptolemy's model of the planetary motion.

A.2.2   Abstraction

Physical theories are often quite abstract and mathematical. By abstract we mean essential features devoid of details. For example, Newton's three laws explain all kinds of motion (except relativistic and quantum). These laws capture the essential features of motion and are called abstract laws.

We can contrast the above approach with the detailed approach often encountered in engineering problems. An engineer needs to worry about many details about the problem at hand. For example, the design of an aeroplane requires a completely different approach; one needs to worry about wing design, controls, safety, cost etc. A physicist trained in abstraction finds these details quite difficult. On the other hand, a physicist is perfectly at home worrying about the motion of stars and galaxies by treating them as point particles. An engineer will find these assumptions absurd.

The above examples provide some idea about the methods of engineering and scientific research. We must be clear about the strength and weaknesses of the two approaches, and choose the one which is more appropriate at a given time. Note however that both the approaches are essential for civilization.

A.2.3   Universality

Physicists are always in search of a universal theory. Newton discovered a universal law of gravitation which could explain the motion of planets and moons, and the motion of objects on the surface of the Earth. To a very good approximation, Newton's law of gravity also describes forces between galaxies, stars, and other astrophysical objects. The search for a universal theory is one of the main thrust of physics research. At present, scientists believe that all the forces in the universe can be derived from four fundamental forces: gravity, electromagnetism, weak nuclear interactions, and strong nuclear interactions. Physicists have been able to show that the last three forces become one force at very high energies. This is called unification of forces. A similar attempt to unify gravity has not been that successful so far.

There are other universal theories in physics that explain varied phenomena using a common framework. One such theory is the band structure of matter that explains the essential features of metals, insulators, and semiconductors. Universality is one of the most powerful concepts in physics.

Let us compare this approach with the methods used in other disciplines. To make a stark contrast, let us take the profession of a chemical engineer specialising in soap manufacturing. If we say that all soaps are the same, the engineer would be extremely upset. For him/her every chemical, perfume, liquid is special. This approach is opposite to the search for universal theories. An universalist would attempt to find a core feature among all chemicals, while the chemical engineer will look for different features in chemicals.

We should however be cautious. Along with the search for specialised solutions for specialized problems, every discipline has certain universal theories. Some examples are Kirchhoff's law in circuit theory, thermodynamics, theory of evolution in biology etc. These theories attempt to provide a general framework to address varied phenomena or processes. Yet it is safe to say that among all scientists, physicists and mathematicians explore universality the maximum. For our present discussion we make a passing remark that abstract mathematical theories need not be connected to the physical world, but physical theories are supposed to model some physical phenomena.

Strangely, physics theories are quite mathematical involving a range of mathematical concepts: simple ones like vectors, differential equations, and complex concepts like topology, differential geometry, number theory etc. We cannot do physics without mathematics; in some sense, physical theories are written in the language of mathematics. Why this is the case, is a profound philosophical question which is beyond the scope of this book.

These are some of the most important characteristics of physical theories. After discussing these characteristics we study some of the basic assumptions made in physics.

A.3   Basic assumptions in physics

A.3.1   Mechanical deterministic world

According to Newton's laws of motion, the future of a system is uniquely determined once the initial condition of the system is specified. Newton and his followers believed that the universe is like a massive clock-work built by God, and set into motion by providing the initial conditions. These ideas were reinforced by the enormous success Newton's law had in explaining the motion of planets, moon, comets, tides, projectiles, etc.

One of the biggest propounder of determinism was Laplace (1749-1827), who wrote in his *Philosophical Essay on Probabilities* (1814): "An intellect which at any given moment knew all the forces that animate Nature and the mutual positions of the beings that comprise it, if this intellect were vast enough to submit its data to analysis, could condense into a single formula the movement of the greatest bodies of the universe and that of the lightest atom: for such an intellect nothing could be uncertain; and the future just like the past would be present before our eyes." Laplace believed in determinism so much so that he claimed that there is no free will in the universe. Our fate is completely determined by the present state. There is a famous conversation between Napoleon and Laplace. Napoleon is supposed to have remarked, "You gave me a copy of your great work, *the Mecanique Celeste*, and I found that, in this massive volume about the universe, there is not a single mention of God, its creator", to which Laplace replied, "Sire, I had no need of that hypothesis."

Newton's assumption of a deterministic world has been challenged by the discovery of quantum mechanics, chaos, and complexity theory. In quantum mechanics, one cannot accurately specify the initial position and velocity simultaneously, and the outcome of a measurement is only probabilistic. In chaos theory, the future of a system depends very sensitively on initial conditions, hence the future of a complex system cannot always be predicted. A classic example of a chaotic system is weather. Note that we cannot predict weather beyond a week even with best available computers. It has been shown that the long-term prediction of weather is simply impossible due to the inherent chaotic nature of the weather system. There goes the predictability of the universe!

Newton's mechanistic–deterministic world view has been applied to economics, history, politics, and ethics (human character). If society, financial markets etc. are some sort of mechanical systems, then they could be analysed, and possibly manipulated for better results. This system of thought is called *deism*.

Is this paradigm correct? There are two issues that have to be cleared here: one, is mechanical modelling of social and psychological process correct? second, what kind of analysis is possible if mechanical modelling of these systems is valid? The first issue is very debatable, and we will not even get into it. We may make a few remarks regarding the second issue. We now know many physical systems that show very complex chaotic behaviour, and predictions of these systems is impossible. Complex social processes involve many strongly interacting components, which may make its analysis very complex. A long-term prediction of these systems may be impossible. We must however add that economics is full of mathematical models that supposedly work quite well in many complex situations. So we are far from the final answer on this issue.

A.3.2   Reductionist and atomistic approach

A basic assumption in physics is that we can understand a system by analysing and understanding its part. For example, Newtonian dynamics starts from the abstraction of a point particle; a rigid body is considered to be a collection of point particles etc. Many physical and chemical properties of a material could be derived from the nature of its constituent atoms and molecules.

There is another method called the holistic approach. According to this paradigm, many properties of a system arise due to the collective behaviour of the system. It is claimed that the colour of a cloth cannot be derived solely from the properties of its constituents (atoms and molecules). Which of the two viewpoints is correct? This is an unresolved issue, yet it is safe to say that a majority of the physical theories are based on reductionist ideas. It is impossible to discuss these profound issues in just a few pages.

The reductionist approach in biology has been very fruitful. Genes determine various traits of an individual. Some scientists believe that the psychological behaviour of an individual may be governed by genetic properties. Since psychological behaviour has not been understood satisfactorily so far, the connection of psychology to genes is not quite clear. Again the debate is between reductionism and holism!

A.3.3   Objective world-view

Typically, it is assumed that the physical laws and the world exists independent of the observer. When we write down the physical laws, we do not need to worry about who is observing the phenomena. The world is assumed to exist whether the observer is present or not.

A.3.4   Space, time, causality

The motion of particles and rigid bodies takes place in space and time. Newton assumed absolute space and absolute time. The property of spacetime in Einstein's special relativity is very different from that of Newton's absolute space and absolute time. The conception of spacetime is more complex in the general theory of relativity; here spacetime and matter affect each other. However in all these theories, the dynamics of a physical system takes place in space and time.

Causality is a very important assumption made in physics. The cause and effect sequence is preserved in all reference frames. In the Newtonian framework, causality is naturally preserved due to the assumption of absolute time. In Einstein's relativity, the events are causally connected in the time-like zone. It can be easily shown that the cause and effect sequence is the same in all inertial frames in the time-like zone.

In this Appendix we have discussed basic assumptions of modern science and physics. These assumptions are very profound, and volumes have been written about them. We introduced them here with the thought that a student should know what are the fundamental assumptions of physical theories. The reader interested in a more thorough discussion may browse Wikipedia and other sources.

Appendix B

Dimensional Analysis and Estimation

Many physical systems are quite complex, and some of them are not well understood. Some such systems are atomic nucleus, black holes, fluid flow, plasma processes etc. You will be surprised to know that even if we do not have a complete understanding of a system, we can get quite far by doing a simple analysis. Some of the techniques used in such situations are

1. Dimensional analysis.

2. Estimation

3. Symmetry arguments

In this chapter we will discuss items 1 and 2. Study of symmetry in physical theories is part of a subject called *group theory*. Simple symmetry arguments have been discussed in Chapter 7 of this book.

B.1 Dimensional analysis

We encounter many physical quantities in physics. Some of these quantities are related to each other, for example, velocity is a ratio of length and time. It turns out that all the quantities we encounter in mechanics can be expressed using mass, length, and time. Therefore, mass, length, and time are called *independent or fundamental quantities*. All other physical quantities are called *derived quantities* because they can be derived from the fundamental quantities.

Every physical quantity has a *dimension*. The meaning of dimension here is a bit abstract; it is not the magnitude or size of that quantity. Dimension also differs from units, which is a standard for measurements. Mass can have several units (like kg, gm, pound etc.), but it has a single dimension. The dimension of mass is denoted by $[M]$, and it is irrespective of the mass content of the body. The dimensions of the other two fundamental quantities, length and time are denoted by $[L]$ and $[T]$ respectively.

The dimensions of derived quantities can be written in terms of $[M]$, $[L]$, and $[T]$. For example, velocity has dimension $[LT^{-1}]$ since it is a ratio of distance over time. I leave it as a simple exercise to determine the dimensions of force, energy, angular momentum etc.

We can say a lot about the relationships between the physical quantities using dimensional analysis. We use a simple fact that we can only compare quantities with the same dimensions. You must have done many exercises in your school using dimensional analysis. Here we will illustrate the procedure using several examples.

EXAMPLE B.1   A pendulum is oscillating in Earth's gravitational field. Derive a formula for its time period using dimensional analysis.

SOLUTION   The time period of the pendulum (T) could depend on the mass of its bob (m), its length (l), and Earth's gravitational acceleration (g). We assume that other effects like air friction have negligible effect on T. Therefore, from dimensional analysis[1]

$$T = m^a l^b g^c.$$

In terms of dimensions

$$[T] = [M]^a [L]^b [LT^{-2}]^c.$$

Matching dimensions of both sides, we obtain $a = 0$, $c = -1/2$, and $b = 1/2$. Hence

$$T = \sqrt{l/g}.$$

A big limitation of dimensional analysis is that we cannot determine the dimensionless factor 2π of the above formula. Other than that we can get a pretty good picture of the problem using dimensional analysis.

EXAMPLE B.2   Estimate the time period of a planet of mass M_P that is going around a star of mass M_S in a circular orbit of radius R.

SOLUTION   The time period T of a planet's orbital motion could depend on gravitational constant G, mass of the Sun M_s, mass of the planet M_p, and the radius of the orbit R. Therefore,

$$T = G^a R^b M_s^c M_p^d.$$

Using $[G] = M^{-1} L^3 T^{-2}$ and matching the dimensions of M, L and T we obtain $a = -1/2$, $b = 3/2$, $c + d = -1/2$. Hence,

$$T \propto R^{3/2}.$$

Thus we can derive Kepler's third law from dimensional analysis. Note that dimensional analysis cannot resolve the dependence of M_s and M_p. It has to come from theory, experiments, or observations.

---

[1]The form $T = m^a l^b g^c$ is chosen because of the symmetry of systems under scaling transformation. This symmetry property is proved from the equation of motion. In Section 8.3.4 we showed that Kepler's orbits are symmetric under scaling. In this Appendix we will assume that the systems at hand are symmetric under scaling transformation.

These examples and many others illustrate the power of dimensional analysis. Many physical systems are not well understood. Using dimensional analysis, we can say quite a lot about these systems. For example, we can derive a formula for the frictional force on a body moving in a fluid, even though the physics of fluid flow is not completely understood (Section B.5).

Mass, length, and time are the fundamental quantities in mechanics. The situation is however different in electrodynamics; in the CGS system, mass, length, and time are the fundamental quantities, but in the SI system, mass, length, time, and charge are the fundamental quantities. We illustrate this difference using the following examples.

In the CGS system, the potential energy of two charges q_1 and q_2 separated by a distance d is

$$U = \frac{q_1 q_2}{d}.$$

From this definition, the dimension of Q is

$$[Q] = [M]^{1/2}[L]^{3/2}[T]^{-1}.$$

Hence charge is a derived quantity in CGS system. This is not the case in the SI system; here charge is a fundamental quantity. Recall that in an SI system, the potential energy of two charge systems is

$$U = \frac{q_1 q_2}{4\pi\epsilon_0 d}$$

where ϵ_0 is permittivity in free space. The new quantity ϵ_0 has the dimension

$$[\epsilon_0] = [M][L]^3[T]^{-2}[Q]^{-2}.$$

This difference came about because the unit current and charge in CGS are too small to be useful in practical situations. In the SI system, ampere was introduced as a unit of current, which brought in new constants μ_0 and ϵ_0. Note that particle physicists deal with small particles (e.g., electron) for which the CGS system is still appropriate. So many physicists use the CGS system, while engineers often use the SI system.

After the above discussion on dimensional analysis, we will introduce another useful technique called estimation.

B.2 Estimation

Estimation is a technique that is used often in science and engineering. In fact, the first stage of any engineering or science project involves estimation. For example, to construct a one-km bridge, first we need to get an idea of the maximum load the bridge has to bear, strength of the trusses, cost estimate etc. A good engineer must be able to quickly see the feasibility of his/her design. Hence estimation is an indispensable tool for an engineer and a scientist. Unfortunately there is no prescribed procedure for estimation. It is an art which comes through experience.

Dimensional analysis and estimations go hand in hand. Typically we derive a rough formula using dimensional analysis, and then plug in numbers to get an estimate of various quantities related to the problem. These tools are specially useful in situations where we do not have a good understanding of the system.

We suggest the following tips to be kept in mind while estimating:

1. Use units in the expressions and cancel them.

2. Round off numbers, e.g, replace 4/3 by 1.

3. Do not use a calculator for estimates. It defeats the whole purpose.

4. Never give your estimate with more than one or two significant digits.

We will illustrate the above tricks by several examples.

EXAMPLE B.3   Estimate the thermal speed of gas molecules at room temperature.

SOLUTION   We assume that the gas molecules are in equilibrium with the ambient heat bath. Forgetting factors like 1/2 etc., we obtain

$$\frac{1}{2}mv^2 = k_B T,$$

where $m = 14 \times 1.6 \times 10^{-27}$ kg is the mass of nitrogen molecules, $T = 300$ K, and $k_B = 1.38 \times 10^{-23} \text{JK}^{-1}$. Therefore,

$$v \approx \sqrt{k_B T/m} \approx 430 \text{ m/s}.$$

Note that $v \approx \sqrt{N k_B T/(Nm)} \approx \sqrt{P/\rho}$, which is close to the sound speed. Note that the above estimate is the average rms speed.

EXAMPLE B.4   Estimate the number of molecules in a glass of water.

SOLUTION   A glass contains roughly 250 gms of water. The mass of a water molecule is $18 \times 1.67 \times 10^{-27}$ kg. Therefore, the glass contains roughly $0.25/\left(18 \times 1.67 \times 10^{-27}\right) \approx 10^{25}$ molecules.

B.3   Various scales of a system

Understanding of a system involves a fairly good idea about its size, duration of its typical processes, energy of the system etc. These quantities are called length-scale, time-scale, and energy-scale of the system. We get a significant idea of the system by estimating these quantities. We will illustrate these concepts by several examples.

EXAMPLE B.5   What is the size of an atom?

SOLUTION   You know from your school physics that the size of an atom is around an armstrong. Can we get this number from dimensional analysis? Let us estimate the size of a hydrogen atom (the Bohr radius). The Bohr radius could depend on the charge of the electron (e), Planck's constant (h), mass of the electron (m_e), and permittivity of free space (ϵ_0). That is

$$r_a = h^\alpha m_e^\beta e^\gamma \epsilon_0^\delta.$$

The dimension of ϵ_0 is $T^2 Q^2 / M L^3$, and that of h is $M L^2 T^{-1}$. Note that we have four unknown, and four equations obtained by matching the dimensions of M, L, T, Q. Solving these equations yields $\alpha = 2, \beta = -1, \gamma = -2, \delta = 1$. Hence,

$$r_a = \frac{h^2 \epsilon_0}{m e^2}. \tag{B.1}$$

This is the formula for the Bohr radius apart from π and some numerical factors. Recall that the Bohr radius is computed by solving Schrdinger's equation, but here we derived the above formula with much less labour. The substitution of the values of constants yields

$$r_a \approx \frac{6^2 \times 10^{-68} \times 10^{-11}}{10^{-30} \times 10^{-38}} \approx 10^{-9} \text{m},$$

which is a good estimate of the Bohr radius.

EXAMPLE B.6   What is the time-scale of atomic transitions, i.e., how long does it take for an electron to go from one energy level to another?

SOLUTION   An electron in an atom moves very fast, and its speed is close to that of light. Therefore the time-scale of atomic transition can be estimated as

$$t_a = r_a/c \approx 10^{-9} \text{m}/(3 \times 10^8 \text{m/s}) \approx 10^{-17} \text{s}.$$

This is called *atomic time-scale*.

EXAMPLE B.7   Estimate the energy-scale of an atom.

SOLUTION   It is of the order of potential energy of the electron

$$E_a = \frac{e^2}{4\pi\epsilon_0 r_a} \approx \frac{10^{-38} 10^{10}}{10^{-9}} \approx 10^{-19} \text{J} \approx 1 \text{eV}.$$

EXAMPLE B.8   Compute length-scale of quantum gravity.

SOLUTION   The length-scale should depend on fundamental quantities, gravitational constant G, Planck's constant h, and c. Hence

$$L = G^\alpha h^\beta c^\gamma.$$

Dimensional analysis yields $\alpha = \beta = 1/2$, and $\gamma = -3/2$. Therefore, the length-scale is

$$L = \sqrt{\frac{Gh}{c^3}} \approx 10^{-34} \text{m}$$

which is the smallest length-scale in the universe, and is called the *Planck scale*.

EXAMPLE B.9   Estimate the size of the universe.

SOLUTION   The age of the universe is approximately 10 billion years. The Big Bang theory says that the universe is expanding. Astronomical observations indicate that the speed of expansion is of the order of the speed of light c. Hence, the size of the universe

$$L = c \times T \approx 10^{10} \text{ light-year}$$

$$= 3 \times 10^8 \times 10^{10} \times 86400 \times 365 \text{ m}$$

$$\approx 10^{26} \text{ m}$$

We understand a physical system better by thinking about it in its natural scales. For an atomic physicist, the unit of length is the atomic scale, not metre or cm. For a nuclear physicist, the size of a nucleus is his/her unit of length. For an astrophysicist, the size of a star or the size of galaxies is a natural length-scale. Note that our choice of length-scale as metre or cm is because the length of common objects around us vary from cm to several metres. Table B.1 describes length-scales of various objects in the universe.

Table B.1   *Length-scales of various objects in the universe*

Objects	length-scale
Planck scale	10^{-34}m
Nucleus	10^{-15}m
Atoms/molecules	10^{-10}m
Bacteria	10^{-6}m
animals	1 m
Earth	10^7m
Solar system	10^{12}m
Galaxy	10^5ly $=10^{21}$m
Universe	10^{10}ly $= 10^{26}$m

We could also rewrite the equation of a physical system in terms of its natural units. This topic will be discussed in the next section.

B.4 Non-dimensional equations of physical systems

A typical differential equation describing a physical system contains many parametres. However if we use the scales of the physical system, the equations become rather transparent. We illustrate this idea using several examples.

1. The equation of a forced oscillator is

$$m\frac{d^2x}{dt^2} + kx = F_0 \cos(\omega_f t) \tag{B.2}$$

where m is the mass of the oscillator, k is the spring constant of the spring, F_0 and ω_f are the amplitude and frequency of the external force, and x is the amplitude of oscillation. By dividing the above equation with m and using $k = m\omega_0^2$ we obtain

$$\frac{d^2x}{dt^2} + \omega_0^2 x = \frac{F_0}{m}\cos(\omega_f t).$$

There are two time-scales in the system: $1/\omega_0$ and $1/\omega_f$. We use $1/\omega_0$ to non-dimensionalise time, i.e.,

$$t' = \omega_0 t. \tag{B.3}$$

Clearly t' is dimensionless. The equation of the oscillator in terms of t' is

$$\frac{d^2x}{dt'^2} + x = \frac{F_0}{m\omega_0^2}\cos\left(\frac{\omega_f}{\omega_0}t'\right).$$

An inspection shows that $F_0/(m\omega_0^2)$ provides a length scale to the system. If we use

$$x' = \frac{x}{F_0/(m\omega_0^2)}, \tag{B.4}$$

we obtain

$$\frac{d^2x'}{dt'^2} + x' = \cos\left(\frac{\omega_f}{\omega_0}t'\right). \tag{B.5}$$

The above equation has only one parametre ω_f/ω_0, and the equation can be analysed in terms of this parameter. After we have solved x' as a function of t', we can go back to the original variables x and t by using eqs. (B.3, B.4).

2. The Schrödinger's equation for the hydrogen atom in SI units is

$$-\frac{\hbar^2}{2m_e}\nabla^2\psi - \frac{e^2}{4\pi\epsilon_0 r}\psi = E\psi. \tag{B.6}$$

Using dimensional analysis we obtain the length-scale for the hydrogen atom as (eq. (B.1))

$$r_a = \frac{\epsilon_0\hbar^2}{m_e e^2}$$

If we use the atomic length-scale r_a as our unit of length, that is

$$r = r' r_a,$$

then Schrödinger's equation can be rewritten as

$$-\frac{\hbar^2}{2m_e r_a^2}\nabla'^2\psi - \frac{e^2}{4\pi\epsilon_0 r_a r'}\psi = E\psi.$$

Substitution of r_a yields

$$-\frac{4\pi}{2}\nabla'^2\psi - \frac{1}{r'}\psi = E'\psi, \qquad (B.7)$$

where $E' = E/E_a$ with $E_a = e^2/4\pi\epsilon_0 r_a$ (the energy-scale of a hydrogen atom). The above equation is simpler than the original equation (Eq. (B.6)). When we wish to solve Schrödinger's equation on a computer or analytically, we use Eq. (B.7) rather than Eq. (B.6).

We solve many engineering and scientific problems on a computer. It is better to feed non-dimensional equations rather than equations with dimensions in the computer. One immediate advantage of this idea is that we need to worry about a smaller set of parameters. Secondly, we do not need to substitute parameters that are too small (e.g., Planck's constant) or too large (e.g., size of the universe). The numbers in the non-dimensional equation is of the order one. This way the equations are simpler, and it would be easier for the computers to solve them.

B.5   Dimensional analysis in fluid mechanics

The physics of fluid flow[2] is not fully solved, yet engineers and scientists are able to design aeroplanes that fly. This success is because of a very interesting combination of experiments, simulations, and modelling. Some of the fluid-flow models are based on dimensional analysis. In the following discussion we present one of the dimensional arguments of fluid mechanics. We will estimate the drag force on a ball that is moving in a viscous fluid.

The drag force on a ball could depend on its mass m, its radius r, viscosity of the fluid μ, the density of the fluid ρ, and the velocity of the ball v:

$$F = m^a r^b \mu^c \rho^d v^e. \qquad (B.8)$$

Experiments and experience shows that the drag force is proportional to the cross-section area of the ball. Two balls of the same radius but made of different material experience the same frictional force. So F is independent of m. Hence $a = 0$. We solve the other constants b, c, d, and e using dimensional analysis. Using $[\mu] = ML^{-1}T^{-1}$ and matching the dimensions of both sides, we obtain $a = 0$, $b = e$, $c = 2 - e$, $d = -1 + e$. Therefore,

$$F \propto r^e \mu^{2-e} \rho^{-1+e} v^e.$$

Clearly, there are more unknowns than what can be determined from dimensional analysis. So we resort to experiments and simulations for clues.

---

[2] The equation of fluid flow, called Navier–Stokes equation, is

$$\rho\left[\frac{\partial \mathbf{u}}{\partial t} + (\mathbf{u} \cdot \nabla)\mathbf{u}\right] = -\nabla p + \mu\nabla^2 \mathbf{u},$$

where $\mathbf{u}(\mathbf{x}, t)$ is the velocity field, ρ and p are the density and pressure fields respectively, and μ is the dynamic viscosity.

Experiments and simulations shows that when $vr\rho/\mu$ is small, $e \to 1$. Therefore, for small $vr\rho/\mu$

$$F \propto \mu r v. \tag{B.9}$$

This is called the *viscous limit*, which is valid when the body moves with small velocity. For large $vr\rho/\mu$, experiments and simulations show that $e \to 2$. Hence

$$F \propto r^2 v^2 \rho. \tag{B.10}$$

This is called the *turbulent limit*.

Why are we using a complicated quantity like $vr\rho/\mu$? This is a dimensionless quantity called *Reynolds number*. It makes sense to call a number large or small (compared to 1), but ambiguous to say that velocity is small or large.

The above formulas for the frictional force on the ball is very useful. We could derive these results using dimensional analysis, and some experimental and simulation inputs. Dimensional analysis alone is not enough to solve this problem. For these sorts of calculations, we use all our knowledge of the system.

We can use the above formula to estimate drag on a car or on an airplane. Let us do it for a car that moves with $v = 100$ km/hour ≈ 30 m/s, $\nu = \mu/\rho = 10^{-5}\,\mathrm{m^2/s}$, $r = 1$ m, $\rho = 1$ kg/m^3. For the car, $vr/\nu = 30 \times 1/10^{-5} \approx 10^6 \gg 1$, signalling that the flow is turbulent. Therefore, force $F \approx r^2 v^2 \rho \approx 10^3$ N, and power $P \approx Fv \approx 3 \times 10^4$ watt. A typical car has the power of 300 horse power, which is $300 \times 750 \approx 2.25 \times 10^5$ watt. Hence our estimate is down by a factor of 10, which is quite good considering the complexity of the problem. Equation (B.9) yields a much smaller number for the drag force on the car; note however that Eq. (B.9) is not valid for the car since $vr/\nu \gg 1$ for the car.

We end this Appendix with a suggestion that students should apply these ideas to the problems they encounter in their studies or research.

Exercises

1. Derive dimensions of force, energy, Newton's gravitational constant, moment of inertia, and torque.

2. In the SI system derive dimensions of magnetic field, electric field, magnetic moment, current, dielectric permittivity and permeability of free space, displacement vector, magnetization.

3. Derive the time period of an oscillator as a function of spring constant, mass, and amplitude of oscillation.

4. Construct a dimensionless quantity using electron's charge, Planck's constant, speed of light, and permittivity for free space ϵ_0. Obtain an estimate for this quantity.

5. Obtain time-scales of an *LRC* circuit in terms of its resistance, inductance, and capacitance.

6. Derive the length-scale, time-scale, and velocity scale for molecules in an ideal gas held at temperature T.

7. Non-dimensionalise the following equations:

   (a) classical oscillator (simple, damped, forced)

   (b) quantum oscillator

   (c) Navier–Stokes equation

   (d) RC, LC, LRC circuit

8. Estimate the following quantities:

   (a) How many air molecules are there in your hostel room?

   (b) Estimate the solar energy incident per unit area on the surface of the Earth.

   (c) Estimate the amount of energy lost by the human body per unit time.

   (d) Estimate the number of photons emitted by a 100 watt bulb.

   (e) Estimate the fuel required to take a rocket of 1000 kg outside Earth, say on the way to Mars.

   (f) Estimate air pressure at the surface of the Earth.

   (g) What is the blood pressure inside our body? Estimate the vertical force exerted by air on our body.

   (h) Estimate the power of a human heart.

Appendix C

Numerical Solution of Differential Equations

We encounter differential equations (DEs) in every branch of science and engineering. Only some DEs (e.g., linear DEs) are analytically solvable. Most non-linear DEs do not have analytic solutions, and they are solved numerically using computers.[1] In the following discussion we will introduce two numerical schemes to solve a first-order DE on a computer.

$$\dot{x} = f(x, t) \tag{C.1}$$

C.1 Euler's scheme

Using the definition of derivative $\dot{x} = [x(t + \Delta t) - x(t)]/\Delta t$, Eq. (C.1) is converted to the following difference equation

$$x(t + \Delta t) = x(t) + f(x, t)\Delta t.$$

Given $x(0)$ at $t = 0$, we can use the above difference equation to compute $x(\Delta t)$, $x(2\Delta t)$, ..., $x(n\Delta t)$, where $n\Delta t = t$. Care should be taken to choose a small enough Δt. It can be shown that the maximum error in each step of the difference equation is $max[f''(x, t)](\Delta t)^2/2$. Due to this large error in each time step, Euler's scheme is rarely used.

C.2 Second-order Runge–Kutta scheme

In this scheme we time advance $x(t)$ in two sub-steps:

$$x_{\mathrm{mid}} = x(t) + f(x, t)\frac{\Delta t}{2}$$

$$x(t + \Delta t) = x(t) + f(x_{\mathrm{mid}}, t + \Delta t/2)\Delta t.$$

---

[1]Another popular technique is called approximate methods, discussion of which is beyond the scope of this book.

We again start from $x(0)$, and compute $x(\Delta t)$, $x(2\Delta t)$, ..., $x(n\Delta t)$. This scheme is more accurate than Euler's scheme.

There are many numerical schemes to solve DEs. The interested reader can refer to the texts given in the bibliography.

Appendix D

Matlab and Octave

Matlab is an interactive mathematical package. You can use this package for plotting, solving ODEs etc. Octave is a clone of Matlab, and is available for free on Linux platforms. Typically octave .m files run in Matlab and vice versa.

Matlab is quite easy to learn. Most operations in Matlab are based on vectors and matrices. Some important points to remember about Matlab are

1. The most fundamental and widely used object used in Matlab is an array. Some of the ways to create arrays in Matlab are

   (a) The command x=0:0.1:4*pi; creates a one-dimensional array in the range $[0, 4\pi)$ in steps of 0.1. The elements of the array are $[0, 0.1, 0.2, ..., 12.5]$. You can access the elements by writing x(i). For example the fifth element of x is x(5).

   (b) The command x=[1,2,3;10,20,30]; creates a two-dimensional array whose elements are

   $$\begin{bmatrix} 1 & 2 & 3 \\ 10 & 20 & 30 \end{bmatrix}.$$

2. We can define various functions of an array. For example, y=sin(x); creates an array $y = \sin(x)$ for each element of x defined above. The size of arrays x and y are the same.

3. We can define functions in Matlab. For example

   ```
   function xdot = f(x)
   xdot = [x(1)*x(1)]
   ```

   The function f returns an array xdot(i) that has value x(i)*x(i).

We illustrate its usage by some simple examples. For details refer to the texts given in the bibliography.

D.1   Plotting

1. **Plot** $y = \sin x;\ z = \sin^2 x$

   The commands to plot the above function is given below.

   ```
   x = 0:0.1:4*pi;       % create x as a vector; x = [0, 0.1, ..., 12.5]
   y=sin(x);             % define y=sin(x) of size 126 elements
   z = sin(x).*sin(x);   % define z;
   % Operation .* multiplies two vectors term by term
   plot(x,y);                 % Plots x vs.  y
   xlabel('x'); ylabel('y'); title('Plot of Sin(x) & sin(x)^2');
   hold on;
   plot(x,z,'-.');
   % All the statements after % are comments.
   % Experiment with 'grid on', 'hold on';
   ```

   The resultant plot is shown in Fig.D. 1.

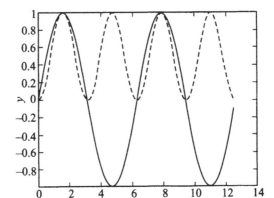

Figure D.1   *A plot of functions* $\sin(x)$ *and* $\sin^2(x)$ *using Matlab. The plot is defined for the range of* $x = [0, 4\pi]$.

2. **Polar plots**
   ```
   %plot r = exp(0.1 * φ)
   phi=0:0.1:6*pi;        % define phi=[0,0.1,...,18.8]
   r=exp(0.1*phi);
   polar(phi,r);          % plots r(phi)
   %
   phi=0:0.01:2*pi;
   e=0.5;
   r=(1+e*cos(phi)).^(-1);
   polar(phi,r,'-.');
   %
   ```

```
hold on;
%
N=size(phi);
r=ones(N);
polar(phi,r,':');
%
e=1;
r=(1+e*cos(phi)).^(-1);
polar(phi,r,'--');
%
e=3;
phic=acos(-1/e);
phi=-phic:0.001:phic;
r=(1+e*cos(phi)).^(-1);
polar(phi,r);
```

The plots obtained using the above commands are displayed in Fig. D.2 and Fig. 8.9.

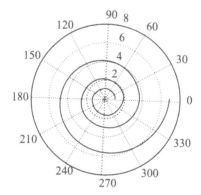

Figure D.2   *A polar plot of a function* $r = \exp(\phi/10)$ *using Matlab.*

Figure D.2 is a spiral $r = \exp(\phi/10)$, while Fig. 8.9 displays the curves of the conic sections $r = p/(1 + e\cos\phi)$ for $e = 0, 0.7, 1$ and 3.

D.2   Differential equation solver

We can solve differential equations in Matlab or Octave. We illustrate the procedure using simple examples.

1. Solve $\dot{x} = 5(x + x^2)$ with initial condition $x(0) = 1$.

   We create a .m file named *f.m* and define a function named *f* whose inputs are independent variable t and a vector $[x(1)]$. *xdot* is the output vector of the function.

The equation is

$$xdot(1) = \frac{d}{dt}x(1)$$

$xdot(1) = d\,x(1)/dt$ contains the definition of a differential equation. The function is defined below.

```
function xdot = f (x, t)       % define dot(x)=f(x,t)
xdot= [5*x(1)* (1+x(1))];
```

Use the following Matlab commands to solve the differential equation.

```
tspan=[0 2*pi];
initcond=[1 ];
[t,x]=ode45(@f, tspan, initcond);
plot(t,x);
```

The function *ode45* solves the differential equation with the initial condition *initcond*. The RHS function is given by *f*. The time range is given by *tspan*, and the output is stored in *x*. For other differential equation solvers refer to the Matlab manual.

2. Solve $\dot{x} = y$; $\dot{y} = -\omega^2 y$ for $\omega = 0.1$ and with initial condition $[x(0), \dot{x}(0)] = [1, 0]$.

We apply the same procedure as above. Here $[x(1), x(2)]$ are arrays with two components. The equations are

$$xdot(1) = \frac{d}{dt}x(1)$$

$$xdot(2) = \frac{d}{dt}x(2)$$

In this example we take $x(2) = dx(1)/dt$ to convert the second-order differential equation to two first-order differential equations. The function definitions g is

```
function xdot = g (x, t)       % define dot(x)=f(x,t)
omega=2.0;
xdot= [x(2); -omega^2*x(1)];
```

After function definitions, we use the following Matlab commands to solve ODE.

```
tspan=[0 2*pi];
initcond=[1 0 ];
[t,x]=ode45(@g, tspan, initcond);
plot(t,x);
```

$x = [x(1), x(2)]$ contains the result as a function of time.

Exercises

1. Plot the following functions using Matlab/Octave.

   (a) $y = \exp(x)$
   (b) $x = \cos(t)$ and $y = \sin(2t)$
   (c) $r = 1/(1 + 0.1 * \cos\theta)$
   (d) $r = 1/(1 + 3 * \cos\theta)$

2. Solve the following equations using Matlab/Octave.

   (a) $\dot{x} = -3x$
   (b) $\ddot{x} = -9x$
   (c) $\ddot{x} + 2\dot{x} + 9x = \cos(t)$
   (d) $\ddot{x} + 2\dot{x} + 9x = t$

Appendix E

Vectors and Tensors

In school books, vectors are defined as physical quantities having both magnitude and direction. It turns out that vectors in physics have a very specific meaning. Thy are defined using the symmetry property of the physical laws (see Section 7.4). In Newtonian physics we define vectors using the operation of rotation.

E.1 Vector

While studying a physical system, we choose a coordinate system, and measure components of the physical quantities in that coordinate system. If we had chosen another coordinate system that is rotated wrt the earlier one, we would have measured a different set of numbers for their components. We define vectors in terms of transformation rules between these components.

DEFINITION   Any physical quantity whose components transform in the same manner as the components (x, y, z) is called a vector.

In Section 7.4 we showed that the transformation rule for the coordinates of a point in the two coordinate systems that make an angle θ between them is

$$x' = x \cos \theta + y \sin \theta$$

$$y' = -x \sin \theta + y \cos \theta.$$

A vector $d\mathbf{r}$ has components (dx, dy, dz) and (dx', dy', dz') in the two coordinate systems, and they transform in the same manner as given above, hence $d\mathbf{r}$ is a vector. A division of $d\mathbf{r}$ by dt yields $d\mathbf{r}/dt$, hence the velocity $\mathbf{v} = d\mathbf{r}/dt$ is also a vector. Similar arguments are used to prove that the acceleration, which is a time derivative of velocity, is a vector as well. Other examples of vectors are force, momentum, angular velocity etc.

E.2   Scalar

DEFINITION   Physical quantities that do not change under the rotation of coordinate systems are called *scalars*.

In Section 7.4 we showed that the dot product of two vectors is a scalar, i.e., $\mathbf{A} \cdot \mathbf{B}$ is a scalar, when $\mathbf{A}$ and $\mathbf{B}$ are vectors. Using this result we deduce that $\mathbf{v} \cdot \mathbf{v} = v^2$ is a scalar. Similarly arguments show that $\omega \cdot \omega$ (where ω is the angular velocity), kinetic energy etc. are scalars.

E.3   Tensor

In physics we encounter physical quantities that have more than 3 components. For example, in electrostatics, the relationship between the electric field and the displacement vector in a linear dielectric is given by

$$D_1 = \kappa_{11}E_1 + \kappa_{12}E_2 + \kappa_{13}E_3,$$

$$D_2 = \kappa_{21}E_1 + \kappa_{22}E_2 + \kappa_{23}E_3,$$

$$D_3 = \kappa_{31}E_1 + \kappa_{32}E_2 + \kappa_{33}E_3.$$

The quantity κ_{ij} $(i, j = 1, 2, 3)$, called *dielectric constant*, is a nine component object. In your school books, all the diagonal components of dielectric constant are assumed to be equal, and the nondiagonal components as zeros. This is valid only for isotropic material like air, water, etc., but not for material with directionality (e.g., a salt crystal). For isotropic material

$$\mathbf{D} = \kappa\mathbf{E}$$

where all the diagonal components of κ_{ij} are equal to κ.

κ_{ij} is a nine component object, and it is called tensor. Since it has two indices i and j, it is a second-rank tensor. Rank is the number of indices of the tensor. According to the above notation, vector is a first-rank tensor, and scalar is a zero-rank tensor.

The transformation rule for a second-rank tensor is somewhat more complex than that for the first rank tensor. To express this relationship we introduce a notation

$$R_{11} = \cos\theta; R_{12} = \sin\theta; R_{13} = 0;$$

$$R_{21} = -\sin\theta; R_{22} = \cos\theta; R_{23} = 0;$$

$$R_{31} = 0; R_{32} = 0; R_{33} = 1;$$

Using R_{ij} we can define the transformation rule for second-rank tensors.

DEFINITION   A second-rank tensor is a physical quantity whose components κ_{ij} transform as

$$\kappa'_{ij} = \sum_a \sum_b R_{ia} R_{jb} \kappa_{ab}. \qquad (E.1)$$

The dielectric constant is a second-rank tensor, and we use the above transformation rules to compare the components of the dielectric constant in two coordinate systems.

Another important example of a second-rank tensor is the *Kronecker delta function* which is defined as

$$\delta_{ij} = \begin{cases} 1 & \text{if } i = j \\ 0 & \text{if } i \neq j \end{cases} \qquad (E.2)$$

We use eq. (E.1) to compute components of the Kronecker delta tensor in the rotated coordinate system, which yields

$$\delta'_{ij} = \sum_a \sum_b R_{ia} R_{jb} \delta_{ab} \qquad (E.3)$$

Let us compute δ'_{11} which is

$$\delta'_{11} = R_{11}(R_{11}\delta_{11} + R_{12}\delta_{12} + R_{13}\delta_{13})$$

$$+ R_{12}(R_{11}\delta_{21} + R_{12}\delta_{22} + R_{23}\delta_{23})$$

$$+ R_{13}(R_{11}\delta_{31} + R_{12}\delta_{32} + R_{23}\delta_{33})$$

$$= R_{11}^2 + R_{12}^2$$

$$= \cos^2 \theta + \sin^2 \theta = 1.$$

We can easily show that $\delta'_{12} = \delta'_{21} = \delta'_{31} = \delta'_{13} = \delta'_{23} = \delta'_{32} = 0$ and $\delta'_{22} = \delta'_{33} = 1$. Hence the components of Kronecker delta function remain the same in both the coordinate systems. This special tensor is very useful in physics as we will see later.

Let us define another quantity with two indices in two dimensions

$$\epsilon_{ij} = \begin{cases} 0 & \text{if } i = j \\ 1 & \text{if } i = 1, j = 2 \\ -1 & \text{if } i = 2, j = 1 \end{cases}$$

Under rotation, the components of the tensor transforms as

$$\epsilon'_{ij} = \sum_a \sum_b R_{ia} R_{jb} \epsilon_{ab} \qquad (E.4)$$

ϵ'_{ij} term by term is

$$\epsilon'_{11} = R_{11}(R_{11}\epsilon_{11} + R_{12}\epsilon_{12}) + R_{12}(R_{11}\epsilon_{21} + R_{12}\epsilon_{22})$$

$$= \cos\theta\sin\theta - \sin\theta\cos\theta = 0$$

$$\epsilon'_{12} = R_{11}(R_{21}\epsilon_{11} + R_{22}\epsilon_{12}) + R_{12}(R_{21}\epsilon_{21} + R_{22}\epsilon_{22})$$

$$= \cos^2\theta + \sin^2\theta = 1$$

Similarly we can show that $\epsilon'_{21} = -1$ and $\epsilon'_{22} = 0$. Hence ϵ_{ij} is a constant second-rank tensor like δ_{ij}.

Another useful tensor which does not change under rotation is Levi-Civita tensor, which is defined as

$$\epsilon_{ijk} = \begin{cases} 0 & \text{if any two indices match} \\ 1 & \text{for cyclic permutation of indices} \\ -1 & \text{for anticylic permutation of indices} \end{cases} \tag{E.5}$$

where i, j, k take values $1, 2, 3$. It can be shown that the components of ϵ_{ijk} are the same in both the coordinate systems. You can check that the ith component of $\mathbf{A} \times \mathbf{B}$ is

$$(\mathbf{A} \times \mathbf{B})_i = \sum_j \sum_k \epsilon_{ijk} A_j B_k.$$

We state a very useful identity on ϵ_{ijk} without proof:

$$\epsilon_{ijk}\epsilon_{iab} = \delta_{ja}\delta_{kb} - \delta_{jb}\delta_{ka}.$$

E.4   Moment of inertia

Moment of inertia is an important property of a rigid body. It is encountered in rotation. We defined moment of inertia in Chapter 17 by brute force algebra of the cross-product. Here we will derive the same results using tensor analysis.

In Section 17.7 we defined the rotational kinetic energy of a rigid body as $\sum m_a(\mathbf{\Omega} \times \mathbf{r}_a)^2/2$, where m_a is the mass of the ath constituent particle located at $\mathbf{r}_a$, and $\mathbf{\Omega}$ is the angular velocity of the rigid body. We can obtain an expression for the rotational kinetic energy using the following algebra:

$$T_{\text{rot}} = \frac{1}{2}\sum m_a(\mathbf{\Omega} \times \mathbf{r}_a)^2$$

$$= \frac{1}{2}\sum m_a\epsilon_{i\alpha\beta}\Omega_\alpha r_{a,\beta}\epsilon_{i\gamma\delta}\Omega_\gamma r_{a,\delta}$$

$$= \frac{1}{2}\left[\sum m_a(r_a^2\delta_{il} - r_{a,i}r_{a,l})\right]\Omega_i\Omega_l$$

$$= \frac{1}{2}I_{il}\Omega_i\Omega_l,$$

where

$$I_{il} = \sum m_a (r_a^2 \delta_{il} - r_{a,i} r_{a,l})$$

is called the moment of inertia tensor of the rigid body. Here $r_{a,\beta}$ denotes the βth component of the position vector of the ath particle. A simple inspection shows that the moment of inertia depends only on the distribution of the mass in the rigid body. The moment of inertia is a nine-component object just like κ_{ij} and δ_{ij}, and it is a second-rank tensor. It is easy to verify that the components of the moment of inertia tensor has a property that $I_{12} = I_{21}$, $I_{32} = I_{23}$, $I_{31} = I_{13}$.

There is a mathematical theorem according to which, we can always choose a coordinate system in which the cross terms I_{12}, I_{13}, and I_{23} are zeros. The axes of the new coordinate system are called *principal axes*.

We can derive an expression for the angular momentum of the rigid body in terms of moment of inertia of the rigid body. The derivation is

$$L_i = \left[\sum \mathbf{r}_a \times m_a (\boldsymbol{\Omega} \times \mathbf{r}_a) \right]_i$$

$$= \sum_a \sum_{jklm} \epsilon_{ijk} r_{a,j} \epsilon_{klm} m_a \Omega_l r_{a,m}$$

$$= \left[\sum_a m_a (\delta_{il}\delta_{jm} - \delta_{im}\delta_{jl}) r_{a,j} r_{a,m} \right] \Omega_l$$

$$= \left[\sum_a m_a (r_a^2 \delta_{il} - r_{a,i} r_{a,l}) \right] \Omega_l$$

$$= I_{il}\Omega_l$$

This is how we use tensor algebra to derive kinetic energy and angular momentum of a rigid body. If you are familiar with the tensor properties of δ_{ij}, ϵ_{ijk}, and sum rules, you will find the above algebra much simpler than the ones discussed in Chapter 17. Note that the quantities I_{xx}, I_{yy}, I_{zz}, I_{xy}, I_{xz}, I_{yz} of Chapter 17 correspond to I_{11}, I_{22}, I_{33}, I_{12}, I_{13}, I_{23} of the present appendix.

The discussion in this appendix shows that tensors are very useful objects in physics. As discussed in Chapter 7, all the laws of physics must be written in tensor form because of symmetry properties. Tensors are also very useful in writing the physical quantities in a convenient form.

We also remark that the definition of a vector and tensor is coupled to the symmetry operation. In the above discussion we used rotation as the symmetry operation. In special relativity however we use Lorentz transformation as the symmetry operation and define vectors and tensors accordingly. In Chapter 19 we introduced relativistic vectors and scalars. More complete discussions on these objects would be covered in your course on relativity.

Appendix F

Vector Operations on Vector and Scalar Fields

A vector field is a physical quantity for which we associate a vector at every point in its domain space. Examples of vector fields are electric field, gravitation field, etc. Scalar fields are quantities for which we associate a scalar at every point in its domain space. Examples are temperature field, density field etc.

There are some very important physical operations on vector and scalar fields. We define a gradient on a scalar field U as (see Section 11.6)

$$\nabla U = \frac{\partial U}{\partial x}\hat{\mathbf{x}} + \frac{\partial U}{\partial y}\hat{\mathbf{y}} + \frac{\partial U}{\partial z}\hat{\mathbf{z}}.$$

By using the definition of vectors, we can show that ∇U is a vector. The proof is as follows: At point $\mathbf{r}$, the components of gradient are $(\partial U/\partial x, \partial U/\partial y, \partial U/\partial z)$ in one coordinate system. Now let us take another coordinate system which is rotated about the z-axis by an angle θ wrt the earlier coordinate system. Note that the origin of both the coordinate systems are at point $\mathbf{r}$. Using elementary calculus, we derive

$$\frac{\partial U}{\partial x'} = \frac{\partial U}{\partial x}\frac{\partial x}{\partial x'} + \frac{\partial U}{\partial y}\frac{\partial y}{\partial x'}$$

$$= \frac{\partial U}{\partial x}\cos\theta + \frac{\partial U}{\partial y}\sin\theta.$$

Similar analysis shows that

$$\frac{\partial U}{\partial y'} = -\frac{\partial U}{\partial x}\sin\theta + \frac{\partial U}{\partial y}\cos\theta$$

The third component of the gradient remains unchanged, i.e.,

$$\frac{\partial U}{\partial z'} = \frac{\partial U}{\partial z}.$$

The above transformation rules are the same as those for vectors. Hence, the *operator* $(\partial_x, \partial_y, \partial_z)$ transforms like a vector.

There is another important operation on vector field called *curl*. Curl of a vector field **A** is defined as

$$\nabla \times \mathbf{A} = (\partial_y A_z - \partial_z A_y)\hat{\mathbf{x}} + (\partial_z A_x - \partial_x A_z)\hat{\mathbf{y}} + (\partial_x A_y - \partial_y A_x)\hat{\mathbf{z}}.$$

It can be shown using similar algebra as above that $\nabla \times \mathbf{A}$ is a vector.

Another operation called *divergence* of a vector field is defined as

$$\nabla \cdot \mathbf{A} = \partial_x A_x + \partial_y A_y + \partial_z A_z.$$

It can be shown that $\nabla \cdot \mathbf{A}$ is a scalar.

The above operations are very useful when we are dealing with fields. For example, in electrostatics the relationship between electric field (**E**) and electric potential (U) is $\mathbf{E} = -\nabla U$. If you look at any book on electrodynamics, you will find the above symbols all over the place. These operations will be discussed in more detail in your future courses like electrodynamics, vector calculus etc.

Appendix G

Important Astronomical Data

G.1 Properties of the Sun, the Earth, and the Moon

Property	Units	Sun	Earth	Moon
Mass	kg	1.99×10^{30}	5.98×10^{24}	7.35×10^{22}
Mean radius	m	6.96×10^8	6.37×10^6	1.74×10^6
Rotation period	d	35 at poles, 25 at equator	23h 56m	27.3d
Average density	kg/m^3	1410	5515	3346
Surface temperature	K	5778	287	220(equator)
Core temperature	MK	5	–	–
Radiation power	W	3.9×10^{26}	–	–

G.2 Properties of planets[1]

Name	Orbital radius (AU)	Orbital period(y)	Eccentricity	Mass	Equatorial diameter	Days
Mercury	0.39	0.24	0.206	0.06	0.38	58.64
Venus	0.72	0.62	0.007	0.82	0.95	243.02
Earth	1.00	1.00	0.017	1.00	1.00	1.00
Mars	1.52	1.88	0.093	0.11	0.53	1.03
Juptier	5.20	11.86	0.048	317.8	11.21	0.41
Saturn	9.54	29.46	0.054	95.2	9.45	0.43
Uranus	19.22	84.01	0.047	14.6	4.00	0.72
Neptune	30.06	164.8	0.009	17.2	3.88	0.67

---

[1]The quantities are given in units of the values for the Earth.

Appendix H

Important Physical Constants

Constant	Symbol	Numerical value	SI units
Gravitational constant	G	6.67×10^{-11}	m^3/s^2kg
Speed of light in vacuum	c	3.00×10^8	m/s
Charge of an electron	e	1.60×10^{-19}	C
Electron mass	m_e	9.11×10^{-31}	kg
Proton mass	m_p	1.67×10^{-27}	kg
	m_p/m_e	1840	
Hydrogen atom mass		$1.0078u$	kg
Permittivity constant	ϵ_0	8.85×10^{-12}	F/m
Permeability constant	μ_0	1.26×10^{-6}	H/m
Boltzmann constant	k_B	1.38×10^{-23}	J/K
Planck's constant	h	6.63×10^{-34}	J s
Stefan–Boltzmann constant	σ	5.67×10^{-8}	W/m^2K^4
Universal gas constant	R	8.31	J/mol.K
Avogadro number	N_A	6.02×10^{23}	mol^{-1}
Bohr radius		5.29×10^{-11}	m

u = Atomic mass unit = $1.66054 \times 10^{-27}kg$

Selected References

Introductory books on mechanics

The following books are at the level of the present book.

1. J. M. Knudsen and P. G. Hjorth. *Elements of Newtonian Mechanics, Volume I.* 3rd edition. New York: Springer, 2000.

2. *D. Kleppner and R. J. Kolenkow. *An Introduction to Mechanics.* Ohio: McGraw-Hill, 1973.

3. *R. P. Feynman, R. B. Leighton and M. Sands. *The Feynman Lectures on Physics.* Boston: Addison Wesley, 1963.

4. *A. P. French. *Newtonian Mechanics.* New York: W. W. Norton & Co., 1971.

5. *C. Kittel, W.D. Knight, M.A. Ruderman. et al. *Mechanics.* Berkeley Physics Course Volume I. New York: McGraw-Hall, 1973.

Introductory books on mechanics with an engineering flavour

1. *J. L. Meriam and L. G. Kraige. *Engineering Mechanics.* 5th edition. New York: John Wiley and Sons, 2002.

2. * F. P. Beer and E. R. Johnston Jr. *Vector Mechanics for Engineers.* 3rd edition. New Delhi: Tata McGraw-Hill, 1999.

Advanced level books on mechanics for advanced undergraduate courses

1. J. R. Taylor. *Classical Mechanics.* Virginia: University Science Books, 2005.

2. R. D. Gregory. *Classical Mechanics.* Cambridge: Cambridge University Press, 2006.

Advanced level books on mechanics for graduate courses

1. L. D. Landau and E. M. Lifsitz. *Mechanics, Courses of Theoretical Physics, Volume I.* 3rd edition. Oxford: Pergamon, 1976.

2. *H. Goldstein, C. Poole and J. Safko. *Classical Mechanics.* 3rd edition. Boston: Addison-Wesley, 2002.

3. V. I. Arnold. *Mathematical Methods of Classical Mechanics.* New York: Springer, 1978.

Books on history of mechanics and science

1. T. Ferris. *Coming of Age in the Milky Way*. New York: Harper Collins, 1988.

2. R. Spangenburg and D. K. Moser. *History of Science, Volume I*. New York: Facts on File, 1993.

Books on relativity

1. E. F. Taylor and J. A. Wheeler. *Spacetime Physics*. 2nd edition. New York: W. H. Freeman & Co.,1992.

2. *R. Resnick. *Introduction to Special Relativity*. New York: Wiley, 1968.

Books on non-linear dynamics and chaos theory

1. J. Gleick. *Chaos, Making a New Science*. New York: Viking-Penguin, 1987.

2. G. Baker and J. Gollub. *Chaotic Dynamics: An Introduction*. 2nd edition. Cambridge: Cambridge University Press, 1996.

3. S. H. Strogatz. *Nonlinear Dynamics and Chaos*. Boston: Addison-Wesley, 1994.

4. R. Hilborn. *Chaos and Nonlinear Dynamics*. 2nd edition. Oxford: Oxford University Press, 2000.

Books on mathematics relevant to this book

1. *G. B. Thomas and R. L. Finney. *Calculus and Analytic Geometry*. 9th edition. Boston: Addison-Wesley, 1996.

2. *E. Kreyszig. *Advanced Engineering Mathematics*. 9th edition. New York: Wiley, 2005.

Books and reference on Matlab and Octave

1. *Octave*. www.octave.org (Contains online manual)

2. *Matlab*. www.mathworks.com (Contains online manual)

3. D. Higham and N. J. Higham. *Matlab Guide*. Philadelphia: Siam, 2005.

Miscellaneous references

1. S. Chandrashekhar. *Newton's Principia for the Common Reader*. Oxford: Oxford University Press, 2003.

2. R. Malhotra, M. Holman and T. Ito. Chaos and Stability of the Solar System. *PNAS* 98: 12342.2001.

3. *D. Halliday, R. Resnick and J. Walker. *Fundamentals of Physics*. 7th edition. New York: Wiley, 2004.

Internet resources

1. *Wikipedia.* http://www.wikipedia.org

2. *Mathworld.* http://mathworld.wolfram.com

3. T. Kanamaru, J. Michael and T. Thompson. *Introduction to Chaos and Nonlinear Dynamics.* http://brain.cc.kogakuin.ac.jp/~kanamaru/Chaos/e/ contains simulations and animations of several chaotic systems including double pendulum.

4. http://members.tripod.com/~gravitee/ contains parts of Newton's *Principia.*

5. *Stanford Encyclopedia of Philosophy.* http://plato.stanford.edu contains certain philosophical issues on Newton's Principia.

6. http://dynamical-systems.org contains very nice animations of many complex mechanical systems including the three-body problem.

7. http://www.gyroscopes.org for various kinds of gyroscopes.

All the books marked with * are available in Indian low-priced editions.

Index

Printed and bound by CPI Group (UK) Ltd, Croydon, CR0 4YY

21/10/2024

01777040-0001